Oxide Nanostructures

edited by
Avanish Kumar Srivastava

Oxide Nanostructures

Growth, Microstructures, and Properties

Pan Stanford Publishing

Published by

Pan Stanford Publishing Pte. Ltd.
Penthouse Level, Suntec Tower 3
8 Temasek Boulevard
Singapore 038988

Email: editorial@panstanford.com
Web: www.panstanford.com

British Library Cataloguing-in-Publication Data
A catalogue record for this book is available from the British Library.

Oxide Nanostructures: Growth, Microstructures, and Properties

ISBN 978-981-4411-35-6 (Hardcover)
ISBN 978-981-4411-36-3 (eBook)

Printed in the USA

Contents

Foreword by Prof. C. N. R. Rao

While not so long ago it seemed to the casual observer that almost everything related to chemical science was understood, discovered, established and was just waiting to be applied for the greater advancement of civilization, things have certainly changed with the arrival of nanoscience and nanotechnology. This branch of materials science, which deals with materials that have at least one of their characteristic dimensions in the nanometer realm, has been in existence for practical purposes since time immemorial but has gained greater importance in the last couple of decades due to the availability of sophisticated techniques. Nanoscience and nanotechnology are fast becoming one of the burgeoning fields of research for scientists and engineers alike. Predictions have been made about nanomaterials becoming the basis of remarkably powerful computers and new medicinal products that could save millions of lives and perhaps bring about the next industrial revolution. Various materials are being investigated within their nano-realm and raw information is being acquired frequently such that updated ensembles of fresh research activities encompassing an entire branch of nanomaterials are becoming a necessity. This book, titled *Oxide Nanostructures: Growth, Microstructures, and Properties*, focuses on metal oxides of the family of nanomaterials.

Metal oxides represent a family of materials that have a wide range of properties with potential applications. The significance of metal oxides to fields such as information storage, energy storage and energy conversion, medicinal implications, heterogeneous catalysis, and humidity and gas sensing have spurred research aiming not only to develop facile synthetic pathways to nanostructured metal oxides but also significant advancement in the characterization methods aiming to obtain a comprehensive understanding of their various properties. The critical challenges that need our attention involve devising inexpensive, greener ways of manufacturing these oxides and obtaining a greater degree of mastery over the manipulation of the shape of the nanostructured metal oxides, thus influencing their performances in practical applications. A thorough understanding of the defect structures of metal oxides is another aspect or relevance. This book deals with some of the important issues dealing with both the conventional and unconventional synthetic strategies of metal oxide nanostructures and their applications. Additionally, there are chapters that discuss some of the contemporary issues in nanostructured metal oxide research.

Designed purposefully to provide with an overview of the present-day research on metal oxide nanomaterials to practitioners, graduate students and engineers, this book treats the subject using terms familiar to materials scientists and engineers. The book contains eleven accounts of various topics written by people who possess significant research experience in this field. Apart from the topics mentioned above, the compilation includes chapters on applications such as sensing, photovoltaics, bioimaging and biomplants. While the book has been compiled keeping in mind specialists working in the field of nanostructured metal oxides, it could be useful to all those interested in nanoscience.

Prof. C. N. R. Rao

Honorary President,
Jawaharlal Nehru Centre for Advanced Scientific Research, India

Foreword by Prof. Ramesh Chandra Budhani

Numerous technological achievements and conceptual excellence have been realized in the past couple of decades due to global research leading to skills in manipulating materials over length scale of few nanometers. While many classes of materials have been investigated, including nano-dimensional polymorphic forms of carbon, metal oxides and their composites placed broadly in the category of ceramics constitute a large group of industrially important nanomaterials.

The metals are able to form a wide variety of oxide compounds with differing crystal symmetries and electronic structures making them insulators, conductors or semiconductors. Many of them also show exotic long-range electronic orders like superconductivity and magnetism. In terms of technological applications, these metal oxides are being used in microelectronic circuits, sensors, piezoelectric devices, fuel cells and coatings for the passivation of surfaces against corrosion, and as heterogeneous catalysts. Additionally, the theoretical studies on various transition metal and 4f metal oxides and their complexes keep providing fresh insights of their physicochemical characteristics transforming the family of metal oxides into one of the most vibrant and relevant topics of research today.

Tackling such a vast branch of materials within the scope of a single compilation of pertinent chapters and to be able to do justice is difficult for obvious reasons. This book makes an honest attempt to address a lot of the intriguing cornerstones of current nano-metal oxide research activities. From the first chapter, which is depicted as an elaborate review of conventional synthesis and characterization methods applied, coupled with the extensive technological and scientific applications available, to chapters corresponding to more specific areas of synthesis and subsequent property studies, e.g. pulsed laser deposition of nanostructured metal oxides, synthesis of porous and hollow metal oxide nanostructures, application of metal oxides in photovoltaics and bioimaging, the book may have the curiosity of the followers of this particular field of nanoscience. Furthermore, discussions focused on more specialized studies on specific metal oxides such as the defect structures within zinc oxide nanostructures, paramagnetic defects of crystalline quartz, plasticity in amorphous conjugate metal oxides shed light on some of the ongoing exclusive research works. Theoretical investigations on the band structures of metal oxides based on density functional theory add to the overall all round compactness of the compilation.

Though an immense volume of research works have already been conducted on materials such as ZnO and TiO_2, the field of metal oxides is still considered to be young and rapidly developing owing to the largeness of the family where a lot of materials are yet to be thoroughly studied. This is a very opportune time to publish a book on this very alive topic of metal oxides. While it is a challenging task to cover all aspects of oxide nanomaterials in a single book, the authors of this monograph have made a serious effort and one can expect the readers to find this an engrossing and useful book.

Prof. R. C. Budhani

Director, CSIR—National Physical Laboratory, India

Foreword by Prof. Anand Mohan

The field of nanoscience and nanotechnology has completely revolutionized the design and generation of novel materials whose properties can be tailored to suit a targeted application. It has shown a great impact on the production of new knowledge-driven products by industries which are quite relevant for socio-economic development. New avenues of nanotechnology have been initiated by physicists, chemists, molecular biologists, material scientists and technologists to coordinate their research strategies into synergistic approaches towards the exploitation of nanotechnology for viable solutions for mankind.

Oxide Nanostructures: Growth, Microstructures, and Properties covers the detailed discussion on metal oxide nanomaterials with physical growth and their applications and attempts to provide in-depth information on the synthesis of undoped and doped metals oxides nanostructures. It contains physical synthesis processes for thin films, porous and hollow nanostructures, paramagnetic lattice defects in crystal quartz, defect structure, space charge depletion layer and core shell model with wide applications in the field of science and technology.

I believe this book will serve as a valuable reference for students, scientists, engineers and specialists in both academia and industry concerned with the fundamental and technological/industrial applications of metal oxide nanostructures.

Prof. Anand Mohan,

Director, National Institute of Technology, Kurukshetra, India

Foreword by Dr. Simon J. Holland

Natural nanotechnology applications have always been around us, and synthetic nanomaterials are being developed across the globe. As advances in characterization techniques are made, we continue to learn about the properties of materials at the nanoscale, which in general terms is a particle size of approximately 100 nm in diameter or less, or some four times smaller than the wavelength of violet light. Here in 2014 it is significant to note that a large proportion of research in the fields of physics, chemistry and medicine is being devoted to developing an understanding of the physicochemical properties of existing and manufactured nanomaterials to enable the development of applications for use in everyday life.

One important family of nanomaterials are the metal oxides. Although applications for nano-sized titanium and zinc oxide are widely known in the scientific community and these initial developments are leading amongst many other applications to the provision of tooth-whitening agents and sunscreens as consumer products, a deep understanding of the structure–property relationships of metal oxides at the nanoscale is essential to enable us to utilize the entire family. This information is provided by this book that you are about to read, and the contents are a good reference for students, researchers and industries alike.

In Chapter 1, the surprisingly broad range of applications of nanoparticulate metal oxides is discussed. These include applications in electronics, catalysis and sensors. They also extend to medical applications such as imaging agents where there are already marketed products, and we are introduced to new materials beyond the commonly referred-to oxides of zinc and titanium to those of vanadium and cerium. The field of engineering is also explored with nanofluids of suspended iron oxides being developed as efficient heat exchanging agents. In addition, there is one area that relates to my own experience in the pharmaceutical industry, namely, the opportunity to improve separation sciences through the use of finely divided zirconium oxide as chromatography column packing agents.

Methods for the manufacture of nanoparticulate metal oxides are described, as are methods for characterizing these materials. This is a subject close to my heart in the field of nanotechnology, as in order to develop robust products we need discriminatory analytical methodologies to effect manufacturing controls for these materials and also to establish that the product remains in its nanoparticulate form throughout the life of the application.

Chapter 2 provides specific details on the manufacture of metal oxides, in particular methods for controlling their morphology, and Chapter 3 also focusses on the control of nanoparticle morphology, which is important for engineering materials.

Chapter 4 focusses on the control of material porosity and also the preparation of hollow structures to exploit potential sensor, catalysis, electronic and medical applications, and Chapter 5 describes developments in gas sensor materials.

Chapter 6 covers the use of coatings to improve biocompatability of surgical implants, which requires a complex mix of morphological considerations to provide materials that are fit for purpose.

Chapter 7 describes how metal oxide films can be utilised to harness energy from light sources. This is an area where a growing number of countries have shown a real interest as basic energy supply costs continue to increase, and we are reminded that the use of semiconductor films to generate electricity at source obviates expensive distribution networks.

Chapter 8 introduces us to the exciting field of nanobiotechnology, specifically to bio-imaging probes. It has been clear for some time that with cell activities taking place at the 200–300 nm scale, it was inevitable that the fields of nanotechnology and biotechnology would eventually coalesce. To assist with standardisation in this new field, the International Organization for Standardization Committee on nanotechnologies, ISO TC229, has created a Study Group on Nanotechnology and Biological Systems, and in the next five years, I am expecting this to be a challenging and also a productive area within ISO and also the other standardisation bodies.

Chapter 9 shows how the fundamental structures of established titanium and zinc oxide nanomaterials can be evaluated. This is an important area given the move to the formation of materials comprising very few layers of atoms.

Chapter 10 introduces us to electron spin resonance analysis for the dating of materials such as quartz through defect measurement. This has utility in geology and climate change studies, once again demonstrating that metal oxides are key to high-impact areas in the modern world.

Finally, Chapter 11 outlines how surface morphology of zinc oxide can be studied through electron paramagnetic resonance spectroscopy to elucidate the defect structures that control the semiconductor properties of this material. So, a whole, this volume provides the reader with the tools to manufacture and characterise nanoparticulate metal oxides for a plethora of applications.

It is pleasing to learn that further understanding of metal oxides and their potential applications in their nano-sized form has been achieved. Opportunities for this family of materials exceed those already established for zinc and titanium, and we are offered a clear view at how potentially useful nanoparticulate forms of metal oxides are.

Through careful development of methods to characterise the physical properties of these materials and also the establishment of standard protocols to evaluate their safety, a wide raft of new applications will become available for the human race. All we need to do is to marry scientific expertise with the public inquisitiveness about utilising new nanotechnology applications. Provided we can also meet the requirements of regulatory scrutiny by working directly with these agencies during the development cycle, this will help us exploit these versatile materials.

Dr. Simon J. Holland

Chairperson,
International Organization for Standardization Committee on
Nanotechnologies, ISO TC229

Preface

Nanomaterials, their synthesis, and their property studies have been an obsession with modern current physicists, chemists, and materials scientists for their vast array of technological implications and the remarkable way their properties are modified or enhanced when the size dimensions are reduced to the realm of nanometers. Although nanomaterials, for a lot of practical purposes, have been in existence since the remotest past of civilization, it is only in the last few decades that the field has gained the attention that it deserves from the scientific and industrial fraternity. A lot of this has to do with the immense improvements we made in tools to study and characterize these materials. Metal oxides have been one of the well-documented and hottest branches of nanomaterials revolution with oxides such as TiO_2, ZnO, CuO, Fe_3O_4, Cr_2O_3, Co_3O_4, MnO_2, and many more being an integral part to a variety of technological advancements and industrial applications. From green power issues such as photovoltaic cells to rechargeable batteries, from drug delivery agents to antimicrobial and cosmetic products, from superconductor materials to semiconductors and insulators, metal oxides have been omnipresent in terms of both commercial prerogatives and research highlights. This book is solely devoted to this special section of nanomaterials with an aim to partially access the science pertaining to the oxides of metals. Quite aptly, the book opens with an introductory chapter that overviews the research activities in this field with its mood inclined toward both the beginners and experienced metal oxide researchers. The following chapters encompass the various corners of metal oxides such as specific synthesis methodologies (e.g., pulsed laser deposition (Chapter 2)), specific morphology processing (e.g., porous and hollow metal oxide nanostructures (Chapter 4)), specific property studies (e.g., phase selection and plasticity (Chapter 3), defect studies (Chapters 10 and 11)), various application purposes (e.g., gas detection (Chapter 5), bio-implants (Chapter 6), photovoltaic applications (Chapter 7), bio-imaging

(Chapter 8)), and theoretical studies (e.g., band energy (Chapter 9)). The chapters not only are dominated by the status of the contemporary research works related to metal oxides but also try to envision the future directions of this field.

It is our sincere belief that this book will prove to be a formidable source of instructive essays for scientists, technologists, teachers, and students from the corresponding fields.

I would love to take this opportunity to express my deepest gratitude to people without whose contributions this book would not have materialized. I acknowledge Director General, Council of Scientific & Industrial Research, and Director, National Physical Laboratory (NPL), for providing all scientific, financial and infrastructural support during the writing and subsequent compilation of the scientific contents of the book. I am grateful to all the authors of the book who have put their best efforts in contributing their respective chapters. I am indebted to my mentors and collaborators, who provided me with tremendous moral support toward this exhaustive endeavor, and my NPL colleagues and students, who are always enthusiastic for various discussion purposes. Last but not the least, I thank my family members and relatives for their constant generous well wishes and extraordinary level of understanding in always making me comfortable so that the best output could be generated at the work.

Dr. Avanish Kumar Srivastava

CSIR—National Physical Laboratory, India

Chapter 1

Metal Oxide Nanomaterials: An Overview

Kajal Kumar Dey and Avanish Kumar Srivastava

CSIR—National Physical Laboratory,
Council of Scientific and Industrial Research,
Dr. K. S. Krishnan Road, New Delhi 110012, India

aks@nplindia.org

This chapter is solely intended for those fascinated by the world of nanoscaled objects and particularly besotted by the metal oxides. It attempts to acquaint its readers with the basics of metal oxide nanomaterials—its various synthesis procedures, characterization methods and applications. The authors have tried to feed as much information as possible, although due to the limited scope, an extremely elaborative depiction of the various topics has never been their objective.

1.1 Initiation

We are in the midst of an exciting and visionary journey toward technological resplendence: an expedition into very small space,

Oxide Nanostructures: Growth, Microstructures, and Properties
Edited by Avanish Kumar Srivastava

ISBN 978-981-4411-35-6 (Hardcover), 978-981-4411-36-3 (eBook)
www.panstanford.com

led by savvy scientists and virtuoso engineers from various disciplines who are pooling their talents to establish a new field—"nanotechnology." We still have an obscured vision about the final outcome of this journey, whether it is going to lead us to a land of myriad opportunities or to a landscape of despair. However, so far, the journey has produced hopes of materializing dreams, which a few decades ago would have been stuffs for the science fiction literatures. No wonder the "nanotechnology mania" has been sweeping all fields of science and engineering and making more people take note of the prophecy made by Christopher Altman about an "omni-linked world populated with intelligent artifacts will bring sweeping changes to virtually every facet of modern life—from science and education to industry and commerce—leaving no segment of society unaffected by its advance [1]." The term nanotechnology was first applied in 1974 by Prof. Norio Taniguchi of Tokyo Science University, describing traditional silicon machining down to regions below one micron. The ability to work at the molecular level, atom by atom, and construct large structures with fundamentally new properties and functions are at the core of nanoscaled science, engineering, and nanotechnology [2]. Although the use of nanomaterials can be traced back to the Roman period, it was not until 1959 that a lecture by physicist Richard P. Feynman at the annual meeting of the American Physical Society held at the California institute of Technology caused the inception of the concept of nanoscience as we know today. In his introduction to the topic of his lecture, he said "I would like to describe a field in which very little has been done, but in which an enormous amount can be done in principle. This field ... will not tell us much of fundamental physics but ... it might tell us much of great interest about the strange phenomena that occur in complex situations. What I want to talk about is the problem of manipulating and controlling things on a small scale. It is staggeringly small world that is below. In the year 2000, when they look back, they will wonder why it was not until the year 1960 that anybody began to move to this direction" [3]. However, that lecture was by and large forgotten by the immediate scientific fraternity and even Feynman's own biography does not mention the "room at the bottom" talk. It was not until 1986, when K. Eric Drexler published *The Engines of Creation: The Coming Era of Nanotechnology*, in which he discussed his ideas about nanotechnology in a way that was comprehensible to the general public [2]. This was the first attempt by anyone to define nanotechnology as an independent field of science.

Today nanotechnology is still in a formative phase—not unlike the condition of computer science in the 1960s or biotechnology in the 1980s. Yet it is maturing rapidly. Between 1997 and 2005, investment in nanotech research and development by governments around the world soared from $432 million to about $4.1 billion, and corresponding industry investment exceeded that of governments by 2005. By 2015, products incorporating nanotech will contribute approximately $1 trillion to the global economy. About two million workers will be employed in nanotech industries, and three times that many will have supporting jobs—not for nothing has it been referred as the little big science [4]. Nanomaterials and devices based on them-the nanodevices, are the fundamental units of nanotechnology revolution. Nanomaterials research has now reached a well-deserved maturity. Ample summaries, reviews, and books document the accumulated information on various types of nanomaterials, their potential and existing applications, synthetic strategies, structural analysis are easily available. This brings us to the basic question: What are these nanomaterials and what makes them such a distinguished breed of materials compared with the macroscopic bulk materials?

1.2 Orientation with the Nanomaterials

Nanomaterials (Greek *Nanos* = dwarf, although in reality nano-particles are infinitesimally smaller than the dwarfs) are functional materials with at least one of the characteristic dimensions measured within 100 nm. However, there are still some discussions about the size of nanoparticles. It has been told that the actual size of nanoparticles should be less than 100 nm. On the other hand, it has also been said that the size must be less than 10 nm. Moreover, some sources classify the size difference as first and second-generation nanoparticles. It is accepted that the first generation nanoparticle size is less than 100 nm, second-generation size is less than 10 nm, the lower limit for the size of nanoparticles has been told to be approximately 1 nm. Just to have a conceptual feeling regarding how small these nanoparticles are it is useful to compare nanoparticles sizes with other minuscule entities in nature, such as bacteria, viruses, and even the human hair. Bacteria are huge when compared to the nanoparticles and it is helpful to note that the equivalent volume of some bacteria can accommodate a million of 5 nm. The human hair is about five magnitudes larger than the regular nanoscaled objects.

The simple presence of nanoscale structure alone is not sufficient to define a nanomaterial, since most if not all materials have structure in this range. The ability to control the structure at this scale is essential. A typical nanomaterial is more often formed of several small "primary particles." The size of the nanoparticle usually refers to the secondary particle size. Primary particles agglomerate into these secondary particles. The primary particles have a very high surface-to-volume ratio. The relative number of particles on the surface and at the grain boundary is a significant parameter for the nanoparticles for they dominate the change these particles display in their respective electronic, electrical, magnetic, optical, mechanical and chemical properties with respect to the bulk materials. The smaller the dimensions of the primary particles, the more the proportion of these particles reside on the surface and at the grain boundaries. Consider for example, a 3 nm iron particle has 50% of its total atoms residing on the surface. A 10 nm iron particle has 20% of the total atoms residing on the surface, whereas a 30 nm iron particle has only 5%. Even the surface energies and surface morphologies are dependent on the size and consequently translate to enhanced intrinsic surface activities. In fact, the size range of the nanoparticles may be more accurately defined from 1 nm to the largest nanometer dimension where the particle ratio on the surface and at the grain boundaries to the center is large enough to cause the significant deviations in the aforementioned properties compared to their bulk counterpart. Thus, even though the popular upper limit for nanoparticles dimension has been defined as 100 nm, in a more accurate definition this limit may well vary depending upon the sizes of the atoms and molecules in question. From the electronic band structure point of view, it is clear that the energies of the electronic states depend on $1/l^2$ where l is the dimension of the system in that particular direction; the spacing between successive energy levels also varies as $1/l^2$. This behavior is also clear from the description of a solid as a giant molecule: as the number of atoms in the molecule increases, the molecular orbitals (MOs) gradually move closer together. Thus, if the number of atoms in a system, and hence the length scale, is substantially different from that in a normal bulk material, the energies and energy separations of the individual electronic states will be very different. Although in principle the Fermi level would not be expected to change since the free electron density should remain constant, there may be associated modifications in structure, which will change this quantity. Furthermore, as the system size decreases, the allowed energy bands become substantially narrower than in an infinite solid. The normal collective (i.e., delocalized) electronic properties of a solid become severely distorted and the electrons in a reduced-dimensional system tend to behave more like the "particle in a box" quantum system; this is the phenomenon of quantum confinement. In other words, the electronic states are more like those found in localized molecular bonds rather than those in a macroscopic solid. From the standpoint of basic sciences, chemistry deals with atoms and molecules, a region of matter with dimensions generally below 1 nm. Condensed matter physics deals with solid materials comprising of an infinite array of bound atoms and molecules of dimensions commonly above 100 nm. Nanomaterials reside somewhere in a region in between where neither the quantum chemistry nor the laws of classical physics hold. In this realm of dimensions a host of new properties evolve and some of the properties vary markedly from their bulk counterpart (e.g., bandgap, melting point, magnetic and optical properties, etc.).

Based on the nature of confinement, nanomaterials can be classified as quasi 1D (system confined in two dimensions, e.g., nanowires, nanorods, nanotubes, nanobelts, nanofilaments, etc.), quasi 2D (system confined in one dimension, e.g., nanoplatelets, nanodiscs, thin films, etc.) and quasi 0D (system confined in all three dimensions, e.g., quantum dots or nanodots). They can be crystalline or amorphous; they can be metallic, organic, polymeric, composites, or any other pure inorganic materials such as silicates, sulfides, and oxides.

Speaking of oxide nanomaterials, the metal oxide nanomaterials, particularly those of transition metals are highly sought after for their widest and most fascinating range of properties among any single class materials. For example, oxide nanomaterials, unlike the other class of nanomaterials, display an increase in lattice parameters in comparison with the bulk. The metal elements with their varying valencies can form a large diversity of oxide compounds processed through suitable synthesis methodologies. Their availability in various shapes and sizes and applications in diversified fields of industrial and technological necessities have channelized the metal oxide nanomaterials to be as much a hot field of research today as carbon nanotubes (CNTs) and graphene. In fact, the surge in the interest level in metal oxides can be well documented by the following plot, which gives a relative idea about the number of papers published on two of the most important metal oxides (TiO_2 and ZnO) during the last 15 years.

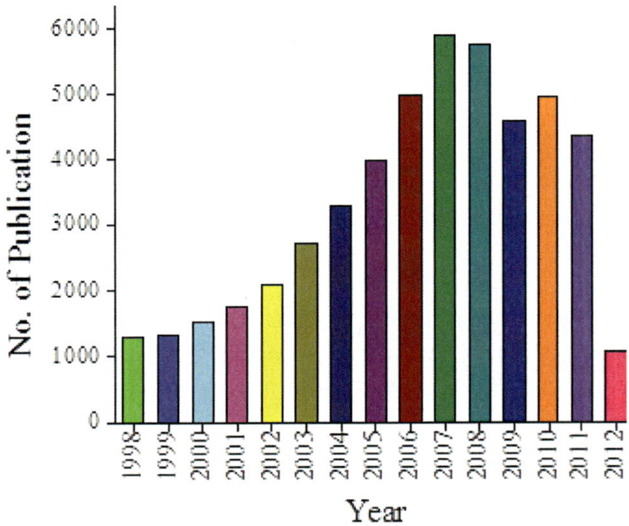

Number of publications of ZnO and TiO_2 combined from 1998 to March 2012 (Data retrieved from ISI Web of Science).

Metal oxides exhibit metallic, semiconducting, or insulating character due to the electronic structure difference. Normally, oxides formed by metals residing at the left or right of the periodic table are generally insulating in nature (e.g., MgO, CaO, Al_2O_3, SiO_2), whereas those formed by the metals residing at the middle of the periodic table are semiconductors or metallic in nature (e.g., ZnO, NiO, TiO_2, CuO, Fe_2O_3, Cr_2O_3). A chart pointing out the nature of various metal oxides in terms of conductivity property is provided in Table 1.1. The variety of attributes of oxides enables the wide applications in the fabrication of microelectronic circuits, sensors, piezoelectric devices, fuel cells, coatings against corrosion, and as catalysts. For example, almost all catalysts involve an oxide as active phase, promoter (or support), which allows the active components to disperse on. In the chemical and petrochemical industries, products worth billions of dollars are generated every year through processes that use oxide and metal/oxide catalysts. For the control of environmental pollution, catalysts or sorbents that contain oxides are employed to remove the CO, NOx, and SOx species formed during the combustion of fossil-derived fuels. Furthermore, the most active areas of the semiconductor industry involve the use of oxides. Thus, most of the chips used in computers contain an oxide component. Research in this field has been immensely served by the fact that a lot of these materials had already been studied vividly as their bulk form before interest in the nanoscaled structures took off and thus paved a way for comparative data between the bulk and nano forms.

Table 1.1 Some metal oxides, their position in the periodic table, electronic bandgap, and conducting nature

Name	Position of the metal in the periodic table and nature	Bandgap (/eV)	Classification
MgO	Group 2 (III); alkaline earth	7.8	Insulator
CaO	Group 2 (IV); alkaline earth	7.7	Insulator
CeO_2	Group 3 (VI); lanthanide	3.2	Semiconductor (n-type at low pressure; p-type at high pressure)
TiO_2	Group 4 (IV); transition metal	3.2 (anatase), 3.0 (rutile), 2.96 (brookite)	Semiconductor (n-type)
ZrO_2	Group 4 (V); transition metal	5.2–5.8	Insulator
VO_2	Group 5 (IV); transition metal	0.5–0.7	Semiconductor (n-type)
V_2O_5	Group 5 (IV); transition metal	2.2–2.3	Semiconductor (n-type)
Cr_2O_3	Group 6 (IV); transition metal	3.3–3.4	Semiconductor (p-type)

(Continued)

Table 1.1 (*Continued*)

Name	Position of the metal in the periodic table and nature	Bandgap (/eV)	Classification
WO_3	Group 6 (VI); transition metal	2.6–3.1	Semiconductor (n type)
MnO	Group 7 (IV); transition metal	4.1	Semiconductor (n-type)
Mn_3O_4	Group 7 (IV); transition metal	2.5	Semiconductor (p-type)
MnO_2	Group 7 (IV); transition metal	β 0.26 γ 0.58–0.7	Semiconductor (n-type)
FeO	Group 8 (IV); transition metal	2.4–2.5	Semiconductor (p-type)
Fe_3O_4	Group 8 (IV); transition metal	0.1	Metallic
Fe_2O_3	Group 8 (IV); transition metal	α 2.2 γ 2.0	α = Semiconductor (n-type) γ = Semiconductor (n-type)
RuO_2	Group 8 (V); transition metal	2.2	Semiconductor (Amphoteric)
CoO	Group 9 (IV); transition metal	2.4	Semiconductor (p-type)
Co_3O_4	Group 9 (IV); transition metal	2.0–0.2	Semiconductor (p-type)
NiO	Group 10 (IV); transition metal	3.6–4.2	Semiconductor (p type)
CuO	Group 11 (IV); transition metal	1.2–1.8	Semiconductor (p type)
Cu_2O	Group 11(IV); transition metal	2.1–2.2	Semiconductor (p-type)
ZnO	Group 12 (IV); poor metal	3.3–3.4	Semiconductor (n type)
CdO	Group 12 (V); poor metal	2.2–2.9	Semiconductor (n type)
Al_2O_3	Group 13 (III); poor metal	6.0–8.8	Insulator
Ga_2O_3	Group 13 (IV); poor metal	β 4.7–4.9	Semiconductor (n-type)

Name	Position of the metal in the periodic table and nature	Bandgap (/eV)	Classification
In_2O_3	Group 13 (V); poor metal	3.6 (direct), 2.6 (indirect)	Semiconductor (n-type)
SnO_2	Group 14 (V); poor metal	3.6–3.9	Semiconductor (n-type)
ITO	Group 13, 14 (V)	4.0	Semiconductor (n-type)

In this chapter, we propose to discuss briefly some structural properties, various synthesis strategies for synthesizing the metal oxide nanomaterials and their subsequent characterization techniques. However, first let us take a look at why this particular species of material has become such an integral part of modern nanomaterials research.

1.3 Metal Oxide Nanomaterials: Why Have They Become Indispensable?

Metal oxide nanomaterials have become an integral part of technological advancement through the sheer magnitude of fundamental value they generate. From medical science to chemistry, from material science to physics or be it hardcore industrial applications, these oxide nanomaterials have become indispensable from every aspect of consideration. Below are some of the applications of these materials that have become popular across different industrial and technological and biological fraternities.

1.3.1 Photocatalytic Activity

The principle of the semiconductor photocatalytic reaction is straightforward. Upon absorption of photons with energy larger than the bandgap of the corresponding catalyst material, electrons are excited from the valence band to the conduction band, creating electron–hole pairs. These charge carriers migrate to the surface and react with the chemicals adsorbed on the surface to decompose these chemicals. A large surface area with a constant surface density of adsorbents leads to faster surface photocatalytic reaction rates.

In this sense, the larger the specific surface area, the higher the photocatalytic activity is. Naturally, nanomaterials with their high surface-to-volume ratio have been highly explored as photocatalyst. Since Fujishima and Honda reported the evolution of oxygen and hydrogen from titanium oxide electrode in the electrolyte cell under irradiation of light [5], photocatalysis has been regarded as one of the most effective ways to solve these problems. Various photocatalysts, especially metal oxide photocatalysts are promising materials for degradation of organic pollutants by utilizing UV or solar light. Examples of some of the most effective metal oxide photocatalysts are as follows:

TiO_2: TiO_2 is regarded as the most efficient and environmentally benign photocatalyst and it has been widely used in the photodegradation of various organic pollutants [6–10]. Not only pure TiO_2 but the doped (metal and nonmetal) TiO_2 materials have also been found to be extremely responsive as photocatalyst [11–13]. Various structures such as mesoporous nanorods and nanotubes have been found particularly suitable for photocatalysis [14,15].

ZnO: As to photocatalysis, ZnO is more efficient than TiO_2, the most intensively studied photocatalyst, in the photodegradation of some organic compounds [16–20]. However, the application of ZnO as photocatalyst was limited due to its photoinstability in aqueous solution. Research works are going on to improve its aqueous photostability without compromising on its efficiency [21].

WO_3: Platinized WO_3 can be used as an environmental photocatalyst. Studies have shown this material to be an effective visible live photocatalyst for organic pollutants degradation [22,23]. WO_3 has also been reported to photo catalytically generate H_2 gas by the reformation of methanol [24]. Cs modified WO_3 surface has shown excellent results in photocatalytic O_2 evolution and Fe^{3+} reduction [25].

1.3.2 Photovoltaic Application

In photosensitive devices such as dye-sensitized solar cells (DSSCs), thick and porous nanostructured material layers are also required to provide sufficient interaction volume for impinging photons. Thick film layers facilitate large surface area for the adsorption of dye sensitizer, assisting in the fabrication of the DSSCs. In producing DSSCs, researchers have reported nanostructured metal oxides being used as the photoanode material. In such types of solar cells, photons with sufficient energy excite dye molecules, and this is followed by a rapid injection of electrons into the conduction band of the metal oxide semiconductor. So far, the most efficient in these regard has been TiO_2, although research on other alternative metal oxides are going on. Below are some of the examples.

TiO_2: TiO_2 can absorb light in the visible region and can convert solar energy into electrical energy for photovoltaic applications. An overall solar to current conversion efficiency of 10.6% was reached by the group led by Grätzel with DSSC technology [26]. Zukalova et al. found that ordered mesoporous TiO_2 nanocrystalline films showed enhanced solar conversion efficiency by about 50% compared to traditional films of the same thickness made from randomly oriented anatase nanocrystals [27]. Various other shapes of titania such as nanotubes, core shell structures along with hybrid materials they have all been reported to exhibit promising solar cell electrode capability [28–30].

Cu_2O: Cuprous oxide has been one of the earliest metal oxide ceramics to be utilized in solar energy conversion. Various types of solar cells such as Schottky barrier solar cell (prepared by evaporating a metal on Cu_2O at high vacuum; e.g., Cu/Cu_2O) [31], heterojunction solar cell (obtained by depositing a p-type semiconductor material on Cu_2O; e.g., ZnO/Cu_2O) [32], homojunction solar cells (p–n homojunction of Cu_2O solar cells) [33,34] have been obtained from cuprous oxide and found to be quite efficient for photovoltaic purposes.

WO_3: Nanostructured thin films fabricated using WO_3 nanoplatelets have been reported to be employed in developing DSSCs [35,36].

ZnO: ZnO is a candidate for use in the photoanode of DSSCs due to its direct wide-bandgap of 3.37 eV and high electron mobility [37,38]. The bandgap and the conduction band energies of ZnO are similar to TiO_2, the common photoanode material, but it has a higher mobility than TiO_2.

SnO_2: SnO_2 possesses certain inherent advantages such as higher electron mobility (100–200 $cm^2\ V^{-1}\ S^{-1}$) and more positive conduction band edge position than TiO_2. The high electron mobility leads to faster transport of photoinjected electrons to a transparent conductive oxide current collector while the more positive conduction band edge position facilitates electron injection from photoexcited dye molecules, especially in low bandgap organic sensitizers [39,40]. Furthermore, in the case of SnO_2, its larger bandgap (3.6 eV), as compared with that of TiO_2 (3.2 eV), creates fewer oxidative holes in the valence band under UV illumination than are created in the case of TiO_2, thereby minimizing the dye degradation rate and improving the long-term stability of DSSCs [41]. Surprisingly, the performance of DSSCs based on SnO_2 photoanodes is much less than those consisting of TiO_2. The inferior photovoltaic properties of a SnO_2 photoanode DSSC are attributed to the faster electron recombination kinetics resulting from a 300 mV positive shift in the conduction band edge and poor dye uptake associated with the low isoelectric point [42]. These issues have been recently overcome by engineering the SnO_2 surface with conformal barrier layers using materials such as TiO_2 or Al_2O_3 [43].

NiO: There have been reports of NiO being used in solar cells both in organic photovoltaic (OPV) and in DSSCs [44,45]. Yang et al. reported that NiO/Graphene composite could be used as photocathode in p-type DSSCs [46].

1.3.3 Catalysis

The term catalysis was introduced by Berzelius way back in 1836 to categorize certain chemical reactions whose progress can be fastened or facilitated by the addition of certain materials that are not part of the reaction products [47]. Heterogeneous catalysis, which is a surface adsorption process, requires high concentration of catalytic sites. In nanostructured materials because of their high surface-to-volume ratio and intrinsic surface defects catalytic activity is normally much higher compared to the bulk materials. Various metals and metal oxides in the nanoregime have been well exploited as either a catalyst or a catalyst support for various important reaction and transformations. Some of the most studied metal oxides in this regard are as follows:

CeO_2: CeO_2 plays a very important role in heterogeneous catalyst, both as active material and support. First, ceria exhibits a particularly low formation energy for oxygen vacancies, which renders lattice oxygen available in chemical reactions and also, ceria is a good electron acceptor, as the initially empty Ce 4f states can be filled up with electrons, thereby reducing the charge state of respective Ce ions from +4 to +3. The latter effect facilitates electron transfer from suitable adspecies into the oxide surface, promoting the chemical activity of the material [48–52].

Ga_2O_3: Gallium oxides have been reported to be a catalyst or cocatalyst for a number of heterogeneous catalytic reactions, which involve the formation and breakage of H-containing bonds, such as water splitting [53–55], steam reforming of methane [53], water gas shift reaction [56], reduction of carbon dioxide by methane [57], and decomposition of volatile aromatic compounds [58].

Co_3O_4: This material has been extensively explored as a potential heterogeneous catalyst for CO oxidation [59,60], Oxygen reduction reaction catalyst when coupled with graphene nanosheet [61], methane combustion [62], conversion of NO to NO_2 [63] and Oxygen evolution reaction [64].

ZrO_2: Zirconia has been widely exploited as a catalyst support material for various metallic catalysts. Wang et al. studied the CO oxidation reaction on Au/ZrO_2 catalyst and the effect of the crystal phase of Zirconia (tetragonal and monoclinic) [65]. Song et al. reported Zirconia as a support for Ni catalyst for the partial oxidation of methane [66]. Zhang et al. observed a remarkable nano-size effect of zirconia as a catalyst support in CO oxidation by Au [67].

RuO_2: Ruthenia has been demonstrated to be a really good oxidation catalyst. Ruthenia-based nanomaterials have been explored as a catalyst or catalyst support material for catalytic oxidation of ethanol [68], oxidation of CO and methanol [69–71].

MnO_2: MnO_2 is an excellent catalyst for volatile organic compound oxidation. Sui et al. reported both α and β MnO_2 nanowires showing excellent catalytic properties in Fenton-like reaction [72]. Liang and coworkers correlated crystallographic phase of MnO_2 nanoparticles with their catalytic activity in terms of CO oxidation [73]. Wang et al. utilized β MnO_2 as a catalyst support for Au for the aerobic alcohol oxidation [74]. Yu et al. reported oxidative decomposition of formaldehyde using Pt/MnO2 nanostructured catalyst [75]. Cheng et al. reported electrochemical oxygen reduction reaction in alkaline media using MnO_2 nanostructured catalysts [76].

Fe_2O_3: The catalytic properties of Fe(III) oxides are widely used in chemical industries. Iron oxides are one of the catalytic components in the manufacture of styrene by the dehydrogenation of ethylbenzene [77,78], in Fischer–Tropsch synthesis of hydrocarbons [79,80] or in Super Claus catalysis [80]. They have been proved effective catalysts for the selective oxidation of polyaromatic hydrocarbons [81], catalytic burning of fuels [82], coal liquefaction [83], and vapor-phase oxidation of benzoic acid to phenol [84].

Cu_2O: Cu_2O is an efficient catalyst in the partial oxidation of olefin compounds. It can convert propylene into acrolein [85].

CuO: CuO can be used as a catalyst in the complete oxidation of olefin compounds. It oxidizes propylene to CO_x and other by products [86].

NiO: reported NiO being employed as a catalyst in the pyrolysis or gasification of biomass [87,88].

WO_3: This material has been reported to be successful as a catalyst in producing H_2 through the steam reformation of methanol [89].

1.3.4 Sensing Applications

Semiconductor transition metal oxide gas sensors are being widely used nowadays to detect various air pollutants. Transition metal oxides are the most promising candidates for gas-sensing materials because of their low preparation cost, high stability, and high compatibility with microelectronic processing. The gas sensing generally involving the gas transport in the materials and the subsequent surface reaction between the gas molecules and the sensing particles suggests that principally the structures that suit gas transport such as hollow structures, porous structures, and hierarchical structures exhibit good gas-sensing properties. Their sensing property is mostly based on the simple principle of change of resistivity when the gas interacts with the surface of the material. One of the drawbacks of using these metal oxide materials as gas sensors has been their lack of selectivity in most cases. Examples of some of the most widely used metal oxide as sensing purposes are as follows:

SnO_2: Tin dioxide is one of the most promising and widely explored gas-sensing materials [90–92]. Song et al. reported good sensing properties of hierarchical SnO_2 nanostructures for reducing gases such as ethanol and acetone [93]. Huang et al. used nanotubular SnO_2 to detect H_2, CO and ethylene oxide [94].

ZnO: Extensive efforts have been made to fabricate various kinds of chemical and biochemical sensors based on ZnO nanostructures, such as fluorescent biosensors with nanoscale

ZnO platforms [95,96], H_2S gas sensor with single ZnO nanowire [97], intracellular pH sensor with ZnO nanorod [98] and ethanol sensor with flowerlike ZnO nanostructure [99]. ZnO nanotube arrays have been shown to be biosensors for glucose [100].

TiO_2: TiO_2 nanocrystalline films have been widely studied as sensors for various gases. Grimes et al. conducted a series of excellent studies on sensing using TiO_2 nanotubes [101,102]. They found that TiO_2 nanotubes were excellent room-temperature hydrogen sensors not only with a high sensitivity of 10^4 but also with an ability to self-clean photoactively after environmental contamination [103]. Oxygen sensors based on TiO_2 nanomaterials include TiO_{2-x} [104], TiO_2–Nb_2O_5 [105], CeO_2–TiO_2 [106]. TiO_2 nanomaterials are promising candidates for CO sensing [107] and for methanol and ethanol sensing [108,109].

CuO: There have been numerous reports about CuO nanostructures, of various shapes, both porous and non-porous being utilized as a high performing gas-sensing material for hazardous gases like H_2S, CO, NO_2, glucose, aldehydes, organic solvents like ethanol, acetone [110–116].

In_2O_3: This material has been widely exploited as a sensing material for hazardous gases like CO, formaldehyde, NH_3, H_2S and NO_2 [117–121].

WO_3: WO_3 nanocrystalline films have been established as one of the best gas sensors for the reducing gases such as NO_2, H_2, and CO [122]. ε-WO_3 has been reported to possess acetone detection ability [123]. Porous WO_3 nanospheres have shown impressive ability to sense ethanol, CS_2, benzene, ether, etc. [124–127].

VO_2: Based on its metal to insulator transition property vanadium dioxide has been explored as an effective and selective sensing material for H_2 [128,129].

ZrO_2: Zr/ZrO_2 electrode is a suitable pH sensor for the hydrothermal systems at midocean ridge or other geothermal systems with the high-temperature environment [130]. Zirconia-based composites have been explored as ethanol sensors based on their chemiluminescent properties [131].

1.3.5 Li-Ion Batteries

Rechargeable Li-ion batteries have been a primary component of portable entertainment, computing and telecommunication devices for today's information rich mobile society. Researchers for quite some time now have been looking for suitable replacements for graphite, the normally used anode material in these batteries because of the low gravimetric capacity (~370 mAh g^{-1}) of carbon. Back in 2000, Poizot et al. had reported that transition metal oxide nanoparticles could be an excellent substitute for constructing the negative electrodes of these batteries [132]. They reported that electrodes made of metal oxide nanoparticles (MO, where M = Co, Ni, Cu or Fe) demonstrated electrochemical capacities of 700 mAh g^{-1}, with 100% capacity retention for up to 100 cycles and high recharging rates. Since then several metal oxide nanoparticles have been reportedly explored as options to be used as anode materials. Following are some of them:

VO_2: VO_2 is an attractive material for Li-ion batteries. VO_2 (B) exhibits a maximum reversible capacity 320 $mAhg^{-1}$ in the range 4 to 1 V in Li cells [133,134]. This "B" phase has been exploited as the cathode material in Li-ion batteries [135,136].

CuO: Various reports confirm CuO nanoparticles to be precious as the anode material in Li-ion batteries with high capacity and excellent retention [137–139].

Co_3O_4: Li et al. reported Co_3O_4 nanotubes, nanoparticles, and nanorods to show a capacity of above 800 mAh g^{-1} for all of them with the nanotubes having the best result [140]. Lin et al. reported a first discharge capacity of ~1100 mAh g^{-1} for mesoporous cobaltite nanoparticles and the discharge capacity was found to be improving with the lowering particle sizes [141].

1.3.6 Capacitors

The new-age lightweight, flexible, and even wearable electronics require lightweight and highly efficient energy management technology. These demands are often not met by the conventional charge storage devices such as batteries, which have limitations like short cycle life and relatively slow charging/discharging rates. Electrochemical capacitors (ECs), another state of-the-art charge storage technique, also known as supercapacitors, have attracted much attention due to their higher power density, fast charging/discharging rate, sustainable cycling life (millions of cycle), longer life cycles than batteries, and higher energy density than dielectric capacitors. Much effort has been dedicated toward the fabrication of carbon-based supercapacitors and it has been well established that decorating their surface with pseudocapacitors such as conducting polymers and several metal oxides can further enhance their electrochemical performance. A number of transition metal oxides have been investigated in this regard. Following are some of them:

MnO_2: Manganese dioxide nanomaterials have found widespread research focus as a material suitable for supercapacitor application [142–147]. The theoretical specific capacitance of MnO_2 is 1370 F/g [148]. However, due to its poor electric conductivity (10^{-5} to 10^{-6} S/cm) such high theoretical capacitance has not been achieved in experiments. Some high-performance results have been reported only from MnO_2 nanometer-thick thin films and/or nano-sized particles. Hou et al. reported a ternary nanocomposites film of MnO_2, CNT and conducting polymer to show a high specific capacitance of 427 F/g [149]. Li et al. reported mesoporous MnO_2 and carbon aerogel composite to show a specific capacitance of 515.5 F/g [150]. Devraj et al. reported that the capacitance depended strongly on the crystallographic nature of MnO_2 with the α phase being the best and the β phase the lowest [151].

Co_3O_4: Electrodes developed with Co_3O_4 nanonetwork structure have reportedly shown high specific capacitance for supercapacitor [152]. Zhu et al. showed Co_3O_4 nanoneedle incorporated electrodes could have a pseudo capacitance of 111 F g^{-1} [153].

RuO_2: $RuO_2 \cdot xH_2O$ is the most promising electrode material for electrochemical supercapacitors due to its intrinsic reversibility of surface redox couples [154]. The theoretical value of $RuO_2 \cdot xH_2O$ nanoparticulates was estimated to range from ca. 1300–2200 F g^{-1} and the maximum utilization of $RuO_2 \cdot xH_2O$ within carbon–RuO_2 composites was found to reach as high as 1580 F g^{-1} [155,156]. Hu et al. reported ultrahigh specific capacitance of ca. 1300 F g^{-1} and excellent charge/discharge behavior at 1000 mV s^{-1} obtained from $RuO_2 \cdot xH_2O$ nanotubes [157]. Zang et al. prepared cone-shaped polypyrrole composite of RuO_2 and its specific capacitance was reported to be 302 G g^{-1} [158]. Nanocomposite of CNT and RuO_2 nanoparticles (with a mass ratio of 7:6) was reported to show a specific capacitance of 953 F g^{-1} [159]. However, it is not widely used yet due to the high processing cost, narrow operating voltage (~1 V) rarity of Ru and the toxic nature of RuO_2, despite being the best supercapacitor material in terms of high specific capacitance (1580 F g^{-1}) and reversibility [157].

Mn_3O_4: Manganese oxide thin films show excellent pseudocapacitive behavior with large specific capacitance. Dubal et al. reported Mn_3O_4 electrode to show specific and interfacial capacitances of Mn_3O_4 electrode were 193 and 0.11 F cm^{-2} [160].

NiO: NiO is a versatile material for supercapacitor applications due to its higher surface area, good redox and charge storage property, controllable size, shape, and structural characteristics. There are several reports of NiO nanomaterials as a potential pseudocapacitance material [161–163].

1.3.7 Biophysical Functionalities

Due to their special electrical, optical, and magnetic properties, metal oxides less than 100 nm in size are very promising for biosensors, bio-separation, and drug delivery. Following are some examples:

Iron Oxides: Iron oxides are perhaps the most explored nanoparticles for bio-related research activities due to their special magnetic, structural, and electrical properties and non-toxic nature. Superparamagnetic iron oxide nanoparticles with appropriate surface chemistry can be used for numerous in vivo applications, such as MRI contrast enhancement, tissue repair, immune-assay, detoxification of biological fluids, hyperthermia, drug delivery, and cell separation. Molecular imaging is of course one of the most promising applications of targeted iron oxide nanoparticles [164]. Various applications using targeted iron oxide nanoparticles have been evaluated in Vitro and in animal experiments. The first cellular imaging studies were performed with unfunctionalized iron oxide nanoparticles for labeling leukocytes, lymphocytes, etc. [165–167]. Another successful application of iron oxide nanoparticles in MRI is specific cell tracking. The ability to load enough magnetic particles (micromolar Fe concentration) in cell culture via cell-permeable peptide or transfection agents in combination with the negatively charged surface of magnetic particles has provided a useful technique to label and track cells in vivo by MRI [168,169]. Another important kind of application of iron oxide nanoparticles is the functionalization for in Vitro protein or cell separation [170]. Kohler et al. have reported the development of a biostable methotrexate-immobilized iron oxide nanoparticles drug carrier that may potentially be used for real-time monitoring of drug delivery through magnetic resonance imaging [171]. Ferrofluids are not only a very powerful material for diagnosis by MRI, but they can be used also for therapeutic purposes [172]. New approach for localized thermotherapy induced by a magnetic fluid is already suitable for both hyperthermia and thermo-ablation. Evaluation of the feasibility and survival benefit of this new hyperthermia approach is in progress on animals, and first clinical trials have been started recently [173,174].

ZnO: Although applications of ZnO nanostructures for DNA-related experiments are still rare, functionalized tetrapod-like ZnO nanostructures have been reported (1) as a solid-phase adsorbent for plasmid DNA purification, (2) as improving reagents for polymerase chain reaction (PCR), and (3) as novel carriers for mammalian cell transfections [175,176]. The availability of ZnO nanoparticles coupled with its facile biomodification and vary low cytotoxicity makes it a promising bio-application material.

ZrO_2: ZrO_2 toughening ceramics are used as the alternate materials of thighbone, oral planting materials [177].

CeO_2: Nano-ceria has attracted a lot of attention due to its SOD (superoxide dismutase) mimetic activity and their reversibility and auto-generative properties, which can be helpful of protecting the cells from oxidative damage [178].

1.3.8 Nanofluid

In the development of efficient heat transfer equipment, the thermal conductivity of the heat transfer fluid plays a vital role. Traditional heat transfer liquids such as water, oil, and ethylene glycol are inherently poor heat transfer fluids. Therefore an attempt of improving their heat transfer ability or thermal conductivity is a topic of current industrial interest. Nanoparticles have been dispersed in these fluids (nanofluids) and are shown to vastly improve the thermal conductivity of these liquids. Although so far the majority of the nanoparticles that have been used to prepare nanofluids are metallic or polymeric or composite types, researchers across the world have been trying to utilize metal oxide nanoparticles and some of these attempts have proven to be quite successful. Some important examples are as follows:

Fe_3O_4: The magnetic Fe_3O_4 nanoparticles have been one of the most widely used metal oxides in nanofluids (also known as ferrofluids). Philip et al. reported a thermal conductivity increment of about 300% using ~6.7 nm magnetite nanoparticles coated with oleic acid, at a load of 6 vol% applying an external magnetic field of 82G [179,180].

CuO: CuO has also been quite frequently explored as an option for nanofluidic material. Das et al. reported ~35% thermal conductivity increment over the base fluid water for 4 vol% CuO nanoparticles at a temperature of 51°C [181]. Zhu et al. reported a thermal conductivity increment of ~31% at 5 vol% loading of CuO nanoparticles in water and stabilized with ammonium citrate [182]. Dey et al. reported a 0.1 vol% addition of CuO nanoparticles could improve the thermal conductivity of water to as much as 46% [183].

Al_2O_3: Al_2O_3 nanoparticles have also been utilized in preparing nanofluids to varying degree of success [181].

1.3.9 Transparent Conducting Oxides

Transparent conducting oxides (TCOs) have attracted increasing attention over the last two decades as critical components of flat panel displays, solar cells, and low-emissivity windows. Following are the examples of some of the most popular metal oxides in this regard:

ITO: ITO or Indium Tin Oxide has been the most popular fabricating transparent conducting films. ITO thin films can be used for various applications that require both optical transparency in the visible light region and high electrical conductivity, such as electroluminescent devices, electrochromic systems, liquid crystal display electrodes, solar cell and energy efficient windows [184–188].

CdO: CdO-based TCOs have been of interest due to their relatively simple crystal structures, high carrier mobilities, and sometimes nearly metallic conductivities. Although CdO has a bandgap of ~2.3 eV leading to poor optical transparency in the low wavelength range, this bandgap as well as the optical transparency can be tuned by the doping of aliovalent metals such as Sn, Sc, and In due to its low effective carrier mass [189–191].

ZnO: ZnO films have their advantages due to the cheap value and good stability (in plasma). Thus, they have been exploited in TCOs [192].

1.3.10 Superconductivity

High-temperature superconductors are materials that have a superconducting transition temperature above 30 K. Metal oxides have been integral part of constructing high-temperature superconducting materials.

CuO: CuO-based superconductors have been the most sought after among all the other superconductors showing superconducting ability above liquid nitrogen temperature. There have been numerous existing reports and research ongoing to further improve CuO-based high-temperature superconductors [194–196].

WO_3: Sodium-doped WO_3 ($Na_{0.05}WO_3$) was reported to possess superconductivity at 91 K [197]. Although this work attracted less attention than high-T_c cuprates, it was the first non-cuprate HTS material. Recently Reich's group obtained a possible 2D H_xWO_3 superconductor with a T_c of 120 K, which is very close to the highest T_c (135 K) of HTS material reported so far. [198].

1.3.11 Antimicrobial Agent

Metal oxide nanomaterials are being exploited as antibacterial agents due to their durability, availability in various shapes and robust nature. Following are a few examples:

ZnO: ZnO has been reported to exhibit notable antibacterial quality, which operates through the ROS mechanism [198–200].

CuO: CuO has been explored by some researchers with respect to their antimicrobial nature and was found to be quite impressive [201,202].

1.3.12 Thermochromic Materials

A thermochromic material changes its optical properties reversible when heated and can revert to its original properties when cooled back to its earlier temperature. Thermochromic materials can be used as glazing of windows which has a great advantage in reducing energy consumption compared with other energy-saving glazing because it blocks near infrared solar transmittance (most of the heat from the sun comes in these wavelengths) in summer and allows it in winter. A few metal oxides have been reported to show thermochromic properties such as the following:

VO_2: VO_2 thin film has been used as antireflection thermochromic glazing material for windows [203–207]. It acts as a "spectrally selective" window coating to block infrared transmission and reduce the loss of building interior heat through windows. This material behaves like a semiconductor at lower temperatures, allowing more transmission, and like a conductor at higher temperatures, providing much greater reflectivity.

WO_3: WO_3 is a remarkable thermochromic material. Recent researches show that tungsten trioxide also has potential application in thermochromic devices [208]. The thermochromic behavior of WO_3 has caused a resurgence of interest in the investigations of its thermochromic phase transition [209–211].

1.3.13 Electrochromic Materials

Electrochromism can be defined as the ability of a material to undergo color change upon oxidation or reduction. Electrochromic devices are able to vary their throughput of visible light and solar radiation upon electrical charging and discharging using a low voltage. A small voltage applied to the windows will cause them to darken; reversing the voltage causes them to lighten. Thus, one can regulate the amount of energy entering through a "smart window" so that the need for air conditioning in a cooled building decreases. The energy efficiency inherent in this technology can be large, provided that the control strategy is adequate. Some metal oxides have been widely exploited in fabricating electrochromic devices. Following are some examples:

TiO_2: Titania nanomaterials have been widely explored as electrochromic devices, such as electrochromic windows and displays [212–215]. Two types of electrochromism of nanocrystalline thin film TiO_2 electrodes have been reported. The first type is the electrochromism of nanocrystalline TiO_2 electrodes in Li-containing electrolytes related to the reversible insertion of Li^+ into the anatase lattice of the nanoparticles [216,217]. The second type is the electrochromism of nanocrystalline TiO_2 electrodes modified with viologens and/or anthrachinons equipped with a surface anchoring group [218,219]. Titania can also be utilized in photoelectrochromic smart window devices as demonstrated by Pichot et al. [220].

1.3.14 Piezoelectric Materials

The piezoelectric effect describes the relation between a mechanical stress and an electrical voltage in solids. The materials showing piezoelectric effect are termed as piezoelectric materials and have applications in sensing (Piezoelectric sensor are devices using the piezoelectric effect to measure acceleration, pressure, strain or force and converting them to an electrical signal), ultrasonic transducers (Ultrasonic transducers (10–70 kHz) are used in imaging and high power applications), piezoelectric motor (A piezo motor is based on the change in mechanical shape of a piezoelectric material when a tension is applied), high-voltage sources, frequency standards, etc. Most of the metal oxide materials that display piezoelectric properties are mixed metal oxides with a perovskite-type crystal structure ($BaTiO_3$, lead zirconate titanate (PZT), etc.) ZnO is so far reported to be the only pure phase metallic oxide piezoelectric material.

ZnO: Due to its non-centrosymmetric symmetry, ZnO is a piezoelectric material. This property can be utilized in fabricating electro-mechanical-coupled sensors and transducers. ZnO nanowire (NW)-based piezoelectric nanogenerators have been demonstrated to convert mechanical energy into electricity by utilizing the coupled semiconducting and piezoelectric properties of ZnO [221,222].

1.3.15 Luminescence Materials

Luminescent materials have uses in light-emitting diodes (LED) and as phosphor materials. There are only a few metal oxide nanoparticles that reportedly have decent luminescence properties.

ZnO: ZnO exhibits a direct band-gap of 3.37 eV at room temperature with large exciton energy of 60 meV. The strong exciton binding energy, which is much larger than that of GaN (25 meV), and the thermal energy at room temperature (26 meV) can ensure an efficient exciton emission at room temperature under low excitation energy. As a consequence, ZnO is recognized as a promising photonic material in the blue–UV region [223].

CdO: CdO has been reported to show electro-generated chemiluminescence (ECL) property, although the corresponding intensity was lower than other notable ECL materials such as CdS and CdSe [224].

1.3.16 Field Emitters

Semiconducting transition and rare-earth metal oxides are attracting significant attention as candidates for chemical and environmental sensors, because their electrical conductivity depends sensitively on the nature and concentration of adsorbed species on their surfaces. The key characteristics of these oxides are (a) cations with mixed valence and (b) oxygen vacancies. The latter are responsible for the observed high sensitivity of the electrical properties to the presence of adsorbed molecules and allow the tuning of the conductance of the oxide. By forming metal electrode/nanostructure electrical contacts and capacitively coupling the nanostructure to a nearby gate electrode, a field-effect transistor (FET) is produced that allows the exploration of new aspects of the physical and chemical properties of the nanostructures. Research works have been going on to explore the field emission properties of metal oxide nanoparticles and so far a few of them have proven to be capable. Following are some examples:

ZnO: ZnO nanobelt and nanowires have been reported to be used in fabrication of field-effect transistors [225,226].

WO_3: WO_3 nanorods have been reported as possessing excellent field emission properties [227,228].

VO_2: Both the rutile VO_2 and VO_2 (B) phase and even the metastable VO_2 (A), they all have shown to possess excellent field emission properties [229,230].

Cu_2O: Liao et al. reported using Cu_2O nanowires as conducting channels for FETs [231].

1.3.17 Lasers

As an alternative to an EHP (electron–hole plasma process), excitonic recombination in semiconductors is a more efficient radiative process and can facilitate low-threshold stimulated emission. To achieve efficient excitonic laser action at room temperature, the binding energy of the exciton must be much greater than the thermal energy at room temperature (26 meV). In this regard, ZnO is a good candidate because its exciton binding energy is 60 meV, substantially larger than that of ZnSe (22 meV) and GaN (25 meV). To further lower the threshold, low-dimensional compound semiconductor nanostructures have been fabricated, in which quantum size effects yield a substantial density of states at the band edges and enhance radiative recombination due to carrier confinement. ZnO nanowires can be used as efficient ultraviolet nanoscaled lasers [232].

1.3.18 Switches

Metal oxide nanomaterials can be utilized in switching devices. Some of the examples are as follows:

ZnO: ZnO can have potential application as an optical switch based on its photoconducting properties [233].

VO_2: Rutile VO_2 based on its metal to insulator transition property can be used as an optical switch [234–236] and electrical switch [237].

1.3.19 Memresistor

Memresistors are passive circuit elements that behave as resistors with memory. Because of its typical metal to insulator transition property, VO_2 can be utilized in a memresistive system [238].

1.3.20 Chromatographic Support

An ideal chromatographic support should have a relatively high specific surface and pore volume, narrow particle size distribution and pore size distribution, and favorable pore connectivity, as well as excellent stability. A lot of metal oxide nanomaterials meet these criteria, such as

ZrO_2: Although it is much lower accessibility, variability of specific surface area and pore volume compared to other traditional chromatographic support materials is a concern; its stability within the whole range of pH 1–14 has prompted researchers to explore its applicability as an alternative chromatographic support in HPLC packing [239–241].

1.3.21 Fuel Cells

Solid oxide fuel cells (SOFCs) have several advantages over the other types of fuel cell systems. The use of a solid electrolyte eliminates most problems regarding corrosion and liquid management. Moreover, the high operating temperature facilitates rapid electrode kinetics without expensive noble metal electro-catalysts such as Pt for phosphoric acid fuel cells. Among the metal oxide nanomaterials, yttrium-stabilized zirconia can be used in fuel cells [242].

1.3.22 Optical Recording and Other Information Storage Devices

Raising the density of data transportation is the key subject for compact disk (CD) storage technology at present. The density of storage in optical memory system is inversely proportional to the square of wavelength of the operating lasers. So the shorter the wavelength of the laser the more information the CD can store. The functions of the semiconductor lasers in the violet-ultraviolet (UV) waveband exceeds that in the blue green light waveband. Optical recording is an advanced technology; characterized by high density, reversible information storage and has advanced applications. A few transition metal oxides have been attempted to explore in terms of optical storage capability. The most notable of them is $\mathbf{WO_3}$. Lu et al. [243] showed that an amorphous WO_3 thin film could be colored by a single pulse of KrF excimer laser light at 248 nm and bleached by a single pulse of Nd–Y–Al–garnet laser light at 1.06 μm in air. Research has also revealed that laser light to effect crystal structure change of WO_x films with concomitant changes in optical transmission [244].

1.3.23 Abrasives and Polishing Agents

$\mathbf{Al_2O_3}$: Alumina is one of the hardest metal oxides with a Moh's hardness of 9.0. Thus, it is not surprising that the material is the most frequently used metal oxide in polishing and as abrasives. Kim et al. studied the effects of alumina abrasive particle behavior in magnetorheological polishing for MEMS applications [245]. Hatanaka et al. reported alumina-based ceramic materials, abrasive materials and the manufacture of the same [246].

$\mathbf{CeO_2}$: Ceria has been explored as a potential abrasive material by researchers [247,248].

Fe_2O_3: Owing to their hardness Iron oxides have been used as abrasives and polishing agents. A lightly calcined form of hematite (Jeweler's rouge) is used to polish gold and silver, while a more strongly calcined hematite (crocus) serves to polish brass and steel [250]. Fe_2O_3 has been utilized as high-density coatings for concrete seabed pipelines that bring oil and gas to shore. These coatings stabilize the pipelines on the sea floor and provide protection against physical damage in shallow water [249].

1.3.24 Ultraviolet Filtration

In the UV radiation range reaching the earth's atmosphere there are two sub-range UV type B (290–320 nm) and UV type A (320–400 nm). Among these UV type A has two domains; short UVA (320–340 nm, comprising the most energetic and harmful type of radiation) and long UVA 340–400 nm; responsible for early skin aging). The need for materials able to filter these UV radiations has increased in the field of cosmetic products. Several metal oxide nanoparticles have been incorporated in these cosmetics to prevent damage due to exposure to UV radiation.

CeO_2: With a band-gap of 3.2 eV, good transparency in the visible range, and no known toxicity, nanostructured CeO_2 appears to be a promising inorganic material for use as a UV filter in sunscreen cosmetic products. There have been several reports showing the UV filtration capability of Ceria in the short UVA range [250–252]. The associated photocatalytic activity of CeO_2, which can degrade other compounds present in the cream, can be controlled by doping of heavy metals (Ca, etc.) in ceria.

TiO_2: Titania owing to its ability to block both UVA and UVB radiation has been extensively used as a component material in sunscreen. There have been numerous reports not only on the effectiveness of TiO_2 as a cosmetic product but also its effect on the skin itself [253–256]. There is evidence from isolated (in vitro) cell experiments that ZnO and TiO_2 may induce free radical formation in the presence of light and this free-radical generation may cause cell damage (photogenotoxicity with ZnO). However, recent work suggests that the photogenotoxicity seen in these studies (with ZnO) may be due to an UV-induced experimental artifact in an in vitro assay, rather than the presence of the nanoparticles [257].

Nanomaterial synthesis has occupied a major part of materials science research for the last couple of decades. Nanoparticles have high surface area and a high reactivity as a direct consequence of the unsaturated bonds on their pristine surface area. As a result they tend to react among themselves and agglomerate to form secondary particles. This inflicts utmost challenge for chemists, physicists, and material scientists alike to be able to fabricate identical nanoparticles in an inexpensive and reliable way that can be reproducible. Not only the synthesis of the materials in the desired size regime, but also maintaining a reproducible and consistent size distribution coupled with the exercise of obtaining the nanoparticles in application specific morphology has made the synthesis of nanoscaled materials an absolutely intriguing topic of research to the materials science enthusiasts.

1.4 Various Synthesis Strategies for Metal Oxide Nanomaterials

Numerous reports can be found regarding the synthesis of these metal oxide nanomaterials. The synthetic strategies can also be classified according to the branches of science involved (physical or chemical), the medium of synthesis, the sources of energy, etc.

There are two fundamental strategies for synthesizing the nanomaterials:

(i) Bottom-up approach: Here the nanomaterials are built up by atom by atom or molecule by molecule, more at the will of their creators.

(ii) Top-down approach: This is more traditional nanomaterial-making process. Here the nanomaterials are synthesized by tearing down bulk materials gradually until they get reduced to nano-sized objects. This is essentially a miniaturization technique, one that is unquestionably the most important approach from an economic point of view.

"Bottom-up" preparation methods are of primary interest to chemists and materials scientists because the fundamental building blocks or primary units are atoms. Gaining control over the way these fundamental building blocks come together and form particles are among the most sought-after goals of synthetic chemists. However, physics is not entirely absent from the bottom-up approach.

Materials like fullerenes and CNTs can only be obtained exploiting physical means. Besides, there are the physical assembling methods such as vapor deposition and molecular beam, which can lead to nanostructured materials.

The bottom-up approach has two major branches, physical and chemical, based on which field of science plays the dominant role in the overall outcome of the synthetic method. Chemical synthesis may be carried out in either the solid, liquid, or gaseous state. Solid-state synthesis usually involves an iterative procedure of bringing solid reaction precursors into intimate contact by mixing and grinding and then promoting atomic diffusion processes via heat treatment at high temperatures to form a reaction product. Such elevated temperatures often lead to rapid grain growth and ultimately a final product with a relatively large grain size unless grain growth inhibitors are present. Consequently, true nanoscaled systems are quite difficult to obtain via the solid-state synthesis process. Diffusion in liquid and in gas phase systems is typically many orders of magnitude greater than solid-phase systems, therefore these synthetic methods can be implemented at a much lower temperatures, not allowing unwanted grain growth. Thus, here our focus mainly will be on liquid and gas phase synthesis method.

The overall schematic branched classification of metal oxide nanomaterials synthesis routes has been provided in Fig. 1.1.

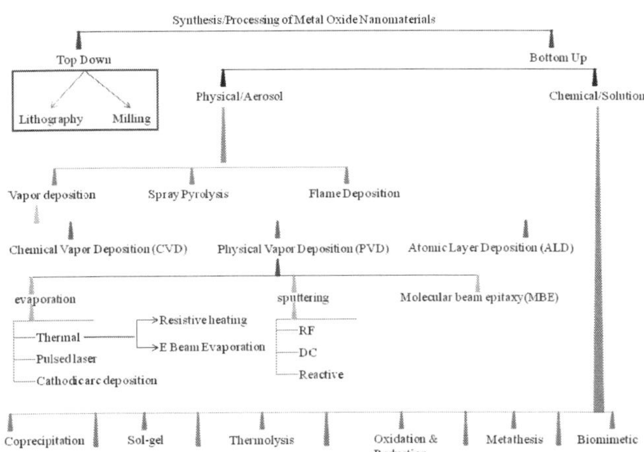

Figure 1.1 Various synthesis methods employed for metal oxide nanomaterials.

The physical or aerosol methods of synthetic procedures include vapor deposition, spray pyrolysis and flame deposition of metal organic precursors. Vapor deposition technique to produce metal oxide nanomaterials, starting from a supersaturated vapor of the metals is among the earliest methods to produce the metal oxide nanoparticles. They generally involve two steps: first, a metallic nanophase powder is condensed under inert convection gas after a supersaturated vapor of the metal is obtained inside a chamber. A high pressure of inert gas is usually needed to achieve supersaturation, then, the powder is oxidized by allowing oxygen into the chamber. This postoxidation is a critical step and very often it becomes necessary for this process is to be performed slowly. Because of the large exothermic reaction, particles heat up for short times (usually less than 1 sec) to temperatures as high as 1000°C, resulting in their agglomeration into large particles by rapid diffusion processes. A subsequent annealing process at higher temperature is often required to complete the oxidation. The system normally consists of a vapor source inside a vacuum chamber containing a mixture of inert gas, usually argon or helium, mixed with another gas that is selected on the basis of the material to be prepared. To produce metal oxides Oxygen is mixed with the inert gas. Nanoparticles are formed when supersaturation is achieved above the vapor source. The vapor deposition can be physical vapor deposition (PVD) or chemical vapor deposition (CVD).

1.4.1 Physical Vapor Deposition

Among the various ways of depositing thin films (a layer of material ranging from fractions of nanometer (monolayer) to several micrometers in thickness), PVD is perhaps the most renowned and highly practiced route. It is a variety of clean, dry, vacuum deposition (Vacuum deposition is a family of processes used to deposit layers atom-by-atom or molecule-by-molecule at sub-atmospheric pressure on a solid surface in which the coating is deposited over the entire object simultaneously, rather than in localized areas). The deposition method involves purely physical processes such as high-temperature vacuum evaporation or plasma sputter bombardment rather than involving a chemical reaction at the surface to be coated as in chemical vapor deposition. PVD methods differ in the means for producing the metal vapor and the details of plasma creation. The material to be deposited is placed in an energetic, entropic environment, so that particles of material escape its surface. Facing this source is a cooler surface which draws energy from these particles as they arrive, allowing them to form a solid layer. The whole system is kept in a vacuum deposition chamber, to allow the particles to travel as freely as possible. Since particles tend to follow a straight path, films deposited by physical means are commonly directional, rather than conformal. The entire PVD process consists of four steps:

(1) Evaporation: During this stage, a target, consisting of the material to be deposited is bombarded by a high-energy source such as a beam of electrons or ions. This dislodges atoms from the surface of the target, "vaporizing" them.
(2) Transport: This process simply consists of the movement of "vaporized" atoms from the target to the substrate to be coated and will generally be a straight-line affair.
(3) Reaction: This is one of the important stages during the PVD processing of metal oxides. In these cases, the target will consist of the metal. The atoms of metal will then react with the appropriate gas (here oxygen) during the transport stage.
(4) Deposition: This is the process of coating build up on the substrate surface.

The first stage, i.e., the target metal atom eviction process, can be either an evaporation process obtained by applying laser, heat, electron beam, etc., or by sputtering obtained by applying plasma. The various evaporation techniques are as follows:

1.4.1.1 Thermal evaporation

Thermal evaporation involves heating a solid material inside a high vacuum chamber, taking it to a temperature that produces some vapor pressure. Inside the vacuum, even a relatively low vapor pressure is sufficient to raise a vapor cloud inside the chamber. This evaporated material now constitutes a vapor stream, which traverses the chamber and hits the substrate, sticking to it as a coating or film. Since, in most cases, the material is heated to its melting point and is liquid, it is usually located in the bottom of the chamber, often in some sort of upright crucible. The vapor then rises above this bottom source, and the substrates are held inverted in appropriate fixtures at the top of the chamber. The surfaces intended to be coated are thus facing down toward the heated source material to receive their coating. There are two primary means of heating the source material. One method, often referred to as filament evaporation, is a simple electrical resistive heat element, or filament. The filament source offers the safety of low voltage, although very high current is required, usually several hundred amps. The other common heat source is an electron beam, or e-beam, and this is generally known as E Beam Evaporation. This is certainly a more "high tech" approach to heating a material up, and involves some dangerous high voltage (usually 10,000 volts), so e-beam systems always include extra safety features. The source itself is an e-beam "gun," where a small and very hot filament boils off electrons, which are then accelerated by the high voltage, forming an electron beam with considerable energy. Accessories such as residual gas analyzers (RGA's), and other custom features and custom automation are also available. Cryogenic pumps are the most popular type of high vacuum pump for Evaporation, but other options are available if desired. The vacuum used in this process is $\sim 10^{-4}$ torr. Electron beam PVD, however, yields a high deposition rate from 0.1 μm/min to 100 μm/min at relatively low substrate temperatures, with very high material utilization efficiency. For example, Hass et al. reported the fabrication of metal oxide coatings utilizing an electron beam directed PVD [258]. In this approach, a transonic helium carrier gas jet has been combined with electron beam evaporation to create and efficiently transport metal vapor to a substrate. Metal oxide coatings were then produced by adding oxygen to the carrier gas. The approach was found to be dependent on the vapor pressures of the individual source components. Results are reported for the reactive deposition of zirconia from a single metal source and reactive deposition of yttria-stabilized zirconia from two metal sources (Y and Zr). Dickey et al. reported the synthesis of ITO nanotube arrays based on e-beam evaporation technique assisted by anodized aluminum oxide (AAO) template [259]. Different chemical states of titanium oxide films were deposited on commercially pure Ti (CP Ti) by electron-beam evaporation at different oxygen flow rates by Lin and coworkers [260].

1.4.1.2 Pulsed laser deposition

Pulsed laser deposition is a well-known method for thin film deposition. This method involves the evaporation of a solid target in an ultra high vacuum chamber by means of short and high-energy laser pulses. In a typical PLD process, a researcher places a ceramic target in a vacuum chamber. A pulsed laser beam vaporizes the surface of the target and the vapor condenses on the substrate followed by subsequent nucleation and growth. The process is generally performed under high vacuum or in the presence of background gases like oxygen, which is used while depositing oxides to fully oxygenate. The main components of a PLD unit are a laser, optics and a vacuum system. Besides being a conceptually simple method, PLD is versatile (many materials can be deposited), cost-effective, and a scalable process. For instance, Irwin et al. reported NiO films grown by pulsed-laser deposition (PLD), using a 248 nm KrF excimer laser with 25 ns duration and a repetition rate of 2–5 Hz. The 230 mJ/pulse beam was focused onto a 1 mm × 2 mm spot on the NiO target [261]. ITO coated glass was used as a substrate here. Han et al. reported the deposition of various metal oxide core–shell synthesis (YBCO–MgO, Fe_3O_4–MgO, etc.). Here the shell oxides were deposited via PLD process on the core MgO nanowires [262].

1.4.1.3 Cathodic arc deposition

This is another potent evaporation PVD method where the cathodic arc is used to evaporate the target material. When an arc discharge is generated under medium and higher vacuum, in general, a cathode spot is formed, but no anode spot is formed. The cathode spot is very active with high temperature (~1500°C) and evaporates the cathode material (the target). At the cathode spot region, very dense plasma is generated, and the evaporated cathode material is ionized and the ions deposit solid film upon reaching the solid surface. Such vacuum arc deposition is a major method in (PVDs) and has the advantage of higher ion energy, compared to the other PVD methods. The ion source of the arc cathode is generally solid. Therefore, no crucible is needed, and the sources can be freely mounted on the wall of the process chamber. In principle, no gas introduction is necessary. However, vacuum arc deposition is suitable for reactive deposition because the ions generated in the vacuum are highly reactive with the gases due to their high energy. Although arc deposition technique is mostly employed in the processing of Nitrides, it can also be used for metal oxide thin film formation. A cathodic arc plasma source with the desired cathode material is operated in an oxygen atmosphere of appropriate pressure, and metal oxides of various stoichiometric compositions can be formed on different substrates. Anders et al. reported a series of experiments on metal oxide formation by cathodic arc deposition for different applications [263]. Black copper oxide was deposited on accelerator components to increase the radiative heat transfer between the parts. Various metal oxides such as tungsten oxide, niobium oxide, nickel oxide, and vanadium oxide were deposited on ITO glass to form electrochromic films for window applications. Tay et al. wrote a review article on the metal oxide films deposited by filtered cathodic vacuum arc technique [264].

1.4.1.4 Sputtering deposition

Sputtering is a mechanism by which atoms are dislodged from the surface of a material as a result of collision with high-energy particles. Thus, PVD by Sputtering is a term used to refer to a PVD technique wherein atoms or molecules are ejected from a target material by high-energy particle bombardment so that the ejected atoms or molecules can condense on a substrate as a thin film. There are three principal sputtering techniques:

(1) DC Sputtering: DC sputtering or direct current sputtering is the simplest sputtering process. To prepare the metal oxides the metal particles prepared by DC sputtering using an argon atmosphere are allowed to react with oxygen, which is the postoxidation process. For example, Ju et al. reported the synthesis of InTaO thin films on glass substrate using DC sputtering method. The sputtering was performed with a fixed Ar gas flow, a variable oxygen gas flow, and a chamber pressure of $\sim 10^{-6}$ torr [265].

(2) RF sputtering: Radio frequency (RF) Sputtering is more adaptable than DC Sputtering. It is not limited to electrically conductive targets like DC Sputtering. It can use targets such as silicon oxides and polymers. Synthesizing metal oxides involve RF sputtering of a metal target using various Ar/O_2 gas mixtures. In this process, oxygen should be continuously supplied to the chamber in order to compensate for the amount consumed. The RF sputtering technique is the more preferred one as it avoids the high-temperature particle agglomeration during the postoxidation process. For example, Cu_2O has been deposited on ITO glass substrates using a reactive RF sputtering technique [266]. Peng et al. reported the deposition

of Fe_3O_4 film on Si and glass substrate from a target consisting of Fe_2O_3 and Fe_3O_4 [267].

(3) Reactive sputtering: The Reactive sputtering is the most complex of the three processes and is where a reactive gas is used along with inert argon to form a plasma. The reactive gas becomes activated and chemically combines with target atoms to form a compound. Two widely used reactive gases are oxygen and nitrogen. This process is used widely for the deposition of dielectrics, resistors, and semiconductors. C. R. Aita reported the reactive sputter deposition of metal oxide nanolaminates on unheated substrates using four archetypical examples: ZrO_2–Al_2O_3, HfO_2–Al_2O_3, ZrO_2–Y_2O_3, and ZrO_2–TiO_2 [268]. TiO_2 films were deposited by reactive sputtering of a titanium (Ti) target, in a mixed argon (Ar) and oxygen (30, 35, and 40%) atmosphere [269].

1.4.1.5 Molecular beam epitaxy

Molecular beam epitaxy (MBE) has evolved into a popular technique for growing compound semiconductor and other materials. MBE can produce high-quality layers with very abrupt interfaces and good control of thickness, doping, and composition. Because of the high degree of control possible with MBE, it is a valuable tool in the development of sophisticated electronic and optoelectronic devices. Molecular beam epitaxy takes place in high vacuum or ultra-high vacuum (10^{-8} Pa). The most important aspect of MBE is the slow deposition rate (typically less than 1000 nm per hour), which allows the films to grow epitaxially. In solid source MBE elements in ultra-pure form, are heated in separate quasi-Knudsen effusion cells until they begin to slowly sublimate. The gaseous elements then condense on the wafer, where they may react with each other. The term "beam" means that evaporated atoms do not interact with each other or vacuum chamber gases until they reach the wafer, due to the long mean free paths of the atoms. During operation, reflection high-energy electron diffraction (RHEED) is often used for monitoring the growth of the crystal layers. The method has been sparsely utilized for the growth metal oxide thin films although there are reports present in the literature. S. A. Chambers reported molecular beam epitaxy coupled with the use of activated oxygen to be a powerful tool for the growth of well-defined, structurally excellent oxide semiconductor films. Several case studies were presented to illustrate some of the physical phenomena that can be investigated; these included Cr- and Co-doped TiO_2 anatase, Ti-doped α-Fe_2O_3 hematite, and N-doped TiO_2 rutile [270]. Holland et al. reported the molecular beam epitaxial deposition of Ga_2O_3 thin films on GaAs (001) substrate [271].

1.4.2 Chemical Vapor Deposition

Chemical vapor deposition or is a generic name for a group of processes that involve depositing a solid material from a gaseous phase and is similar in some respects to PVD. Basically, thin film processes that use gases, evaporating liquids, or chemically gasified solids as source materials are categorized as CVD. Precursor gases (often diluted in carrier gases) are delivered into the reaction chamber at approximately ambient temperatures. As they pass over or come into contact with a heated substrate, they react or decompose forming a solid phase, which is deposited onto the substrate. The substrate temperature is critical and can influence what reactions will take place. Chemical vapor deposition coatings are usually only a few microns thick and are generally deposited at fairly slow rates, usually of the order of a few hundred microns per hour. Precursors for CVD processes must be volatile, but at the same time stable enough to be able to be delivered to the reactor. Generally precursor compounds will only provide a single element to the deposited material, with others being volatilized during the CVD process. However, sometimes precursors may provide more than one. Such materials simplify the delivery system, as they reduce the number of reactants required to produce a given compound. Typical precursor materials include halides, hydrides, alkoxides, carbonyls, etc. Researchers have been using this technique to fabricate a thin film of various metal oxides. For example, metal organic chemical vapor deposition (MOCVD) has been one of the frequently utilized synthesis pathways for preparing transparent conducting films of CdO. Metz et al. employed complex organic Cd-precursor on MgO (111) substrate [191]. Malandrino and coworkers synthesized CuO nanotubes through MOCVD method using $Cu(tmhd)_2$ (Htmhd) 2,2,6,6-tetramethyl-3,5-heptandione) precursor. The deposition temperature was 400°C and the oxygen partial pressure of 1.5 torr. Anodic aluminum Oxide membranes were used as templates [272]. Pan and coworkers grew Tin Oxide nanostructures grown by the decomposition of $Sn(O^tBu)_4$ at 650–750°C in a cold walled quartz CVD reactor [273]. A high frequency field was used to inductively heat the Au coated Alumina substrate. Hao et al. reported In_2O_3 deposition on Si substrate by CVD method from metallic in a temperature range of 950–1350°C [274].

1.4.3 Atomic Layer Deposition

Atomic layer deposition (ALD) is an advanced process for depositing ultra-thin films one atomic layer at a time. The technique offers unprecedented control over thickness, uniformity, quality, and material properties of monolayer films that are very "thin." It is a chemical gas phase thin film deposition method based on alternating saturative surface reactions. In ALD the source vapors are pulsed into the reactor alternately, one at a time, separated by purging, or evacuation periods. Each precursor exposure step saturates the surface with a monolayer of that precursor. Thin film growth occurs in a cyclic manner in ALD; the cycle in the simplest case consists of four steps. In the first step a precursor is exposed to the surface, which reacts with the surface in a saturating manner until a monolayer has been chemisorped and no further adsorption takes place. Then the reaction stops, so the process is called "self-limiting." Any excess precursor is then evacuated or purged out (step 2). A second vapor from the next precursor then reacts with the surface in a second self-limiting reaction, thus depositing a second layer of atoms onto the first; the second reaction also must return the surface to a state in which it is ready to react with the first reactant. This is followed by another evacuation (step 4). The cycle can be repeated to build up a binary compound atomic layer by layer. Most of the ALD processes reported are based on the above exchange reactions between molecular precursors. Another possible reaction type is additive with elemental precursors but practical applications are limited due to the low volatility of metals. The third reaction type, also quite rare, involves a self-limiting adsorption of a precursor followed by its decomposition by an appropriate energy or by its reduction using an appropriate reductant. In a majority of the ALD processes, the reactions are activated only thermally under isothermal conditions. Atomic layer deposition method has been used for the successful deposition of high dielectric constant oxides (e.g., Al_2O_3, ZrO_2, TiO_2, etc.). Wang et al. reported ALD of alumina on pristine Graphene surface using precursors trimethyl aluminum and water at ~100°C [275]. Elam et al. reported a new method for preparing thin films of SnO_2 by ALD using alternating exposures to tetrakis (dimethylamin) tin and hydrogen peroxide. Tin oxide films were successfully deposited on a variety of substrates using deposition temperatures of 50–300°C at an average growth rate of 1.2 Å/cycle [276]. Willinger and coworkers coated CNTs with V_2O_4 through ALD using vanadium n-propoxide and acetic acid [277].

Vapor deposition methods to prepare nanoparticles directly from supersaturated vapor have many advantages over other techniques including versatility, ease in performance and analysis, and high-purity products. They can also be employed to produce films and coatings. In spite of the success of these methods, the drawback lies in the high production cost because of low yields and the difficulty in scaling-up. Heating techniques have other disadvantages that include the possibility of reactions between the metal vapors and the heating source materials. Furthermore, the operating temperature is limited by the choice of the source material, and because of that, they cannot be used to make a wide variety of materials.

1.4.4 Spray Pyrolysis

Spray pyrolysis (SP) is a powerful technique to synthesize a wide variety of high purity and chemically homogenous metal oxide powders. Large quantities of oxide powders with homogenous particle sizes and crystallite sizes within the nano-regime may be synthesized through this process. In this process a water-based precursor solution is fed into a rotating furnace through a nozzle. The precursor can be metallic salts (usually chlorides or nitrates) in solution or suspension. Sometimes appropriate organic compounds are added in order to make a stable and uniform suspension. The process involves the generation of aerosol (suspension of solid or liquid particles in gas [278]) droplets by atomization (atomization is the production of droplets and their dispersion into the gas [278]) of the starting solution. The commonly used atomizers in SP are pressure, nebulizer, ultrasonic and electrostatic [279], although to obtain nano-sized droplets nebulizer is known to be the most effective atomizer. The droplet is evolved through following processes:

- Evaporation: evaporation of solvent from the surface, diffusion of solvent vapor away from droplet, change in droplet temperature, diffusion of solute toward the center of the droplet, change in droplet size;
- Precipitation/drying: involves volume precipitation or surface precipitation of the solute, followed by the evaporation of the solvent through the nanoporous crust;
- Thermal decomposition or pyrolysis: forms a nanoporous structure;
- Sintering: involves the adhesion/solidification of the crystallites.

The morphologies and the size distribution can be controlled by the cautious choice of precursors, concentration of the solution, residence time in the furnace, etc. Its advantages are that the method does not require high quality targets or substrates and continuously produces the material. The chemical reaction occurs within the created micron-to sub-micron sized liquid droplets—a microcapsule reactor. Example: Prathap et al. reported the spray pyrolysis deposition of In_2O_3 thin film on glass substrate using $InCl_4$ as the precursor. The substrate temperature was maintained at 300–400°C and compressed purified air was used as the carrier gas [280].

1.4.5 Thermochemical or Flame Deposition of Metal Organic Precursors

This is another type of gas condensation technique, with the precursor being a liquid chemical material. The precursor is vaporized followed by a combustion process using a fuel-oxidant mixture such as propane-oxygen or methane-air [281]. The process combines the rapid thermal decomposition of a precursor-carrier gas stream in a reduced pressure environment with thermophoretically driven deposition of the rapidly condensed product particles on a cold substrate [282]. The flame usually provides a high temperature (1200–3000 K), which promotes rapid gas phase reactions [283].

Several types of flame reactors have been used by researchers [281–286], e.g., diffusion flame reactor, low-pressure flat flame combustion reactor, etc. In the first one the combustion process are mainly determined by the rate of inerdiffusion of the oxidant and the fuel. The maximum flame temperature usually occurs at the tip of the flame where agglomerates fuse and leave the flame. As they leave the flame, the flame temperature drops quickly and particles continue to coagulate without sintering, resulting in large agglomerates of primary particles. For low-pressure flat flame combustion reactor an increase in production rate and the fact that pyrolysis and condensation occur in the gas phase, which prevents impurity entrainment from the reactor walls.

Among the precursor metal chlorides are the most widely used because of the high vapor pressure of chlorides and the fact that they can be safely stored and handled. The disadvantages of using chloride precursors are the formation of acidic gases and the contamination of the products with halide residues. Flame processes are industrially exploited to produce commercial quantities of metal oxide powders, but the control of particle size, morphology, phase, and composition is difficult and limited. For instance, Ding et al. reported the synthesis of titania thin films were prepared by aerosol flame deposition process via the pyrolysis of titanium tetra-isopropoxide (TTIP) precursor [287].

Although physical methods are better suited for thin film deposition, Chemists have risen rather successfully to the challenge of making nanoparticles. It is well within their means to prepare dispersions of any desired simple or complex nanoparticles in a high degree of monodispersity. Furthermore, it is often possible to isolate the nanoparticles as solid powders and redisperse them without affecting their sizes and size distributions. Besides the experimental setups required for obtaining nanoparticles through chemical or solution methods is well within the financial constraints that are often the biggest obstacles while dealing with the much more sophisticated instruments required in the physics dominated world of PVDs or CVDs. Thus, researchers across the globe have really dug into the chemical methods of nanoparticles synthesis and as a result myriads of literatures are available today describing nanomaterials synthesis via various chemical methodologies. Some of the most relevant of those methods for the synthesis of metal oxide nanoparticles are described below with appropriate examples.

1.4.6 Chemical/Solution Approach

1.4.6.1 Coprecipitation

Metal oxides can be synthesized by precipitation of the sparingly soluble products (formed under the condition of supersaturation) from aqueous solutions followed by thermal decomposition of those products to oxides. Coprecipitation reactions involve the simultaneous occurrence of nucleation, growth, coarsening, and/or agglomeration processes. Due to the difficulties in isolating each process for independent study, the fundamental mechanisms of coprecipitation are still not thoroughly understood. The supersaturation condition demands that nucleation be a critical step in the synthesis method as a large number of particles will form simultaneously, which can subsequently undergo secondary processes such as Ostwald ripening, aggregation, etc. These factors will dramatically influence the size and morphology. Anything that influences the chemical reactions leading to the supersaturation (e.g., rate of addition the precursors, stirring, etc.) can have a far-reaching effect on the eventual particle characteristics. When the product contains only one or two elements (e.g., a metal, binary oxide, etc.), precipitation reactions are relatively straightforward. In more complicated ternary and quaternary systems, the process becomes more complex, as multiple species must be precipitated simultaneously (hence, the term *coprecipitation*). Just mere precipitation does not guarantee the formation of nanoparticulate structures. A large number of small crystallites are initially formed (nucleation) which then undergo secondary processes such as aggregation (growth) and form larger and thermodynamically more stable particles. To produce particle in the nano-sized region the nucleation must be a fast process while the growth process must be slow. One way of producing nano-sized particles is to add stabilizers so the initial smaller particles are stabilized and further agglomeration is prohibited. There are generally two approaches to stabilize the particles: (i) causing steric repulsion between the particles by the addition of surfactants, polymers or other organic species bound to the nanoparticle surface (generally termed as capping agents); (ii) electrostatic repulsion caused by the chemisorptions of charged species (H^+ or OH^-) on the particle surface.

Preparation of metal oxide nanoparticles through the coprecipitation method can be obtained in two ways: direct precipitation of metal oxide nanoparticles or production of intermediates that can be converted to metal oxides by further treatments such as calcination or annealing. For monodispersity capping ligands or other surface bound ligands are necessary to prevent agglomeration of the particles. In case of calcinations or annealing some agglomeration may be unavoidable and monodispersity is likely to be absent. Nonaqueous solvents are used to co-precipitate dissimilar metals that cannot be simultaneously precipitated from aqueous solution due the large variation in their pH values necessary to induce the precipitation. The products of co-precipitation reactions especially those produced near or at the room temperature are usually amorphous in nature. In such cases, during the precipitation of hydroxides or carbonates of mixed metals and their subsequent annealing it becomes really difficult to experimentally determine whether the as prepared precursor is a single-phase solid solution or a homogenous mixture of the constituent metal precursors. Co-precipitation is also known as wet chemical solution method.

Many nanoparticulate metal oxides have been synthesized employing the co-precipitation method. Some of the examples are given below.

Spinels and other mixed metal oxides: co-precipitation is arguably the most popular method in synthesizing spinel structured mixed metal oxides. Albuquerque et al. prepared spinel-structured $Ni_{0.5}Zn_{0.5}Fe_2O_4$ by precipitating a mixture of Fe, Ni, and Zn nitrates with NaOH and calcining at 300°C or higher [288]. Particle sizes ranged from 9 to 90 nm, depending on the calcination temperature. $MgFe_2O_4$ [289] and $Sm_{1-x}Sr_xFeO_{3-\ddot{a}}$ [290] have been prepared by nearly identical methods. Fe_3O_4 has been synthesized by co-precipitating the mixed solution of Fe^{2+} and Fe^{3+} induced by NaOH. Here a temperature of 70–100°C is sufficient during the precipitation itself and no postcalcination treatment is necessary [291]. Chinnasamy et al. reported the precipitation of $CoFe_2O_4$ using the solution of a mixture of Fe^{3+} and Co^{2+} induced by NaOH [292]. $LiCoO_2$ can be synthesized by coprecipitating LiOH and $Co(OH)_2$ from the ethanol solution of Li^+ and Co^{2+} and KOH precipitant followed by annealing at 400–700°C [293].

CeO_2: Ceria nanoparticles were synthesized through the reactions of $Ce(NO_3)_2$ and $(NH_4)_2CO_3$ followed by annealing the precipitate at 300°C. Particle sizes were ~6 nm [294]. Traffault et al. reported the synthesis of ceria nanoparticles with precursors $Ce(NO_3)_2$ and NaOH followed by oxidation by H_2O_2 [250].

NiO: NiO has been prepared by precipitating aqueous Ni^{2+} solutions with $(NH_4)_2CO_3$ and calcining the products at 400°C. The particle sizes were in the region of 10–15 nm [295]. NH_4OH has also been used as the precipitant along with CTAB stabilizer [296].

TiO_2: TiO_2 nanoparticles were prepared through the reaction of $TiCl_3$ and NH_4OH under ambient conditions [297]. Here no postthermal treatment or annealing was necessary. The particles were stabilized by poly (methyl methacrylate).

ZnO: ZnO has been synthesized from the solution containing Zn^{2+} ion using NH_4OH as the precipitant and CTAB stabilizer followed by annealing at 500°C [298].

SnO_2: SnO_2 has been reported to be synthesized using $SnCl_4$ solution and NH_4OH and a stabilizer CTAB. The corresponding annealing was performed at 500°C [298].

CuO: Precipitation or the chemical solution method has perhaps been the most popular chemical method of synthesizing nanostructured CuO particles. The basic principle has been to hydrolyze a cupric salt in alkaline medium maintaining a pH ≥ 7.5 followed by thermal dehydration of the precipitate [299–301, 183]. The variation of pH here is an important parameter in controlling the morphology of the particles [183]. The presence of microwave radiation has been reported to significantly decrease the thermal treatment duration [302].

1.4.6.2 Hydrothermal/solvothermal approach

To produce oxide nanomaterials, high temperature associated with refluxing of high boiling solvent is often required. Hydrothermal or solvothermal method is an alternative method to the use of high-temperature reaction solvents, which can be both expensive and toxic. More usual solvents are normally limited by their low boiling points. However, in hydrothermal method the solvents are heated in a sealed vessel where the autogenous pressure far exceeds the ambient pressure and thus the solvent is boiled at a temperature higher than its normal boiling point. The solvent attains a state that is called a supercritical state, which is an in-between the state of liquid and vapor. For instance, the critical point for water lies at 374°C at 218 atm.; above this temperature and pressure, water is said to be supercritical. The solubility to the reactants and transportation ability to the ions in the liquid of such a fluid is much better than that it is in normal liquid solvent. Therefore, some reactions that are impossible to carry on in water in ambient atmosphere can happen at a hydrothermal condition. In hydrothermal systems, the dielectric constant and viscosity of water decrease with rising temperature and increase with rising pressure with the temperature effect predominating. Owing to the changes in the dielectric constant and viscosity of water, the increased temperature within a hydrothermal medium has a significant effect on the speciation, solubility, and transport of solids. Formation of metal oxides through a hydrothermal method should follow such a principal mechanism: the metal ions in the solution react with precipitant ions in the solution and form precipitate, and the precipitate dehydrate in the solution at a high temperature and form crystalline metal oxide nanostructures. In principle this method is performed in an autoclave. The various components to be assembled are mixed with a diluent (generally H_2O). An acidic or basic catalyst is required. Additives which are not directly involved in the final composition of the target material may be introduced to manipulate the morphology, crystallinity, etc. the mixtures are heated from 100°C up to several hundred degrees. Some researchers have explored microwave assisted hydrothermal process revealing a prominent effect on the speed of the reaction. When the diluent or solvent is nonaqueous, the method is termed as solvothermal. Here the temperature can be elevated to much higher than in hydrothermal method since a variety of organic solvents with high boiling point can be used. Besides, the solvothermal method has better control of size and shape distribution than hydrothermal method. Some examples of the hydrothermal processing of metal oxide nanoparticles are provided below.

CuO: CuO has been prepared via the reaction of $Cu(CH_3COO)_2 \cdot H_2O$ with NaOH followed by a hydrothermal treatment at 140°C [303]. Some researchers have adopted hydrothermal method with the inclusion of a surfactant (e.g., sodium dodecylbenzenesulfonate (SDBS), PEG, etc.) [304,305].

TiO_2: Zhang et al. obtained TiO_2 nanorods by hydrothermally treating a dilute $TiCl_4$ solution at 333–423 K for 12 h in the presence of acid or inorganic salts [306,307]. TiO_2 nanowires are obtained by treating TiO_2 white powders in a 10–15 M NaOH aqueous solution at 150–200°C for 24–72 h without stirring within an autoclave [308]. TiO_2 nanotubes can be obtained by treating TiO_2 powders hydrothermally by NaOH solution at 20–110°C for 20 h followed by washing with dilute HCl solution and distilled water [309].

SnO_2: Xi et al. reported the synthesis of SnO_2 from a hydrothermal method starting with $SnCl_4$ precursors and treating it with urea and fuming HCl and a temperature of 90°C [310]. Kar and coworkers synthesized SnO_2 nanorods and nanoparticles by treating $SnCl_4$ solution with NaOH followed by hydrothermal treatment at 180°C for 12 h [311].

V_2O_5: V_2O_5 has been synthesized hydrothermally treating a solution of $V_2O_5 \cdot H_2O$ with H_2O_2. The temperature was varied between 60 and 200°C and the reaction time between 6 and 24 h to obtain different morphologies of the material [312].

VO_2: Hydrothermal method has been the most popular solution method for processing nanostructured VO_2 [313–315]. Liu et al. synthesized VO_2 nanobelts by reducing ammonium metavanadate with formic acid at 180°C under hydrothermal condition [316]. There have been plenty of reports of VO_2 being synthesized hydrothermally reducing V_2O_5 by oxalic acid at a temperature range of 200–300°C [317,318].

Fe_3O_4: Sun et al. have described a high-temperature reaction of iron (III) acetylacetonate with 1,2-hexadecanediol in the presence of oleic acid and oleylamine to obtain monodisperse magnetite nanoparticles. The particle diameter can be tuned from 4 to 20 nm, and the hydrophobic particles can be transformed into hydrophilic ones by adding a bipolar surfactant [319].

Fe_2O_3: Zhu et al. reported the synthesis of α-Fe_2O_3 3D hierarchical nanostructures via the solvothermal route where a $FeCl_3$ solution in isopropyl alcohol was treated with urea in a temperature of about 200°C [320].

1.4.6.3 Sol-gel approach

This is one of the most popular and well-known methods for the preparation of metal oxide nanoparticles and nanomaterials in general. Here reactive metal precursors (normally alkoxides), in an alcoholic solution are hydrolyzed to produce the corresponding hydroxide. This is a nucleophilic reaction and the generalized chemical equation is as follows:

$$M(OR)_n + xH_2O \rightleftharpoons M(OR)_{n-x}(OH)_x + xROH$$

Condensation of the hydroxide molecules with concomitant elimination of water leads to the formation of a network of metal hydroxide. When all the hydroxide species are linked in one network-like structure, gelation is achieved and a dense porous gel is obtained. Gel is a polymeric species consisting of a three-dimensional skeleton surrounding interconnected pores. Removal of the solvents and appropriate drying of the gel results in an ultrafine powder of metal hydroxide. Further heat treatment of the hydroxide leads to the corresponding ultrafine powder of the metal oxide. Since the process begins with a nano-sized unit, and undergoes reactions at the nanometer scale, the final product formed is also in the nano realm. The rate and mode of hydrolysis is an important parameter that can manipulate the final product. Slower and more controlled hydrolysis leads to small particles. The rates of hydrolysis depend on the factors like the electronegativity of the metal atom, its co-ordination number, steric hindrance, etc. More the electronegativity of the metal atom the hydrolysis rate will be generally faster. Steric hindrance plays a very big role as branched or bulky alkoxide groups cause the reaction to be slower. The higher the co-ordination no. around the metal center in the alkoxide the slower the rate of hydrolysis and thus easier to control. Consequently the alkoxides that can form oligomer show the slower rates of hydrolysis. Currently the n-butoxide precursor is preferred than other metal alkoxides because it is the largest alkoxide that does not prevent oligomerization.

The rates of hydrolysis, if needed, can be enhanced by the addition of acid or base catalyst. Acids protonate the alkoxy group, making it a better leaving group, whereas bases provide better nucleophiles in the form of hydroxyl group, which facilitates the hydrolysis reaction. When the drying is achieved under normal condition the gel network shrinks as a result of capillary pressure and the resultant hydroxide powder network is called xerogel. However, if supercritical drying is applied within a high-pressure autoclave reactor at temperatures higher than the critical temperatures of the solvents, less shrinkage of the gel network occurs as there is no capillary pressure and no liquid–vapor interface, which better protects the porous structure. The hydroxide product obtained in this process is referred as aerogel. Aerogel powders exhibit larger surface areas and higher porosities than xerogels.

For the preparation of metal oxide nanoparticles sol-gel processing has been particularly useful because of the present of metal–oxygen bond in the precursor material. Some typical examples are as follows:

ZrO_2: Condensation and supercritical drying of a $Zr(NO_3)_4 \cdot 5H_2O$ or $ZrO(NO_3)_2 \cdot 2H_2O$ precursor in alcohol, for instance, produced amorphous, 1 nm particles of ZrO_2. Calcination of the product at 400°C resulted in 2–3 nm agglomerated particles with a mixed tetragonal/monoclinic phase composition [321,322].

TiO_2: Sol-gel has been the single most utilized chemical mode of synthesis for TiO_2 nanoparticles. There have been numerous reports in the literature about the acid catalyzed hydrolysis of Ti-precursor followed by condensation and subsequent heat treatment. Formation of Ti–O–Ti chains and $Ti(OH)_4$, depend on the quantity of solvent and the rate of hydrolysis [323,324]. Highly crystalline anatase TiO_2 nanoparticles with different shapes and sizes can be obtained with the polycondensation of titanium alkoxide precursor in the presence of tetramethyl ammonium hydroxide [325,326]. By a combination of sol-gel method and anodic alumina membrane (AAM) template TiO_2 nanorods have been successfully fabricated [327,328]. TiO_2 nanotubes can also be obtained using sol-gel method and suitable AAM or other organic template [329,330].

VO_2: Shi and coworkers reported the sol-gel thin film deposition of VO_2 on silicon substrate from molten V_2O_5 precursor by using dip coating method [331].

Al_2O_3: Alumina nanofibers were synthesized via sol-gel methodology from hydrolysis precursor material aluminum isopropoxide and finally cacining at 500°C [332].

1.4.6.4 Microemulsions/micelles approach

This method is one of the newest entrants in the chemical synthesis routes and extremely promising in synthesizing various shapes of nanomaterials that would otherwise seem difficult to obtain. Many research centers have used reverse micelles (water droplets in oil, stabilized by a surfactant) to exercise very accurate control over the size of spherical nano-objects. Surfactants are molecules with a hydrophilic polar head and a hydrophobic hydrocarbon chain.

Such molecules tend to spontaneously position themselves at the interface separating two immiscible media, such as water and oil, to form different aggregates [333]. The very composition of these molecules induces a reduction in surface tension between oil and water. Naturally, if the solvent is capable of simultaneously solubilizing the polar head and the alkyl chains, no aggregate will form. The shape of the surfactant will play an important role in the kind of assembly that eventually forms. If the surfactant has a very bulky polar head with a short chain, the surfactant then has a cone shape; the chains will associate together to form spherical aggregates called micelles. When micelles form at low concentrations, they are spherical and their diameter is determined by the length of the hydrocarbon chain and the polar head. The system is dynamic, although the micelle structure is always conserved. Surfactants in micelles can leave the aggregate and are replaced by others that move freely in the aqueous phase. After a few microseconds, all surfactants making up a micelle have been replaced by others. If the surfactant has a small polar head and a branching (ramified) hydrocarbon chain, the aggregates it forms are spherical droplets of water in oil. These aggregates are commonly called reverse micelles. Both miceller and inverse miceller aggregation have been schematically represented by Figure 1.2. The sizes of these reverse micelles vary with the variation of water content, i.e., the amount of water added to the system.

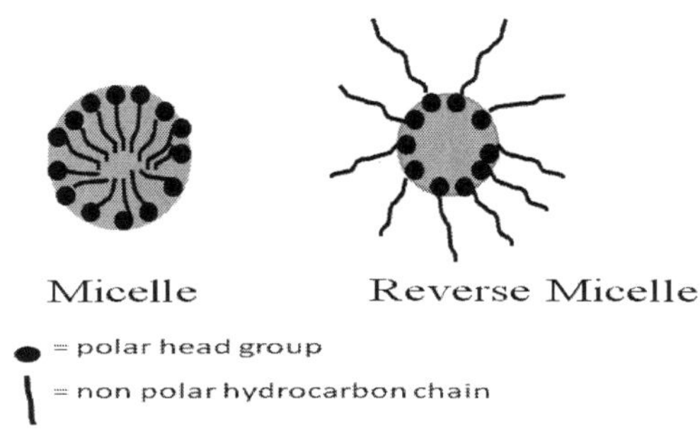

Figure 1.2 Schematic depiction of a miceller and an inverse miceller aggregation.

Reverse micelles or microemulsions are used to prepare nanoparticles through the aqueous solution of reactive precursors, which can be converted to insoluble nanoparticles. Nanomaterial synthesis inside the micelles can be achieved by various methods, such as hydrolysis of reactive precursors, such as alkoxides, precipitation reactions of metal salts, etc., solvent removal, and calcinations leads to the final product. A variety of surfactants can be used in the microemulsion processes, e.g., cetyltrimethylammonium bromide (CTAB), Sodium bis(2-ethylhexyl)sulfosuccinate (generally referred to by its commercial name, Aerosol OT or AOT), decaoxyethylene nonyl phenyl ether (TNT-10), pentadecaoxyethylene nonyl phenyl ether (TNP-35), poly(oxyethylene)$_5$ nonyl phenol ether (NP5), etc. several parameters such as concentration of the reactive precursor in the micelle and the mass percentage of the aqueous phase in the microemulsions, affect the properties, including the particle size, particle size distribution, agglomerate size and phases of the final nanopowders. Some typical examples are provided below:

TiO_2: Li et al. developed TiO_2 nanoparticles with the chemical reactions between $TiCl_4$ solution and ammonia in a reversed microemulsion system consisting of cyclohexane, poly(oxyethylene)$_5$ nonyle phenol ether, and poly(oxyethylene)$_9$ nonyle phenol ether [334]. Shuttle-like crystalline TiO_2 nanoparticles were synthesized by Zhang et al. with hydrolysis of titanium tetrabutoxide in the presence of acids (hydrochloric acid, nitric acid, sulfuric acid, and phosphoric acid) in NP-5 (Igepal CO-520)-cyclohexane reverse micelles at room temperature [335]. In the study carried out by Lim et al. TiO_2 nanoparticles were prepared by the controlled hydrolysis of TTIP in reverse micelles formed in CO_2 with the surfactants ammonium carboxylate perfluoropolyether ($PFPECOO^-NH_4^+$) (MW 587) and poly(dimethyl amino ethyl methacrylate-block-1H,1H,2H, 2H-perfluorooctyl methacrylate) (PDMAEMA-b-PFOMA) [336].

Cu_2O: Cuprous oxide hierarchical double tower-tip like nanostructures were synthesized by mixing two solutions (an aqueous $CuSO_4 \cdot 5H_2O$, EDTA and NaOH mixed solution or a $C_6H_{12}O_6 \cdot H_2O$ solution into an n-octane/cetyl trimethyl ammonium bromide (CTAB)/1 butanol system) and treating them in a Teflon-lined autoclave at 60°C [337].

1.4.6.5 Thermolysis/thermochemical decomposition

This is one of the methods where we can directly synthesize the metal oxide nanomaterials by thermally decomposing a precursor metal compound. It is a chemical decomposition where the precursor material dissociates into the required metal oxide and other by-product. Following are a few examples:

CuO: CuO has been prepared by thermally decomposing Cupric salts [338].

Co_3O_4: Li et al. synthesized nano-sized Co_3O_4 tubes by decomposing $Co(NO_3)_2 \cdot 6H_2O$ at ~500°C using alumina membranes as template [140]. Lin et al. used the same precursor but a different template (SBA-15) and a decomposition temperature of 400°C to obtain mesoporous Co_3O_4 particles of sizes around 4–5 nm [141].

Fe_2O_3: There are plenty of examples in the literatures where researchers have processed Fe_2O_3 by thermally treating Iron (III) bearing precursors such as $Fe_2(CO_4)_3$ [339,340], $Fe_2(SO_4)_3$ [341], $Fe(OH)_3$ [342], $FeCl_3$ [343], $NH_4Fe(C_2O_4)_2$ [344], etc. The nature of the polymorph is dependent on the nature of the precursor and the processing conditions applied.

1.4.6.6 Electrodeposition

It is a very simple method with uncomplicated equipment, controllable experimental conditions, and environment friendly conditions. This method has been utilized to synthesize various types of ceramics and semiconductor type materials. One of the principles to synthesize metal oxide nanomaterials via this method has been to deposit the corresponding metal on a substrate by electrolyzing a solution containing a salt or complex of the metal ion and then let the deposited metal layer to self-oxidize into the desired metal oxide. Utilizing the proper solution precursor coupled with suitable electrochemical parameters the direct precipitation of the metal oxide is also possible. Factors like the choice of electrolytic solutions, applied potential, addition of ligands, etc. can affect the final outcome and can be tailored to obtain the desired morphology. Following are a few examples:

CdO: Wang et al. have reported the synthesis of dendritic CdO nanomaterials through the deposition of Cd on substrate by the electrolysis of $CdCl_2$ solution followed by self-oxidation of the deposited metal layer [224].

CuO: Widmer and coworkers deposited thin film of CuO galvanostatically directly on Au(001) substrate from a solution comprising of 0.2 M tartaric acid (TA), 0.2 M $CuSO_4$, $5H_2O$, and 3 M NaOH in ultrapure water [345].

TiO_2: TiO_2 nanowires have been obtained by deposition Ti from $TiCl_3$ in the pores of AAM followed by annealing at 500°C [346].

Cu_2O: Cu_2O crystals were prepared by cathodic deposition from aqueous solutions of 0.02M $Cu(NO_3)_2{\cdot}6H_2O$. SDS or Sodium sulfate was introduced as additives and the pH was maintained within 3.3 and 4.9 by the addition of NaOH or HCl [347].

MnO_2: Wei et al. used three different precursor solutions (0.3 M $MnSO_4$, 0.3 M $MnSO_4$ + 0.2 M EDTA disodium salt, and 0.3 M $MnSO_4$ + 0.3 M sodium citrate) for the anodic electrodeposition of MnO_2. The deposition current density, electrolyte pH value, and electrolyte temperature were adjusted to be 100 mA/cm^2, 7.0, and 70°C, respectively [348].

1.4.6.7 Oxidation and reduction

In chemistry oxidation reaction implies addition of oxygen or other electronegative species or simply an increment in the oxidation number to the corresponding metal. Reduction reaction implies the decrement in oxidation number. Thus, oxidation reactions can be a metal getting converted into its oxide ($Zn \rightarrow ZnO$) or an oxide getting converted to its higher oxide ($Cu_2O \rightarrow CuO$). A reduction reaction can be conversion of a higher oxide to its lower oxide ($CuO \rightarrow Cu_2O$).Chemical reduction is almost exclusively used in the liquid phase including aqueous and non-aqueous media. Sodium borohydride, aluminohydrides, oxalic acids, tartaric acids are some of the popular reducing agents. Oxidation reactions, on the other hand, can take place in the gas phase as well as in solution phase. Both these principles have been widely exploited in synthesizing various metal oxide nanomaterials. Some typical examples are as follows:

CeO_2: An unusual reaction reported by Inoue et al. is the direct oxidation of Ce metal in 2-methoxyethanol at temperatures between 200°C and 250°C. Most of the product obtained was bulk CeO_2 as a yellow solid, but in addition, they obtained a brown solution of 2 nm CeO_2 nanoparticles [349].

CuO: Chen and coworkers reported the synthesis of CuO nanobelt arrays by oxidizing Cu foil in an aqueous solution of NaOH and oxidizing agent $(NH_4)_2S_2O_8$ [350]. Zhang et al. reported the oxidation of Cu foil into CuO by using naturally occurring oxygen in formamide [351]. Xia and coworkers reported the synthesis of CuO nanowires by thermally oxidizing Cu substrates in air at a temperature of 400–700°C [352].

TiO_2: TiO_2 nanorods can be obtained by the oxidation of Ti metallic plate by H_2O_2 solution. The addition of halide salts can control the crystal phase formation [353]. At high temperature, acetone can be used as the oxygen source [354]. TiO_2 nanotubes can be obtained by the anodic oxidation of Ti plate. Here Ti plate is first anodized followed by annealing at 500°C for 6 h in oxygen [355].

Cu_2O: Liao et al. reported synthesizing Cu_2O nanowires by reducing CuO nanowires in H_2/Ar 20% atmosphere at 200°C for 1 h [231]. Sahoo and coworkers used anodic oxidation of Cu electrode to synthesize Cu_2O nanostructures [356]. Teo et al. reported the synthesis of cubic Cu_2O nanoparticles from a solution containing $Cu(NO_3)_2$ by treating it with DMF as a reducing agent [357]. Monodisperse Cu_2O nanocubes are synthesized by reducing freshly prepared $Cu(OH)_2$ with $N_2H_4 \cdot H_2O$ in water at room temperature [358].

1.4.6.8 Metathesis

In a metathesis reaction two compounds AB and CD exchange their components to produce a new set of compounds AD and CB. Such routes have been explored in the preparation of nanoparticles and metal oxide nanoparticles are no exception. Chen et al. has reported the direct synthesis of metal oxide nanoparticles employing this method and they termed it as "direct liquid phase participation" or DLPP [359]. Here they have synthesized uniformly dispersed metal oxide nanoparticles in non-aqueous medium starting from a metal salt precursor. Through this process all the metallic salts including the lanthanum ones can form their oxides. In a typical synthesis procedure metallic salts (usually chlorides) were dissolved in methanol and that solution was treated with the methanolic solution of equivalent amount of Na_2O. The metal oxide particles are precipitated as a result of the direct reaction between the two precursors. The reaction can be accelerated by sonicating the reaction mixture. A generalized equation of this metathesis mechanism can be represented as follows:

$$MCl_n + \frac{n}{2} Na_2O \rightarrow n NaCl + MO_{n/2} \downarrow \ (n = 1\text{–}6)$$

Chen and his coworkers were able to synthesize a wide variety of metal oxide nanoparticles (Cu_2O, CuO, ZnO, Al_2O_3, Fe_2O_3, Bi_2O_3, TiO_2, SnO_2, CeO_2, Nb_2O_5, WO_3, and $CoFe_2O_4$) through this method [359].

1.4.6.9 Combustion synthesis

Combustion synthesis (CS) has emerged as an advanced technique for the processing of ceramic nanomaterials. In CS the exothermicity of the redox chemical reaction is used to produce important materials. Depending upon the nature of reactants, elements or compounds and the exothermicity, CS is described as self-propagating high-temperature synthesis (SHS); low-temperature combustion synthesis (LCS), solution combustion synthesis (SCS), gel combustion, volume combustion, etc. In SHS, the reactants are pressed into a pellet followed by ignition by an external source (tungsten coil, laser, or microwave) which initiate exothermic reaction ($\Delta H \sim 40$ kcal mol^{-1}). The rate of heat generation should be greater than the rate of heat dissipated. Otherwise the reaction is not going to be self-propagating. For weak exothermic reactions the pellet requires preheating or activation, which can be achieved by heating the pellet uniformly in a controlled manner until the reaction occurs simultaneously through the entire volume. In solution combustion method various oxide materials are synthesized by rapidly heating the aqueous solutions containing a redox system constituted by the nitrate ions of the corresponding metal precursor, acting as an oxidizer and a fuel like urea/hydrazides (carbo hydrazide (CH), oxalyl dihydrazide (ODH), malonic acid dihydrazide (MDH), tetra formal tris azine (TFTA), citric acid (citrate or Pechini method), etc. Upon dehydration, the strongly exothermic redox reaction develops, which is generally self-sustaining and provides the energy for the formation of the oxides. Some of the fuels used can be precursor specific. They are the source of Carbon and Hydrogen, which on combustion form CO_2 and H_2O and liberate heat. Urea has been the popular choice of fuel for most oxide material syntheses due to its ready availability and high exothermicity. A sol-gel variation of the citrate method has also been proposed, actually in the citrate method a viscous gel is already obtained during the heating of the starting solution. The addition of ammonia for adjusting the pH may result in further networking of the precursor molecules, by forced hydrolysis, so justifying the classification as sol–gel combustion method. Although SHS has been successfully used to make non-oxide materials, its application for synthesis of oxide materials was delayed due to economic reasons. Furthermore, it being a solid-state method, phase purity, and particle size control is not possible. Also, due to its high-temperature course it is not suitable for the preparation of nanocrystalline materials. In this context, a low-temperature initiated combustion method [360] developed by Patil et al. at Indian Institute of Science, Bangalore, has carved a niche. This low-temperature initiated self-propagating combustion process is different from the well-known Pechini (citrate process), which uses external heating at high temperatures to burn away the extra carbon. Combustion process is different from pyrolysis since once ignited it does not require external heating. In the synthesis of nanomaterials by soft routes there are two approaches: (i) breaking-down and (ii) building-up processes. Solution combustion synthesis of nanocrystalline oxide materials while appearing to be a breaking-down process is in fact an integrated approach, as the desired oxide products nucleate and grow from the combustion residue. A typical example is **Al_2O_3**. Using hydrous alumina and urea as fuel Zhai et al. synthesized α-alumina powder from LCS method [361]. In fact, there are multiple reports of the synthesis of α-alumina through combustion method with urea as fuel [362,363].

1.4.6.10 Biomimetic approach

Biomimetic method indicates the group of synthesis procedures where a biological entity is exploited as a hard template for structural reproduction or to mimic them by applying advanced nanotechnologies. The biggest advantage of these methods is the exertion of less harsh reaction conditions that is normally associated with nonbiological chemical reaction as accessories. Some typical examples are as follows:

SnO_2: Song and coworkers synthesized 3D hierarchical porous SnO_2 nanostructures by treating an ethanolic solution of $SnCl_4$ with rape pollen grain followed by calcination at temperatures ranging from 600–800°C (sensing) [19].

Ga_2O_3: Lee et al. reported the biomimetic synthesis of β-Ga_2O_3 from $GaCl_3$ solution by treating with NH_4OH and using bolaamphiphile peptide monomer solution as the biotemplate [364].

Apart from the distinct methodologies of chemical processing of metal oxide nanomaterials there are means of influencing a particular reaction methodology by exercising external impetus such as ultrasound effect or microwave heating. These are not any exclusive synthesis procedure, but just accelerating or sometimes even initiating a particular synthetic procedure that would otherwise seem of very low feasibility. Two of the most widely studied of those methods are discussed here.

1.4.6.11 Sonochemical approach

Sonochemistry refers to the field of research where molecules undergo chemical transformation aided by powerful ultrasound radiation (20 kHz to 10 MHz). In a sonochemical reaction the principle reaction mechanism is sequenced by the creation, growth, and collapse of a bubble in the liquid medium. The presence of unseen particles or gas bubbles has been proposed which decreases the intermolecular interaction between the water molecules enabling the formation of the bubble. The growth of the bubble occurs via the diffusion of the solute vapor into the bubble. Consequently when the bubbles reach a critical size they collapse. According to the hot spot mechanism upon the collapse of the bubble a very high temperature is obtained (5000–2500 K) [365]. The collapse being an extremely fast process a very high cooling rate is also obtained and under such a cooling rate the organization and crystallization of the products is hindered. That is why in almost all the cases dealing with volatile precursors, where gas phase reactions are predominant we end up obtaining amorphous nanophase particles. The fast reaction kinetics probably does not allow for the growth of the nuclei containing them in the nano realm. In other cases when the reaction occurs involving a nonvolatile precursor it takes place in a 200 nm ring surrounding the collapsing bubble [366]. In this case the reaction mainly occurs in the liquid phase and the product can be either nanoamorphous or nanocrystalline depending on the ring temperature. Parameters like temperature, frequency of the ultrasound radiation, power, gas under which the reaction takes place, gas pressure, etc. influence the shape and size of the obtained nanoparticles [367]. This method has been employed by a lot of researchers to synthesize metal oxide nanoparticles. Some examples are as follows:

Fe_2O_3: Sonication of a decalin solution of iron penta carbonyl in air yields amorphous Fe_2O_3 [368].

Cr_2O_3: Ultrafine powders of Cr_2O_3 has been obtained by the sonochemical reduction of aqueous solution containing $(NH_4)_2Cr_2O_7$ [369].

TiO_2: Yu et al. reported the synthesis of nano-sized phocatalytic titania by the hydrolysis of $Ti(OPr)_4$ in an ethanolic or aqueous medium under ultrasonic environment. This method gave yield to both anatase and brookite phases [370]. A novel sonochemical method for the direct preparation of anatase nanocrystalline TiO_2 was reported by Guo et al. [371]. Nanocrystalline TiO_2 was synthesized by the hydrolysis of titanium tetrabutyl in the presence of water and ethanol under high-intensity ultrasonic irradiation (20 kHz, 100 W cm^{-2} at 363 K for 3 h). The structure and particle sizes of the product were dependent upon the reaction temperature, the acidity of the medium, and the reaction time.

CuO: Alammar and coworkers synthesized CuO nanorods from copper (II) acetate and NaOH solution with the aid of a room temperature ionic liquid (RTIL) $[C_{4mim}][Tf_2N]$ in ultrasound-assisted method [372].

1.4.6.12 Microwave heating

Microwaves are electromagnetic radiation in the frequency region of 0.3–300 GHz. Only a narrow frequency window centered at 900 MHz and 2.45 GHz are allowed for heating purposes. Microwave acts as high frequency electric field and will heat any material containing mobile electric charges, such as polar molecules in a solvent or ions in a solid. Conventional heating normally utilizes furnace or an oil bath and the walls of the reactor are heated initially through a conduction or convection method and the core sample takes much longer time to attain the desired temperature. Microwaves are able to heat the materials without heating the entire oil bath or the furnace, thus saving time and energy. Microwaves have been used to accelerate chemical reactions for quite a while now, although the solution phase nanoparticles synthesis using microwave irradiation is quite a recent phenomenon. The exact mechanism remains highly speculative as of now, but the transfer of energy from the microwave to the material is believed to occur either through resonance or through relaxation, which results in rapid heating. Recent innovations in microwave reactor technique now allow controlled parallel and automated sequential processing under sealed vessel conditions and the use of continuous or stop-flow reactors for scale-up purposes. In addition, dedicated vessels for solid-phase synthesis, for performing transformations using pre-pressurized conditions and for a variety of other special applications, have been developed.

CuO: Wang et al. synthesized CuO nanosheets and nanowhiskers applying microwave radiation to a solution containing $CuCl_2$, room temperature ionic liquid 1-n-butyl-3-methyl imidazolium tetrafluoroborate ([BMIM][BF_4]) and NaOH [373].

SnO_2: G. Xi reported the surfactant assisted microwave synthesis of microporous SnO_2 nanostructures from the precursor $SnCl_4$ and surfactant CTAB. The reaction vessel was exposed to microwave heating for 10 min at 160°C [374].

ZrO_2: Liang et al. reported the polymer-stabilized tetragonal ZrO_2 nanopowders with average size of ca. 2.0 nm have been prepared by microwave heating in an aqueous solution containing $Zr(NO_3)_4 \cdot 5H_2O$, PVA, and NaOH [375].

Top-down approach: Top-down processes are effective examples of solid-state processing of nanomaterials. The most acknowledged top-down methods are discussed below.

1.4.7 Milling

One nanofabrication process of major industrial importance is high-energy ball milling also known as mechanical alloying or mechanical attrition. Coarse-grained materials in the form of powders are crushed mechanically in rotating drums by hard steel or tungsten carbide balls, usually under controlled atmospheric conditions to prevent unwanted reactions such as oxidation. This repeated deformation can cause large reductions in grain size via the formation and organization of grain boundaries within the powder particles. The grinding works on the principle of critical speed. The critical speed can be understood as that speed after which the steel balls (which are responsible for the grinding of particles) start rotating along the direction of the cylindrical device; thus causing no further grinding. High-quality ball mills are potentially expensive and can grind mixture particles to as small as 5 nm, enormously increasing surface area and reaction rates. A nanometer dispersion of one phase in another can also be achieved. Microstructures and phases produced in this way can often be thermodynamically metastable. The technique can be operated at a large scale, hence the industrial interest. Generally any form of mechanical deformation under shear conditions and high strain rates can lead to the formation of nanostructures, since energy is being continuously pumped into crystalline structures to create lattice defects. The severe plastic deformation that occurs during machining, cold rolling, drawing, cyclic deformation, or sliding wear has also been reported to form nanostructured material. Mechanical milling has been quite widely applied to synthesize metal oxide nanoparticles. Arbain et al. reported the fabrication of hematite nanoparticles through planetary milling [376]. Salah et al. fabricated nanostructured ZnO particles through high-energy ball milling method [377].

1.4.8 Lithography

Conventional lithographic processes are akin to the emulsion-based photographic process and can be used to create nanostructures by the formation of a pattern on a substrate via the creation of a resist on the substrate surface. One lithographic method uses either visible or ultraviolet (UV) light, X-rays, electrons or ions to project an image containing the desired pattern onto a surface coated with a photoresist material; this method requires the prior fabrication of an absorbing mask through which the parallel radiation passes before shadowing onto the photoresist. Alternatively, primary patterning (or direct writing) of the resist is possible using a focused electron, ion or possibly X-ray beam; here either the focused beam or the resist itself is scanned according to the desired pattern design. These different techniques are generally termed photolithography, X-ray lithography, electron beam lithography, or ion beam lithography depending on the radiation employed. The resist material, typically a polymer, metal halide or metal oxide, is chemically changed during irradiation, often altering the solubility or composition of the exposed resist. The pattern transfer processes that utilize the patterned resist may be divided into solution-based wet chemical etching procedures, dry etching in a reactive plasma, doping using ion implantation techniques, or thin film deposition. Dry etching is the collective term for a range of techniques such as reactive ion etching (RIE) and chemically assisted ion beam etching (CAIBE), which are used extensively for high-resolution pattern transfer; both methods produce, either directly or indirectly, reactive ion species which combine with the elements in the substrate material to form volatile reaction products which are removed into a vacuum system. Following are some typical examples of lithographic synthesis of metal oxide nanoparticles:

ZrO_2: Park et al. reported the UV assisted imprinting of ZrO_2 thin films on Si substrate starting from the precursor solution of zirconyl 2-ethylhexanoate [378].

CuO: Standard lithographic techniques have been employed to fabricate CuO single-nanowire devices.

1.5 Nature of Bonding and Defects

An understanding of not only the crystal structure and bonding, but also the local microstructures, which result from defects, is necessary to understand the structure of complex transition metal oxides. Of course, at the nanoscale, the number of defects due to edges, corners, centers, and other surface imperfections is greatly enhanced by the large surface area. Determining the structure and nature of bonding between its components at the atomic level is a difficult task for nanoparticles in general. These particles are made of extremely small crystallites or are amorphous in nature. In spite of these difficulties recent advancement in crystallographic techniques employing XRD, electron and neutron diffraction techniques has helped scientists make noteworthy headway in determining their structure.

For bulk structures five types of crystals can be defined based on the nature of bonding interaction between component units-metallic, ionic, covalent, molecular (van der Waal) and hydrogen bonded. These structural classifications also apply in the nanoscale. In case of ionic bonding the lattice is comprised of highly electropositive and highly electronegative components. In fact, the presence of pure ionic interaction within a lattice of metal oxide is a utopian concept with some degree of covalent interaction accompanying the ionic force. The degree of covalency can be predicted by using Kashimir Fajan's quanticule theory [379], which through its simple, if inexact, approach states the polarizing power of the positive species and the polarizability of the negative species as the deciding factor. According to his theory small size and high charge of the cation and large size and large charge of the anion increases the amount of covalent interaction within the lattice. Except for a few metal oxides like MgO, CaO most of the others are only partly ionic. Van der Waals interaction is normally prevalent in layered structured transition metal oxides. In some oxide hydrates or hydroxyl oxides, hydrogen bonding is found to exist. Some of the transition metal oxides are also found to be metallic in nature.

Simple oxides can be hemioxides (M_2O; e.g., Cu_2O, etc.), monoxides (MO; e.g., ZnO, CuO, MnO, FeO, MgO, CaO, etc.), sesquioxides (M_2O_3; e.g., Al_2O_3, Fe_2O_3, Mn_2O_3, etc.) or dioxides (TiO_2, VO_2, MnO_2, SnO_2, etc.). Among the transition metal oxides, only copper can form hemioxide. Metal Oxide of the formula MO can have the rock salt (B1), CsCl (B2), Zn blend (B3), wurzite (B4), or NiAs (B8) structure [380]. Alkaline earth metal oxides, such as MgO and monoxides of 3D transition metals, as well as Lanthanides and Actinides, such as TiO, NiO, EuO, and NpO, exhibit the rock salt structure with the 6:6 octahedral coordination. One notable exception being CuO, which has a monoclinic crystal structure and square planar co-ordination. Some dioxides such as ZrO_2 and HfO_2 have fluorite structure, whereas TiO_2 conforms to the rutile structure. Table 1.2 lists some important metal oxide nanoparticles and their crystal structure.

Table 1.2 Different metal oxides and their crystal structure, melting and boiling point, hardness, and refractive index

Name	Crystal structure	Melting point	Boiling point	Hardness (Moh)	Refractive index
MgO	Cubic	2852°C	3600°C	5.8	1.736
CaO	Cubic	2572°C	2850°C	3.5	1.84
CeO_2	Cubic	2400°C	3500°C	6.0	2.35
TiO_2	Tetragonal (rutile, anatase), orthorhombic (brookite)	1843°C	2972°C	6.0–6.5	2.5–2.6
ZrO_2	Monoclinic (<1170°C), tetragonal (1170–2370°C), cubic (>2370°C)	2715°C	4300°C	6.5–7.08	2.13
VO_2 (rutile)	Tetragonal (above 68°C), monoclinic (below (58°C))	1967°C	—	—	—
V_2O_5	Orthorhombic	690°C	1750°C	3.0–3.5	2.42
Cr_2O_3	Corundum (rhombohedral)	2435°C	4000°C	8–8.5	2.55

Name	Crystal structure	Melting point	Boiling point	Hardness (Moh)	Refractive index
WO_3	Triclinic (−50 to 17°C), monoclinic (17–330°C), orthorhombic (330–740°C), tetragonal (>740°C)	1473°C	~1700	2.5	2.3
MnO	Cubic	1945°C	—	5.0–6.0	2.16
Mn_3O_4	Cubic (spinel)	1567°C	2847°C	5.5	2.46
MnO_2	Rutile (β, tetragonal) Orthorhombic (γ)	535°C (decomp)	NA	6.0–0.65	2.4
FeO	Cubic (rock salt)	1377°C	3414°C	5.0–5.5	2.23
Fe_3O_4	Cubic (inverse spinel	1537°C	2222°C	5.5–6	2.42
Fe_2O_3	Rhomohedral/ corundum (α), cubic (γ)	1566°C (decomp)	NA	5.5–6.5	2.94–3.22
RuO_2	Tetragonal (rutile)	1200°C (sublimes)	—	—	—
CoO	Cubic (rock salt)	1933°C	3800°C	310–345 Knoop	1.74
Cu_2O	Cubic	1235°C	1800°C	3.5–4.0	2.56
ZnO	Hexagonal (most stable)	1975°C	2360°C	4.5	2.00
CdO	FCC (Cubic rock salt)	900–100°C	1559°C	3.0	2.49
Al_2O_3	Hexagonal (α-alumina), cubic (γ-alumina)	2072°C	2977°C	9.0	1.77
Ga_2O_3	Rhombohedral (α), monoclinic (β)	1900°C 1725°C	—	—	α 1.92–1.95 β 1.84–1.88
In_2O_3	Cubic, rhombohedral	850°C	2000°C	—	2.10
SnO_2	Rutile (tetragonal)	1630°C	1800–1900°C	6.0–7.0	2.006
ITO	Cubic, rhombohedral	1800–2200°C	Sublimes at 982°C	—	1.76

Note: The oxides have been arranged as per their position in the periodic table.
NA, not applicable.

In ionic solids we normally observe point, linear, planar, and volumetric defects. Because of the high surface-to-volume ratio resulting in high percentage of atoms at the surface and limited number of atoms within the lattice, the presence of defect sites has a huge influence on the chemistry and the bonding of oxide nanoparticles. Of course, for nanoscaled materials defects due to edges, centers and other surface imperfections is greatly enhanced due to the large surface area.

Point defects are caused by the absence of one of the constituent atoms (or ions) in the usual lattice site or by the presence of it in the interstitial positions. Foreign atoms or ions present in the lattice represent another type of point defect. These defects cause displacements of neighboring ions or atoms because of polarization in the surrounding region. A cationic vacancy in an ionic solid will have an electronegative charge, causing displacements of neighboring anions [381]. The most usual point defects in ionic solids are Schottky pairs (pairs of cation and anion vacancies) and Frenkel defects (cation or anion interstitial plus a vacancy) [380]. Boundaries between small crystallites (grain boundaries), stacking faults, crystallographic shear planes, twin boundaries, and antiphase boundaries are planar defects. Three-dimensional volumetric defects are a result of segregating point defects. Creation of defects is generally an endothermic process. Thus, the formation energies of vacancies in ionic solids are generally 2 eV or more. Therefore, the intrinsic defect concentration in these solids is extremely low even at high temperatures [380].

In close packed solids, one also often encounters stacking faults. For example, in a solid with cubic close packing, ABC ABC ABC, there can be a fault such as ABC AB ABC. Other types of planar defects include tilt boundary (array of periodically space of edge dislocations), twist boundary (array of screw dislocations), twin boundary (a layer with mirror plane symmetry with respect to the rotation of one part of the crystal, on a specific plane, with respect to another, and antiphase boundary across which the sublattice occupation becomes interchanged) [382]. The highly ionic nature of the metal oxides allows the formation of many stable defect sites including edges, corners, anion, and cation vacancies. In the insulator ceramics the most common type of defects are coordinately unsaturated ions due to the presence of planes, edges, corners, anion/cation vacancies and electron excess centers.

Metal oxides can be both semiconductors and insulators (some metal oxides at certain condition can display properties that resembles metallic characteristics; e.g., rutile VO_2 above 68°C). Table 1.2 incorporates the crystal structure, corresponding melting & boiling point, hardness and refractive index of some of the important metal oxides.

As mentioned before, nanomaterials and nanotechnology in general have existed since a long time back in history. So why is it that only recently the research and activities related to nanomaterials have sparked up? The reason without any doubt lies in the invention of various characterization techniques and analytical procedure that have had a huge impact on the research front of nanomaterials. Nowadays, it is practically possible to view atoms, a development that was unimaginable even a couple of decades back. The characterization of nanoparticles is generally divided in two categories: structure analysis and property measurements. Structure analysis is carried away by various microscopic and spectroscopic methods, whereas property measurements are rather challenging. Here some of the structural characterization tools utilized for metal oxide nanomaterials are briefly discussed.

1.6 Structural Characterization Tools for Metal Oxide Nanomaterials

1.6.1 X-Ray Diffraction

Ninety-five percent of solid materials can be described as crystalline. When X-ray interacts with a crystalline substance it produces a diffraction pattern that is exclusive of that material. The X-ray diffraction (XRD) pattern of a pure substance is, therefore, like a fingerprint of the substance. The powder diffraction method is thus ideally suited for characterization and identification of polycrystalline phases. An electron in an alternating electromagnetic field will oscillate with the same frequency as the field. When an X-ray beam hits an atom, the electrons around the atom start to oscillate with the same frequency as the incoming beam. In almost all directions, we will have destructive interference, that is, the combining waves are out of phase and there is no resultant energy leaving the solid sample. However, the atoms in a crystal are arranged in a regular pattern, and in a very few directions we will have constructive interference. The waves will be in phase and there will be well-defined X-ray beams leaving the sample at various directions. Hence, a diffracted beam may be described as a beam composed of a large number of scattered rays mutually reinforcing one another. In X-ray diffraction work we normally distinguish between single crystal and polycrystalline or powder applications. The single crystal sample is a perfect (all unit cells aligned in a perfect extended pattern) crystal with a cross section of about 0.3 mm. The single crystal diffractometer and associated computer package is used mainly to elucidate the molecular structure of novel compounds, either natural products or man-made molecules. Powder diffraction is mainly used for "finger print identification" of various solid materials, e.g., asbestos, quartz. In powder or polycrystalline diffraction it is important to have a sample with a smooth plane surface. If possible, we normally grind the sample down to particles of about 0.002 mm to 0.005 mm cross section. The ideal sample is homogeneous and the crystallites are randomly distributed. Ideally we now have a random distribution of all possible *h*, *k*, *l* planes. Only crystallites having reflecting planes (*h*, *k*, *l*) parallel to the specimen surface will contribute to the reflected intensities. If we have a truly random sample, each possible reflection from a given set of *h*, *k*, *l* planes will have an equal number of crystallites contributing to it. We only have to rock the sample through the glancing angle THETA in order to produce all possible reflections. A typical diffraction spectrum consists of a plot of reflected intensities versus the detector angle 2-THETA or THETA depending on the goniometer configuration. International Center Diffraction Data (ICDD), formerly known as (JCPDS) Joint Committee on Powder Diffraction Standards, is the organization that maintains the database of inorganic and organic spectra. The database is available from the diffraction equipment manufacturers or from ICDD direct. Currently the database is supplied either on magnetic or optical media. Two database versions are available: the PDF I and the PDF II. The PDF I database contains information on d-spacing, chemical formula, relative intensity, RIR quality information, and routing digit. The information is stored in an ASCII format in a file called PDF1.dat. For search/match purposes most diffraction manufactures are reformatting the file in a more efficient binary format. The PDF II database contains full information on a particular phase including cell parameters.

Scintag's newest search/match and look-up software package is using the PDF II format. Optimized database formats, index files and high-performance PC-computers make PDF II search times extremely efficient. The XRD pattern of a given sample not only assists in identifying the sample, but also in texture analysis through the determination of the preferred orientation of the crystallites in polycrystalline aggregates, and in stress determination of the sample.

1.6.2 Small Angle X-Ray Scattering

Strong diffraction peaks result from constructive interference of X-rays scattered from ordered arrays of atoms and molecules. A lot of information can be obtained from the angular distribution of scattered intensity at low angles. Fluctuations in electron density over lengths on the order of 10 nm or larger can be sufficient to produce an appreciable scattered X-ray intensities at angles $2\theta < 5°$. These variations can be difference in density or difference in composition or due to both. The amount and angular distribution of scattered intensity provides information, such as the size of very small particles or their surface area per unit volume, regardless of whether the sample or particles are crystalline or amorphous.

1.6.3 Scanning Electron Microscopy

This is one of the most widely used techniques for the characterization of metal oxides. Not only does the scanning electron microscopy (SEM) provide topographical information, but also chemical composition information near the surface. It has a resolution approaching a few nanometers and can operate at magnifications that can be adjusted from ~10× to over 500,000×. In a typical SEM, a source of electrons is focused into a beam, with a very fine spot size of ~5 nm and having energy ranging from a few hundred eV to few hundred KeV that is rastered over the surface of the specimen by deflection coils. As the electrons strike and penetrate the surface, a number of interactions occur that result in the emission of electrons and photons from the sample, and SEM images are produced by collecting the emitted electrons on a cathode ray tube (CRT). When a high-energy primary electron interacts with an atom, it undergoes either inelastic scattering with atomic electrons or elastic scattering with the atomic nucleus. In an inelastic collision with an electron, the primary electron transfers part of its energy to the other electron. When the energy transferred is large enough, the other electron will emit from the sample. If the emitted electron has energy of less than 50 eV, it is referred to as a secondary electron. Backscattered electrons are the high-energy electrons that are elastically scattered and essentially possess the same energy as the incident or primary electrons. An additional electron interaction in the SEM is that the primary electron collides with and ejects a core electron from an atom in the sample. The excited atom will decay to its ground state by emitting either a characteristic X-ray photon or an Auger electron, both of which have been used for chemical characterization. Combining with chemical analytical capabilities, SEM not only provides the image of the morphology and microstructures of bulk and nanostructured materials and devices, but can also provide detailed information of chemical composition and distribution. The theoretical limit to an instrument's resolving power is determined by the wavelengths of the electron beam used and the numerical aperture of the system. The resolving power, *R*, of an instrument is defined as:

$$R = \frac{\lambda}{2NA},$$

where λ is the wavelength of electrons used and NA is the numerical aperture, which is engraved on each objective and condenser lens system.

1.6.4 Transmission Electron Microscopy

A high-voltage electron beam is allowed to pass through a very thin sample and the portion of the sample that does not allow the electron beam to pass present an image. Due to the flourished technological advances these microscopes nowadays can have resolution up to 0.1 nm, thus enabling the imaging of heavy atoms is possible in some cases. The sizes of metal oxides particles, shapes, and corresponding crystal structure can all be determined through this technique. Normally, the sample is prepared by dilute dispersion of the nano-sized metal oxide particles in liquid medium followed by dropcasting the dispersion on a carbon coated copper grid. Another alternative is to embed the particle in a solid organic polymer, slicing very thin sections, and passing electron beams through that section. The high magnification or resolution of all transmission electron microscopy (TEM) is a result of the small effective electron wavelengths, λ, which is given by the de Broglie relationship:

$$\lambda = \frac{h}{\sqrt{2mqV}},$$

where m and q are the electron mass and charge, h is Planck's constant, and V is the potential difference through which electrons are accelerated. The higher the operating voltage of a TEM instrument, the greater is its lateral spatial resolution. High-voltage TEM instruments have the additional advantage of greater electron penetration, because high-energy electrons interact less strongly with matter than lower-energy electrons. So it is possible to work with thicker samples on a high-voltage TEM. One shortcoming of TEM is its limited depth resolution. Electron scattering information in a TEM image originates from a three-dimensional sample, but is projected onto a two-dimensional detector. Therefore, structure information along the electron beam direction is superimposed at the image plane. Selected-area diffraction (SAD) offers a unique capability to determine the crystal structure of individual nanomaterials, such as nanocrystals and nanorods, and the crystal structures of different parts of a sample. Selected-area diffraction patterns are often used to determine the Bravais lattices and lattice parameters of crystalline materials by the same procedure used in XRD. Although TEM has no inherent ability to distinguish atomic species, electron scattering is exceedingly sensitive to the target element and various spectroscopies are developed for the chemical composition analysis. Examples include Energy-dispersive X-ray Spectroscopy (EDS) and Electron Energy Loss Spectroscopy (EELS).

1.6.5 Scanning Probe Microscopy

Scanning probe microscopy (SPM) is unique among imaging techniques in that it provides three-dimensional (3-D) real-space images and among other analysis techniques in that it allows spatially localized measurements of structure and properties. Under optimum conditions subatomic spatial resolution is achieved. Scanning probe microscopy is a general term for a family of microscopes depending on the probing forces used. Two major members are scanning tunneling microscopy (STM) and atomic force microscopy (AFM). This technique is based on dragging a very sharp needlelike probe across the sample very close to the sample surface. In a typical STM, a conductive tip is positioned above the surface of a sample. When the tip moves back and forth across the sample surface at very small intervals, the height of the tip is continually adjusted to keep the tunneling current constant. The tip positions are used to construct a topographic map of the surface. An extremely sharp tip usually made of metals or metal alloys, such as tungsten or Pt–Ir alloy is mounted on to a three-dimensional positioning stage made of an array of piezoelectrics. Such a tip would move above the sample surface in three dimensions accurately controlled by the piezoelectric arrays. Typically the distance between the tip and the sample surface falls between 0.2 and 0.6 nm, thus a tunneling current in the scale of 0.1–10 nA is commonly generated. The scanning resolution is about 0.01 nm in the x–y direction and 0.002 nm in the z direction, offering true atomic resolution three-dimensional image. In spite of atomic resolution and other advantages, STM is limited to an electrically conductive surface since it is dependent on monitoring the tunneling current between the sample surface and the tip for conducting samples. For nonconducting samples AFM can be used where the probe is actually in contact with the sample surface. The surface can be mapped through the weak interaction between the probe tip and the sample surface. In the AFM mode the resolution is substantially poorer than it is for the tunneling mode. Near-field scanning optical microscopy (NSOM) can be considered as yet another member of SPM.

1.6.6 Differential Scanning Calorimetry

Heating nanostructured material can lead to crystal growth by amalgamation (exothermic), melting (endothermic) or crystal phase change (exo/endo). When the nanoparticles are ligated (e.g., organic coatings on several metal oxide nanoparticles) the ligands can disappear and other chemical reactions can take place, which can be both exo- or endothermic. By differential scanning calorimetry

(DSC) this transformation can be monitored and the extent of exothermicity or endothermicity can be determined.

1.6.7 Superconducting Quantum Interference Magnetometry

This method is especially useful for magnetic nanomaterials. It is an extremely sensitive instrument that can yield information on blocking temperatures, Néel temperatures, coercivity, saturation magnetization, ferromagnetism, and superparamagnetism. The device is cooled with liquid helium and the sample can be studied starting from the liquid helium temperature up to well above the room temperature.

1.6.8 Ultraviolet-Visible Spectroscopy

Ultraviolet-Visible spectroscopy (UV-Vis) is a widely used absorption or reflectance spectroscopic technique in the ultraviolet-visible region (190–800 nm). Radiation is composed of a stream of photons and atoms and molecules exist in a number of defined energy states or levels and a change of level requires the absorption or emission of an integral number of a unit of energy called a quantum, or in our context, a photon. Absorption in the UV-Vis region corresponds to transitions among the electronic energy levels within the molecule. For ultraviolet and visible wavelengths, one should expect from this discussion that the absorption spectrum of a molecule (i.e., a plot of its degree of absorption against the wavelength of the incident radiation) should show a few very sharp lines. In practice it is found that the ultraviolet and visible spectrum of most molecules consists of a few humps rather than sharp lines. These humps show than the molecule is absorbing radiation over a band of wavelengths. One reason for this band, rather than line absorption is that an electronic level transition is usually accompanied by a simultaneous change between the more numerous vibrational levels. In addition, when molecules are closely packed together as they normally are in solution, they exert influences on each other which slightly disturb the already numerous, and almost infinite energy levels and blur the sharp spectral lines into bands. An UV-Vis absorption peak for a metal oxide nanoparticle can be typical with the electronic transition it undergoes upon absorption of the radiation thus giving insights into the electronic (molecular orbitals, etc.) structure of the species. Besides the absorption peak can be utilized in finding out the optical bandgap of the species through Tauc plot. The square root of the product of the absorption coefficient and photon energy is plotted versus photon energy. The curve should have a section of straight line. If extended to the x-axis, the x-intercept of this line gives the optical bandgap. Besides the method can be used to find out the concentration of the absorbing species in the solution through Beer–Lambert Law ($A = \varepsilon cl$; $A \equiv$ absorbance, $\varepsilon \equiv$ molar absorption coefficient in L mol^{-1} cm^{-1}, $c \equiv$ molar concentration in mol L^{-1}, $l \equiv$ sample path length, length of the sample holder).

1.6.9 Secondary Ion Mass Spectroscopy

Secondary ion mass spectrometry (SIMS) is a surface analysis technique used in materials science and surface science to characterize the surface and the sub-surface region of materials. It effectively employs the principles of mass spectrometry of ionized particles, which are emitted when a solid surface is bombarded with energetic primary particles such as electrons, ions, neutrals, or protons. In SIMS the solid surfaces of the specimen are sputtered by a focused ion beam and the ejected secondary electrons are collected and analyzed to determine the elemental, isotopic, or molecular composition of the surface. SIMS is the most sensitive surface analysis technique, being able to detect elements present in the parts per billion range. The sample chamber requires high vacuum ($\sim 10^{-6}$ torr) to ensure that the secondary ions do not collide with the background gases on their way to the detector. Two types of SIMS are utilized in surface science: (i) static SIMS, which is usually a surface atomic monolayer analysis, associated with a pulsed ion beam and a time of flight mass spectrometer, and (ii) dynamic SIMS, which is a process involved with bulk analysis. Dynamic SIMS mostly utilizes a DC primary ion beam and a magnetic sector or quadrupole mass spectrometer.

1.6.10 Bruner–Emett–Teller Gas Adsorption Surface Area Measurement and Pore Structure Analysis

This method allows the measurement of surface area of metal oxide nanopowders by nitrogen gas adsorption at near liquid nitrogen temperature. Physisorption of a layer of nitrogen gas allows the calculation of the surface area through the plotting of pressure vs. gas uptake. In recent years not only the surface area measurement but also pore size distributions, pore volumes and even fractal volumes can be determined because of the recent developments in this technology.

1.6.11 X-Ray Photoelectron Spectroscopy

It is a quantitative non-destructive spectroscopic technique also known as Electron spectroscopy for chemical analysis or ESCA. This technique is routinely employed in investigating the elemental composition of the sample surface, surface contaminations, chemical state identification of the elements, binding energy determination of the one or more electronic states and the density of electronic states present in the sample. In X-ray photoelectron spectroscopy monoenergetic soft X-rays (highly focused 20 to 200 μm beam of monoenergetic aluminum *K*α X-rays or a broad 10–30 mm beam of non-monochromatic magnesium X-rays) bombard the sample material in ultra high vacuum condition ($\sim 10^{-9}$ torr), causing electrons to be ejected (normally from the top 1–12 nm of the sample being analyzed). Identification of the elements present in the sample can be directly made from the kinetic energies and the numbers of ejected photoelectrons. A typical X-ray photoelectron spectroscopy (XPS) spectrum is a plot of the number of electrons detected (sometimes per unit time) (*y*-axis, ordinate) versus the binding energy (BE) of the electrons detected (*x*-axis, abscissa). Each element produces a characteristic set of XPS peaks at characteristic BE values that directly identify each element that exist in or on the surface of the material being analyzed. These characteristic peaks correspond to the electronic configuration of the electrons within the atoms, e.g., 1s, 2s, 2p, 3s, etc. The number of detected electrons in each of the characteristic peaks is directly related to the amount of element within the area (volume) irradiated. XPS typically detects all the elements having an atomic number above 2 and can detect elements up to parts per million range.

1.6.12 Raman Spectroscopy

Raman spectroscopy is a spectroscopic technique based on inelastic scattering of monochromatic light, usually from a laser source (in the visible, near infrared or near ultraviolet range). Photons of the laser light are absorbed by the sample and then re-emitted. In this inelastic scattering frequency of the re-emitted light can be shifted up or down in comparison with the original monochromatic frequency. This shift (known as the Raman effect) provides valuable information regarding the vibrational, rotational, and other low-frequency modes in a system. Infrared spectroscopy yields similar but complementary information. An intense monochromatic light beam (typically a laser beam) can be considered an oscillating electromagnetic wave. Upon impingement on the samples the beams induces electrical dipole moment in the samples, which deforms the molecules (by distorting the electron clouds associated with the chemical bond in the sample). Because of the periodical deformation, the molecule starts vibrating with a characteristic frequency. Such oscillating dipoles can emit light of three different frequencies. If the excited molecule returns back to the same vibrational state the emitted light has the same frequency of that of the original radiation source. This is called Rayleigh scattering and the molecule has no Raman active modes. If the resulting frequency is diminished then the frequency is called Stokes frequency and if the frequency is enhanced the frequency is called anti-stokes frequency. Both Stokes and anti-Stokes scattering are applicable for molecules with Raman active modes of vibrational/rotational motion. The shift is a direct result of the difference in the energy between the initial and the final energy state of the molecule. The amount of the polarizability change will determine the Raman scattering intensity. The pattern of shifted frequencies is determined by the rotational and vibrational states of the sample. Among the different types of advanced Raman spectroscopy surface enhanced Raman spectroscopy or SERS has become really important tool for material scientists. This is normally performed in a silver or gold colloid or a substrate containing silver or gold. Surface plasmons of silver and gold are excited by the laser, resulting in an increase in the electric fields surrounding the metal. Given that Raman intensities are proportional to the electric field, there is large increase in the measured signal (by up to 10^{11}). Raman is particularly effective with minerals and ceramics as water does not interfere with the Raman spectral analysis. IR results from the absorption of energy by vibrating chemical bonds. Raman scattering results from the same types of transitions, but the selection rules are different so that weak bands in the IR may be strong in the Raman and vice versa.

1.6.13 Fourier Transform Infrared Spectroscopy

Fourier transform infrared spectroscopy (FTIR) spectroscopy is an absorption spectroscopic technique which investigates the vibrational transitions in molecules generated by various stretching and bending motions in the chemical bonds present in the corresponding molecule. An infrared spectrum represents a fingerprint of a sample with absorption peaks which correspond to the frequencies of vibrations between the bonds of the atoms making up the material. Because each different material is a unique combination of atoms, no two compounds produce the exact same infrared spectrum. Therefore, infrared spectroscopy can result in a positive identification (qualitative analysis) of every different kind of material. In addition, the size of the peaks in the spectrum is a direct indication of the amount of material present. With modern software algorithms, infrared is an excellent tool for quantitative analysis. The application of traditional infrared spectroscopy to low concentration measurements, such as ambient air measurements, is limited by several factors. First is the significant presence of water vapor, CO_2, and methane, which strongly absorb in many regions of the infrared (IR) spectrum. Consequently, the spectral regions that can easily be used to search for pollutants are limited to 760–1300 cm^{-1}, 2000–2230 cm^{-1}, and 2390–3000 cm^{-1}. Another problem is that the sensitivity is not enough to detect very small concentrations in the sub-ppm level. Finally, spectral analysis was difficult since subtraction of background spectra had to be carried out manually. The development of FTIR in the early 1970s provided a quantum leap in infrared analytical capabilities for monitoring trace pollutants in ambient air. This technique offered a number of advantages over conventional infrared systems, including sensitivity, speed, and improved data processing. Rather than shining a *monochromatic* beam of light at the sample, this technique shines a beam containing many different frequencies of light at once, and measures how much of that beam is absorbed by the sample. Next, the beam is modified to contain a different combination of frequencies, giving a second data point. This process is repeated many times. Afterwards, a computer takes all these data and works backwards to infer what the absorption is at each wavelength.

1.6.14 Electron Paramagnetic Resonance/Electron Spin Resonance

Electron paramagnetic resonance (EPR), or electron spin resonance (ESR), spectroscopy is a technique for studying chemical species that have one or more unpaired electrons, such as organic and inorganic free radicals or inorganic complexes possessing a transition metal ion. When the molecules of a solid exhibit paramagnetism as a result of the unpaired electron spins, transitions can be induced between spin states by applying a magnetic field and then supplying electromagnetic energy, usually in the microwave range of frequencies. The resulting absorption spectra are described as ESR or EPR. The interaction of an external magnetic field with an electron spin depends upon the magnetic moment associated with the spin, and the nature of an isolated electron spin is such that two and only two orientations are possible. The application of the magnetic field then provides a magnetic potential energy that splits the spin states by an amount proportional to the magnetic field (Zeeman effect), and then radio frequency radiation of the appropriate frequency can cause a transition from one spin state to the other. The energy associated with the transition is expressed in terms of the applied magnetic field B, the electron spin g-factor g, and the constant μB, which is called the Bohr magneton.

Electron paramagnetic resonance spectroscopy is widely used in investigating various defect centers in metal oxide nanoparticles, which can often lead to free radical formation. Also various radiation-induced reactions are best studied using the EPR.

1.6.15 Luminescence Spectroscopy

Light is a form of energy. For the creation of light any other form of energy must be supplied. In luminescence some energy source promotes an electron within a material from a lower energy ground state to a higher energy excited state and the electron release to a lower energy state by releasing the energy in the form of light. The source of energy cannot be heat energy (incandescence); thus making it a form of cold body radiation. This form of emission spectroscopy can be of multiple types depending on the source of radiation. Some of the most frequently used in material science are briefly discussed below.

Chemiluminescence: Chemiluminescence or chemoluminescence is the emission of energy as a result of a chemical reaction. Given reactant A and B react to produce the product C and accompanying radiation, through an excited intermediate I,

$$A + B = [I] \rightarrow C + h\nu \text{ (chemilumilescence)}$$

Chemiluminescence differs from the fluorescence in a way that a chemical reaction rather than the absorption of radiation is producing the electronically excited intermediate state. This can be termed as the antithesis of photochemical reaction.

Electroluminescence: Electroluminescence (EL) is one of the few phenomena where direct conversion of electrical energy into heat energy takes place without the release of heat. This is an optical-electrical phenomenon where light is emitted in response to the passage of an electric current or to a strong electric field. Usually, in a semiconductor material, electroluminescence originates as a consequence of radiative recombination between electron and hole and the process can be both intrinsic of the material or as a result of charge injection. In case of intrinsic electroluminescence thermal activation and the electrical field liberates electron in to the conduction band where the electrons are accelerated by the field until they collide with luminescent centers, ionizing them by ejecting electrons. Photons are generated when these ejected electrons recombines with an ionized atomic center. In the second case, a voltage may be applied to p–n junction causing a current to flow as electrons flow from the n-type material into a p-type material or an electrode contacts a crystal to cause electrons or holes to flow. The luminescence occurs when the electrons lose energy through combining with positive centers or holes. For instance, Sun et al. reported the electroluminescence obtained from n-ZnO/p-Si heterojunction containing an UV and a green band. By controlling the forward bias they obtained stronger UV light than the visible band [383].

Cathodoluminescence: Cathodoluminescence is another example of optical-electrical phenomenon. Here, a beam of electrons generated by an electron gun impinges upon a luminescent material causing the material to emit visible light. CL color and intensity are dependent on the characteristics of the sample and on the working conditions of the electron gun. Cathodoluminescence is employed in scanning electron microscope with specialized optical detector and also in optical cathodoluminescence microscope. Here it assists in investigating the internal structures of ceramic, rocks, or glasses.

Photoluminescence: Photoluminescence (abbreviated as PL) is a process where luminescence originates as a result of absorption of photons (electromagnetic radiation). This is a very fast process. The time between absorption of photons and the re-radiation is extremely small (in the order of 10 nanoseconds). However, this period can be manipulated through expert instrumentation and sample modification and can be extended to a few minutes to several hours. Luminescence process can be affected by factors such as pre- and postfilter, self-absorption, refraction, solvent and temperature. Several forms of photoluminescence processes are present, e.g., fluorescence, photoluminescence excitation (PLE), time-resolved photoluminescence (TRL), phosphorescence, etc. valuable information such the defect states of a metal oxide surface, its bandgap (approximately), etc. van be obtained through photoluminescence spectroscopy.

1.7 The Others (Non-Metal Oxides)

Non-metal oxides are likely to occupy a significant section whenever someone should attempt a discussion on oxide materials. There are non-metal oxide materials that are extremely important looking at from any particular point of view. The single most important oxide material hydrogen monoxide (H_2O) or water is composed of a non-metal and is the basis of evolution of life on this planet. There are oxide materials such as silica (SiO_2), phosphorous oxides (P_2O_5, P_2O_3), nitrogen oxides (NO, NO_2, N_2O_5), carbon oxides (CO, C_3O_2, CO_2), boron oxides (B_2O_3, B_2O, B_6O), sulfur oxides (SO_2, SO_3), and oxides of metalloids (As_2O_3, Sb_2O_3), etc., that are equally if not more important from industrial perspective, than their metallic counterparts. Recently graphene oxide has been a hot shot material due to its effectiveness as sensor, FET (field-effect transistor) material and in biomedical applications among others. Besides, the material can be used as a precursor to synthesize chemically modified graphene—a material with outstanding electronic properties. A few of these oxides have been covered to some degrees with respect to their particular properties in some of the other chapters in this book. As mentioned earlier, this chapter has been solely intended to the nanomaterials of metal oxides and thus the non-metal oxides, although no less important, have been left out of the current discussion.

1.8 Future Prospects for Metal Oxide Nanomaterials

There is little if any doubt that the huge steps taken in the last few decades to attain excellence in various characterization techniques have made the whole nano-branch a much more pursuable area of research. Thus, the structure-property investigation and consequent correlation strategies being set on a firm footing, the principle focus has now turned into the synthesis of various metal oxide nanomaterials in shapes, complementary to specific applications and then, engaging them to actual technological endeavors. So in the coming days of research on the metal oxide nanomaterials we are likely to see more innovative synthesis strategies targeted at specific shapes and phases of these materials. Although materials such as ZnO and TiO_2 have become quite saturated in terms of the synthesis strategies available and the applications they can generate, they will continue to grab researcher's interest because of their industrial and technological utilities. However, going by the current trend, materials like SnO_2, MnO_2, and VO_2 are most probably going to dominate the coming era of metal oxide nanomaterials research. Complex transition metal oxides (perovskite, etc.) are another breed of materials that are gaining in prominence from basic science perspective as well as technological viewpoint. The unique electronic structures and the understanding of the complex interplay between their structural, magnetic, and electronic properties poses a great challenge to the fellows of condensed matter physics. Besides the same interplays leads to a tremendous set of physical properties such as metallic, semiconducting, or insulating behavior, high dielectric constants, piezoelectricity, ferroelectricity, ferromagnetism, metal-insulator transitions, and high-temperature superconductivity, which are extremely attractive for use in modern electronic devices such as nonvolatile memory, integrated circuits, or new types of sensors and actuators, and in addition offer great prospects for the development of future technologies. More and more works are likely to be done for the bio-implications of these materials and the environmental adaptability or toxicity of these materials is going to become a matter of prime interest.

References

1. Altman, C. (2002). "Converging Technologies: The Future of the Global Information Society," First Committee Chair Report to the UN General Assembly. UNISCA.
2. Drexler, K. E. (1986). *Engines of Creation: Coming Era of Nanotechnology*, (Anchor Books, New York)
3. Feynman, R. (1960). *Eng Sci*, **23**, pp. 22–36.
4. Stix, G. (2001). *Sci Am*, **285**, pp. 32–37.
5. Fujishima, A., and Honda, K., (1972). *Nature*, **238**, pp. 37–38.
6. Canle, L., Santaballa, J. A., and Vulliet, E. (2005). *J. Photochem. Photobiol. A*, **175**, pp. 192–200.
7. Chatterjee, D., and Dasgupta, S. (2005). *J. Photochem. Photobiol. C*, **6**, pp. 186–205.
8. Lhomme, L., Brosillon, S., Wolbert, D., and Dussaud, J. (2005). *Appl. Catal. B*, **61**, pp. 227–235.
9. Li, S. X., Zheng, F. Y., Liu, X. L., Wu, F., Deng, N. S., and Yang, J. H. (2005). *Chemosphere*, **61**, pp. 589–594.
10. Bosc, F., Edwards, D., Keller, N., Keller, V., and Ayral, A. (2006). *Thin Solid Films*, **495**, pp. 272–279.
11. Cao, Y., Yang, W., Zhang, W., Liu, G., and Yue, P. (2004). *New J. Chem.*, **28**, pp. 218–222.
12. Tokudome, H., and Miyauchi, M. (2004). *Chem. Lett.*, **33**, pp. 1108–1109.
13. Klosek, S., and Raftery, D. (2001). *J. Phys. Chem. B*, **105**, pp. 2815–2819.
14. Peng, T., Zhao, D., Dai, K., Shi, W., and Hirao, K. (2005). *J. Phys. Chem. B*, **109**, pp. 4947–4952.
15. Beyers, E., Cool, P., and Vansant, E. F. (2005). *J. Phys. Chem. B*, **109**, pp. 10081–10086.
16. Sakthivel, S., Neppolian, B., Shankar, M. V., Arabindoo, B., Palanichamy, M., and Murugesan, V. (2003). *Sol. Energy Mater. Sol. Cells*, **77**, pp. 65–82.

17. Yan, H., Hou, J., Fu, Z., Yang, B., Yang, P., Liu, K., Wen, M., Chen, Y., Fu, S., and Li, F. (2009). *Mater. Res. Bull.*, **44**, pp. 1954–1958.

18. Pardeshi, S. K., and Patil, A. B. (2009). *J. Hazard. Mater.*, **163**, pp. 403–409.

19. Tian, J., Chen, L., Yin, Y., Wang, X., Dai, J., Zhu, Z., Liu, X., and Wu, P. (2009). *Surf. Coat. Technol.*, **204**, pp. 205–214.

20. Pardeshi, S. K., and Patil, A. B. (2009). *J. Mol. Catal. A: Chem.*, **308**, pp. 32–40.

21. Chu, D., Masuda, Y., Ohji, T., and Kato. K. (2010). *Langmuir*, **26**, pp. 2811–2815.

22. Kim, J., Lee, C. W., and Choi, W. (2010). *Environ. Sci. Technol.*, **44**, pp. 6849–6854.

23. Abe, R., Takami, H., Murakami, N., and Ohtani, B. (2008). *J. Am. Chem. Soc.*, **130**, pp. 7780–7781.

24. Gondala, M. A., Hameedb, A., and Yamani Z. H. (2004). *J. Mol. Catal. A: Chem.*, **222**, pp. 259–264.

25. Miseki, Y., Kusama, H., Sugihara, H., and Sayama, K. (2010). *J. Phys. Chem. Lett.*, **1**, pp. 1196–1200.

26. Grätzel, M. (2004). *J. Photochem. Photobiol. A*, **164**, pp. 3–14.

27. Zukalova, M., Zukal, A., Kavan, L., Nazeeruddin, M. K., Liska, P., and Grätzel, M. (2005). *Nano Lett.*, **5**, pp. 1789–1792.

28. Adachi, M., Murata, Y., Okada, I., and Yoshikawa, S. (2003). *J. Electrochem. Soc.*, **150**, pp. G488-G493.

29. Han, H., Zan, L., Zhong, J., and Zhao, X. (2005). *J. Mater. Sci.*, **40**, pp. 4921–4923.

30. Bisquert, J., Garcia-Belmonte, G., and Fabregat-Santiago, F. (1999). *J. Solid State Electrochem.*, **3**, pp. 337–347.

31. Olsen, L. C., Addis F. W., and Bohara, R. C. (1980). *14th IEEE photovoltaic Specialist Conf. Proc.*, IEEE, New York pp. 462.

32. Mittiga, A., Salza, E., Sarto, F., Tucci, M., and Vasanthi, R. (2006). *Appl. Phy. Lett.*, **88**, pp. 163502 (1–2).

33. Longcheng, W., and Meng, T. (2007). *Electrochem. and Solid State Lett.*, **10**, pp. H248–H250.

34. Kunhee, H., and Meng, T. (2009). *Sol. Energ. Mat. Sol. Cells*, **93**, pp. 153–157.

35. Sadek, A. Z., Zheng, H., Breedon, M., Bansal, V., Bhargava, K., Latham, S. K., Yu, Z. Z., Paul Z. L., and Spizzirri, G. (2009). *Langmuir*, **25**, pp. 9545–9551.

36. Zheng, H., Tachibana, Y., and Kalantar-zadeh, K. (2010). *Langmuir*, **26**, pp. 19148–19152.

37. Hsu, Y. F., Xi, Y. Y., Yip, C. T., Djurišić, A. B., and Chan, W. K. (2008). *J. App. Phys.*, **103**, pp. 083114 (1–4).

38. Chiu W.-H., Lee, C.-H., Cheng, H.-M., Lin, H.-F., Liao, S.-C., Wub, J.-M., and Hsieh, W.-F. (2009). *Energy Environ. Sci.*, **2**, pp. 694–698.

39. Snaith, H. J., and Ducati, C. (2010). *Nano Lett.*, **10**, pp. 1259–1265.

40. Mishra, A., Fischer, M. K. R., and Bäuerle, P. (2009). *Angew. Chem. Int. Ed.*, **48**, pp. 2474–2499.

41. Senevirathna, M. K. I., Pitigala, P., Premalal, E. V. A., Tennakone, K., Kumara, G. R. A., and Konno, A. (2007). *Sol. Energy Mater. Sol. Cells*, **91**, pp. 544–547.

42. Green, A. N. M., Palomares, E., Haque, S. A., Kroon, J. M., and Durrant, J. R. (2005). *J. Phys. Chem. B.* **109**, pp. 12525–12533.

43. Ramasamy, E., and Lee, J. (2010). *J. Phys. Chem. C.* **114**, pp. 22032–22037.

44. Lepleux, L., Chavillon, B., Pellegrin, Y., Blart, E., Cario, L., Jobic, S., and Odobel, F. (2009). *Inorg. Chem.*, **48**, pp. 8245–8250.

45. Irwin, M. D., Servaites, J. D., Buchholz, D. B., Leever, B. J., Liu, J., Emery, J. D., Zhang, M., Song, J.-H., Durstock, M. F., Freeman, A. J., Bedzyk, M. J., Hersam, M. C., Chang, R. P. H., Ratner, M. A., and Marks, T. J. (2011). *Chem. Mater.*, **23**, pp. 2218–2226.

46. Yang, H., Guai, G. H., Guo, C., Song, Q., Jiang, S. P., Wang, Y., Zhang, W., and Li, C. M. (2011). *J. Phys. Chem. C.*, **115**, pp. 12209–12215.

47. Berzelius, J. J. (1836). *New Philos. J.*, XXI, p. 223.

48. Trovarelli, A. (2002). *Catalysis by Ceria and Related Materials*, Imperial College Press: London.

49. Fu, Q., Saltsburg, H., and Flytzani-Stephanopoulos, M. (2003). *Science*, **301**, pp. 935–938.

50. Yang, F., Graciani, J., Evans, J., Liu, P., Hrbek, J., Fdez. Sanz, J., and Rodriguez, J. A. (2011). *J. Am. Chem. Soc.*, **133**, pp. 3444–3451.

51. Vayssilov, G. N., Lykhach, Y., Migani, A., Staudt, T., Petrova, G. P., Tsud, N., Skála, T., Bruix, A., Illas, F., and Prince, K. C. (2011). *Nat. Mater.*, **10**, pp. 310–315.

52. Nilius, N., Kozlov, S. M., Jerratsch, J.-F., Baron, M., Shao, X., Vines. F., Shaikhutdinov, S., Neyman, K. M., and H.-J. Freund. (2012). *ACS Nano*, **6**, pp. 1126–1133.

53. Shimura, K., Yoshida, T., and Yoshida, H. (2010). *J. Phys. Chem. C*, **114**, pp. 11466–11474.

54. Yanagida, T., Sakata, Y., and Imamura, H. (2004). *Chem. Lett.*, **33**, pp. 726–727.

55. Sakata, Y., Matsuda, Y., Yanagida, T., Hirata, K., Imamura, H., and Teramura, K. (2008). *Catal. Lett.*, **125**, pp. 22–26.

56. Jochum, W., Penner, S., Kramer, R., Feottinger, K., Rupprechter, G., and Kleotzer, B. (2008). *J. Catal.*, **256**, pp. 278–286.

57. Yuliati, L., Itoh, H., and Yoshida, H. (2008). *Chem. Phys. Lett.*, **452**, pp. 178–182.

58. Hou, Y., Wu, L., Wang, X., Ding, Z., Li, Z., and Fu, X. (2007). *J. Catal.*, **250**, pp. 12–18.

59. Pang, H., Gao, F., Chen, Q., Liua, R., and Lu, Q. (2012). *Dalton Trans.*, **41**, pp. 5862–5868.

60. Xie, X., Li, Y., Liu, Z.-Q., Haruta, M., and Shen, W. (2009). *Nature*, **458**, pp. 746–749.

61. Liang, Y., Li, Y., Wang, H., Zhou, J., Wang, J., Regier, T., and Dai. H. (2011). *Nat. Mater.*, **10**, pp. 780–786.

62. Miao, S., and Deng, Y. (2001). *App. Catal. B: Environ.*, **31**, pp. L1–L4.

63. Ayoub, M., Irfan, M. F., and Abdullah, A. Z. *Environmental Progress & Sustainable Energy* DOI 10.1002/ep.10577.

64. Jiao, F., and Frei, H. (2009). *Angew. Chem. Int. Ed.*, **48**, pp. 1841–1844.

65. Wang, C.-M., Fan, K.-N., and Liu, Z.-P. (2007). *J. Am. Chem. Soc.*, **129**, pp. 2642–2647.

66. Song, Y.-Q., Liu, H.-M., and He, D.-H. (2010). *Energ. Fuel.*, **24**, pp. 2817–2824.

67. Zhang, X., Wang, H., and Xu, B.-Q. (2005). *J. Phys. Chem. B*, **109**, pp. 9678–9683.

68. Zhang, B., Zhang, C., He, H., Yu, Y., Wang, L., and Zhang, J. (2010). *Chem. Mater.*, **22**, pp. 4056–4061.

69. Madhavaram, H., Idriss, H., Wendt, S., Kim, Y. D., Knapp, M., Over, H., Abmann, J., Löffler, E., and Muhlerz, M. (2001). *J. Catal.*, **202**, pp. 296–307.

70. Zhan, B.-Z., White, M. A., Sham, T.-K., Pincock, J. A., Doucet, R. J., Rao, K. V. R., Robertson, K. N., and Cameron, T. S. (2003). *J. Am. Chem. Soc.*, **125**, pp. 2195–2199.

71. Knapp, M., Seitsonen, A. P., Kim, Y. D., and Over, H. (2004). *J. Phys. Chem. B*, **108**, pp. 14392-14397.

72. Sui, N., Duan, Y., Jiao, X., and Chen, D. (2009). *J. Phys. Chem. C*, **113**, pp. 8560–8565.

73. Liang, S., Teng, F., Bulgan, G. R. Zong, G., and Zhu, Y. (2008). *J. Phys. Chem. C*, **112**, pp. 5307–5315.

74. Wang, L.-C., Liu, Y.-M., Chen, M., Cao, Y., He, H.-Y., and Fan, K.-N. (2008). *J. Phys. Chem. C*, **112**, pp. 6981–6987.

75. Yu, X., He, J., Wang, D., Hu, Y., Tian, H., and He, Z. (2012). *J. Phys. Chem. C*, **116**, pp. 851–860.

76. Cheng, F., Su, Y., Liang, J., Tao, Z., and Chen, J. (2010). *Chem. Mater.*, **22**, pp. 898–905.

77. Chang, J. S., Park, S. E., and Park, M. S. (1997). *Chem. Lett.*, **11**, pp. 1123–1124.

78. Jiang, K., Yang, J., Hu, B., Yang, X., Mao, L., Yuan, Y., and Zhang, G. (1998). *Hyperfine Interact.*, **111**, pp. 45–49.

79. Jung, H., and Thomson, W. J. (1993). *J. Catal.*, **139**, pp. 375–382.

80. van der Kraan, A. M., (1998). *Hyperfine Interact.*, **111**, pp. 23–24.

81. Baerns, M., Borchert, H., Kalthoff, R., Kässner, P., and Majunke, F. (1992). *Stud. Surf. Sci. Catal.*, **72**, pp. 57–70.

82. Sazonov, V. A., Prokudina, N. A., and Gavrilov, V. Y. (1997). *Russ. J. Appl. Chem.*, **70**, pp. 87.

83. Kotanigawa, T., Yokoyama, S., and Yamamoto, M. (1989). *Fuel*, **68**, pp. 618–621.

84. Miki, J., Asanuma, M., and Shikada, T. (1994). *Chem. Commun.* **14**, pp. 1685–1686.

85. Adams, C. R., and Jennings, T. J. (1964). *J. Catal.*, **3**, pp. 549–558.

86. Holbrook, L., and Wise, H. (1971). *J. Catal.*, **20**, pp. 367–373.

87. Li, J., Yan, R., Xiao, B., Liang, D. T., and Lee, D. H. (2008). *Energ. Fuel.*, **22**, pp. 16–23.

88. Li, J., Yan, R., Xiao, B., Liang, D. T., and Du, L. (2008). *Environ. Sci. Technol.*, **42**, pp. 6224–6229.

89. Gondala, M. A., Hameedb, A., and Yamani, Z. H. (2004). *J. Mol. Catal A: Chem.*, **222**, pp. 259–264.

90. Wang, Y., Jiang, X., and Xia, Y. (2003). *J. Am. Chem. Soc.*, **125**, pp. 16176–16177.

91. Liu, Y., Koep, E., and Liu, M. (2005). *Chem. Mater.*, **17**, pp. 3997–4000.

92. Chen, Y. J., Nie, L., Xue, X. Y., Wang, Y. G., and Wang, T. H. (2006). *Appl. Phys. Lett.*, **88**, pp. 083105 (1–3).

93. Song, F., Su, H., Chen, J., Moon, W.-J., Laucd, W. M., and Zhang, D. (2012). *J. Mater. Chem.*, **22**, pp. 1121–1126.

94. Huang, J., Matsunaga, N., Shimanoe, K., Yamazoe, N., and Kunitake, T. (2005). *Chem. Mater.*, **17**, pp. 3513–3518.

95. Dorfman, A., Kumar, N., and Hahm, J. (2006). *Langmuir*, **22**, pp. 4890–4895.

96. Dorfman, A., Kumar, N., and Hahm, J. (2006). *Adv. Mater.*, **18**, pp. 2685–2690.

97. Liao, L., Lu, H. B., Li, J. C., Liu, C., Fu, D. J., and Liu, Y. L. (2007). *Appl. Phys. Lett.* **91**, pp. 173110/1–173110/3.

98. Al-Hilli, S. M., Al-Mofarji, R. T., and Willander, M. (2006). *Appl. Phys. Lett.*, **89**, pp. 173119–173121.

99. Feng, P., Wan, Q., and Wang, T. H. (2005). *Appl. Phys. Lett.*, **87**, pp. 213111/1–213111/3.

100. Yang, K., She, G.-W., Wang, H., Ou, X.-M., Zhang, X.-H., Lee, C.-S., and Lee, S.-T. (2009). *J. Phys. Chem. C*, **113**, pp. 20169–20172.

101. Varghese, O. K., Gong, D., Paulose, M., Ong, K. G., and Grimes, C. A. (2003). *Sens. Actuators B.*, **B93**, pp. 338–344.

102. Varghese, O. K., and Grimes, C. A. (2003). *J. Nanosci. Nanotechnol.*, **3**, pp. 277–293.

103. Mor, G. K., Carvalho, M. A., Varghese, O. K., Pishko, M. V., Grimes, C. A. *J. Mater. Res.*, **2004**, 19, pp. 628–634.

104. Zheng, L., Xu, M., and Xu, T. (2000). *Sens. Actuators B*, **66**, pp. 28–30.

105. Demarne, V., Balkanova, S., Grisel, A., Rosenfeld, D., and Levy, F. (1993). *Sens. Actuators B*, **14**, pp. 497–498.

106. Trinchi, A., Li, Y. X., Wlodarski, W., Kaciulis, S., Pandolfi, L., Viticoli, S., Comini, E., and Sberveglieri, G. (2003). *Sens. Actuators B*, **B95**, pp. 145–150.

107. Savage, N. O., Akbar, S. A., and Dutta, P. K. (2001). *Sens. Actuators B*, **B72**, pp. 239–248.

108. Tan, O. K., Cao, W., Zhu, W., Chai, J. W., and Pan, J. S. (2003). *Sens. Actuators B*, **B93**, pp. 396–401.

109. Tang, H., Prasad, K., Sanjines, R., and Levy, F. (1995). *Sens. Actuators B*, **B26**, pp. 71–75.

110. Gou, X., Wang, G., Yang, J., Park, J., and Wexler, D. (2008). *J. Mater. Chem.*, **18**, pp. 965–969.

111. Zhang, X., Wang, G., Liu, X., Wu, J., Li, M., Gu, J., Liu, H., and Fang, B. (2008). *J. Phys. Chem. C*, **112**, pp. 16845–16849.

112. Li, J.-Y., Xiong, S., Pan, J., and Qian, Y. (2010). *J. Phys. Chem. C*, **114**, pp. 9645–9650.

113. Zhao, Y., Zhao, J., Li, Y., Ma, D., Hou, S., Li, L., Hao, X., and Wang, Z. (2011). *Nanotechnology*, **22**, pp. 115604 (1–9).

114. Zhu, G., Xu, H., Xiao, Y., Liu, Y., Yuan, A., and Shen, X. (2012). *ACS Appl. Mater. Interfaces*, **4**, pp. 744–751.

115. Liao, L., Zhang, Z., Yan, B., Zheng, Z., Bao, Q. L., Wu, T., Li C. M., Shen, Z. X., Zhang, J. X., Gong, H., Li, J. C., and Yu, T. (2009). *Nanotechnology*, **20**, pp. 085203 (1–6).

116. Chen, J., Wang, K., Hartman, L., and Zhou, W. (2008). *J. Phys. Chem. C*, **112**, pp. 16017–16021.

117. Lai, X., Wang, D., Han, N., Du, J., Li, J., Xing, C., Chen, Y., and Li, X. (2010). *Chem. Mater.*, **22**, pp. 3033–3042.

118. Singh, N., Gupta, R. K., and Lee, P. S. (2011). *ACS Appl. Mater. Interfaces*, **3**, pp. 2246–2252.

119. Li, C., Zhang, D., Lei, B., Han, S., Liu, X., and Zhou, C. (2003). *J. Phys. Chem. B*, **107**, pp. 12451–12455.

120. Yao, K., Caruntu, D., Zeng, Z., Chen, J., O'Connor, C. J., and Zhou, W. (2009). *J. Phys. Chem. C*, **113**, pp. 14812–14817.

121. Elouali, S., Bloor, G., Binions, R., Parkin, I. P., Carmalt, C. J., and Darr, J. A. (2012). *Langmuir*, **28**, pp. 1879–1885.

122. Zhang, D., Liu, Z., Li, C., Tang, T., Liu, X., Han, S., Lei, B., and Zhou, C. (2004). *Nano Lett.*, **4**, pp. 1919–1924.

123. Wang, L., Teleki, A., Pratsinis, S. E., and Gouma, P. I. (2008). *Chem. Mater.*, **20**, pp. 4794–4796.

124. Li, X.-L., Lou, T.-J., Sun, X.-M., and Li, Y.-D. (2004). *Inorg. Chem.*, **43**, pp. 5442–5449.

125. Ma, J., Zhang, J., Wang, S., Wang, T., Lian, J., Duan, X., and Zheng, W. (2011). *J. Phys. Chem. C*, **115**, pp. 18157–18163.

126. Lee, D.-S., Nam, K.-H., and Lee, D.-D. (2000). *Thin Solid Films*, **375**, pp. 142–146.

127. Ionescu, R., Llobet, E., Brezmes, J., Vilanova, X., and Correig, X. (2003). *Sens. Actuators B-Chem.*, **95**, pp. 177–182.

128. Byon, J. W., Kim, M.-B., Kim, M. H., Kim, S. Y., Lee, S. H., Lee, B. C., and Baik, J. M. (2012). *J. Phys. Chem. C*, **116**, pp. 226–230.

129. Baik, J. M., Kim, M. H., Larson, C., Yavuz, C. T., Stucky, G. D., Wodtke, A. M., and Moskovits, M. (2009). *Nano Lett.*, **9**, pp. 3980–3984.

130. Zhang, R. H. Zhang, X. T., and Hu, S. M. (2008). *Anal. Chem.*, **80**, pp. 2982–2987.

131. Sun, Z., Zhang, X., Na, N., Liu, Z., Han, B., and An, G. (2006). *J. Phys. Chem. B*, **110**, pp. 13410–13414.

132. Poizot, P., Laruelle, S., Grugeon, S., Dupont, L., and Tarascon, J.-M. (2000). *Nature*, **407**, pp. 496–499.

133. Li, W., Dahn, J. R., and Wainwright, D. S. (1994). *Science*, **264**, pp. 1115–1118.

134. Tsang, C., and Manthiram, A. (1997). *J. Electrochem. Soc.*, **144**, pp. 520–524.

135. Li, H., He, P., Wang, Y., Hosono, E., and Zhou, H. (2011). *J. Mater. Chem.*, **21**, pp. 10999–11009.

136. Gao, S., Yang, S., Shu, J., Zhang, S., Li, Z., and Jiang, K. (2008). *J. Phys. Chem. C*, **112**, pp. 19324–19328.

137. Wang, Z., Su, F., Madhavibc, S., and Lou, X. W. (2011). *Nanoscale*, **3**, pp. 1618–1623.

138. Gao, S., Yang, S., Shu, J., Zhang, S., Li, Z., and Jiang, K. (2008). *J. Phys. Chem. C*, **112**, pp. 19324–19328.

139. Dar, M. A., Nam, S. H., Kim, Y. S., and Kim, W. B. (2010). *J Solid State Electrochem.*, **14**, pp. 1719–1726.

140. Li, W.-Y., Xu, L.-N., and Chen, J. (2005). *Adv. Funct. Mater.*, **15**, pp. 851–857.

141. Lin, Z., Yue, W., Huang, D., Hu, J., Zhang, X., Yuanb, Z.-Y., and Yang, X. (2012). *RSC Adv.*, **2**, pp. 1794–1797.

142. Yuan, L., Lu, X.-H., Xiao, X., Zhai, T., Dai, J., Zhang, F., Hu, B., Wang, X., Gong, L., Chen, J., Hu, C., Tong, Y., Zhou, J., and Wang, Z. L. (2012). *ACS Nano*, **6**, pp. 656–661.

143. Bao, L., Zang, J., and Li, X. (2011). *Nano Lett.*, **11**, pp. 1215–1220.

144. Qu, Q., Zhang, P., Wang, B., Chen, Y., Tian, S., Wu, Y., and Holze, R. (2009). *J. Phys. Chem.* C, **112**, pp. 4406–4417.

145. Zhang, G., Zheng, L., Zhang, M., Guo, S., Liu, Z.-H., Yang, Z., and Wang. Z. (2012). *Energy Fuels*, **26**, pp. 618–623.

146. Patel. M. N., Wang, X., Slanac, D. A., Ferrer, D. A., Dai, S., Johnston, K. P., and Stevenson, K. J. (2012). *J. Mater. Chem.*, **22**, pp. 3160–3169.

147. Ghodbane, O., Pascal, J.-L., and Favier, F. (2009). *ACS Appl. Mater. Interfaces*, **1**, pp. 1130–1139.

148. Toupin, M., Brousse, T., and Belanger, D. (2004). *Chem. Mater.*, **16**, pp. 3184–3190.

149. Hou, Y., Cheng, Y., Hobson, T., and Liu, J. (2010). *Nano Lett.*, **10**, pp. 2727–2733.

150. Li, G.-R., Feng, Z.-P., Ou, Y.-N., Wu, D., Fu, R., and Tong, Y.-X. (2010). *Langmuir*, **26**, pp. 2209–2213.

151. Devraj, S., and Munichandraiah, N. (2008). *J. Phys. Chem. C*, **112**, pp. 4406–4417.

152. Yang, L., Cheng, S., Ding, Y., Zhu, X., Wand, Z. L., and Liu, M. (2012). *Nano Lett.*, **12**, pp. 321–325.

153. Zhu, T., Chen, J. S., and Lou, X. W. (2010). *J. Mater. Chem.*, **20**, pp. 7015–7020.

154. Conway, B. E. (1999). *Electrochemical Supercapacitors*; Kluwer Academic/Plenum: New York.

155. Hu, C. C., Chen, W. C., and Chang, K. H. (2004). *J. Electrochem. Soc.*, **151**, pp. A281–A290.

156. Hu, C. C., and Chen, W. C. (2004). *Electrochim. Acta*, **49**, pp. 3469–3477.

157. Hu, C.-C., Chang, K.-H., Lin, M.-C., and Wu, Y.-T. (2006). *Nano Lett.*, **6**, pp. 2690–2695.

158. Zang, J., Bao, S.-J., Li, C. M., Bian, H., Cui, X., Bao, Q., Sun, C. Q., Guo, J., and Lian, K. (2008). *J. Phys. Chem. C*, **112**, pp. 14843–14847.

159. Bi, R.-R., Wu, X.-L., Cao, F.-F., Jiang, L.-Y., Guo, Y.-G., and Wan, L.-J. (2010). *J. Phys. Chem. C*, **114**, pp. 2448–2451.

160. Dubala, D. P., Dhawale, D. S., Salunkhea, R. R., Pawar, S. M., Fulari, V. J., and Lokhande, C. D. (2009). *J. Alloy. Compd.*, **484**, pp. 218–221.

161. Meher, S. K., Justin, P., and Rao, G. R. (2011). *ACS Appl. Mater. Interfaces*, **3**, pp. 2063–2073.

162. Justin, P., Meher, S. K., and Rao, G. R. (2010). *J. Phys. Chem. C*, **114**, pp. 5203–5210.

163. Sun, X., wang, G., Hwangb, J.-Y., and Lian, J. (2011). *J. Mater. Chem.*, **21**, pp. 16581–16588.

164. Ferrucci, J. T., and Stark, D. D. (1990). *Am. J. Roentgenol.*, **155**, pp. 943–950.

165. Kalish, H., Arbab, A. S., Miller, B. R., Lewis, B. K., Zywicke, H. A., Bulté, J. W. M., Bryan, L. H., and Frank, J. A. (2003). *Magn. Reson. Med.*, **50**, pp. 275–282.

166. Schulze, E., Ferrucci, J. T., Poss, K., Lapointe, L., Bogdanova, A., and Weissleder, R. (1995). *InVest. Radiol.*, **30**, pp. 604–610.

167. Moore, A., Marecos, E., Bogdanov, A., and Weissleder, R. (2000). *Radiology*, **214**, pp. 568–574.

168. Arbab, A. S., Yocum, G. T., Kalish, H., Jordan, E. K., Anderson, S. A., Khakoo, A. Y., Read, E. J., and Frank, J. A. (2004). *Blood*, **104**, pp. 1217–1223.

169. Bulte, J. W. M., Douglas, T., Witwer, B., Zhang, S.-C., Strable, E., Lewis, B. K., Zywicke, H., Miller, B., van Gelderen, P., Moskowitz, B. M., Duncan, I. D., and Frank, J. A. (2001). *Nat. Biotechnol.*, **19**, pp. 1141–1147.

170. Gao, Y. (2005). *Nanotechnologies for the Life Scienc*es Vol. 1, ed. Kumar, C. Chapter 3 "Biofunctionalization of Nanomaterials," (Wiley: New York).

171. Kohler, N., Sun, C., Fichtenholtz, A., Gunn, J., Fang, C., and Zhang, M. (2006). *Small*, **2**, pp. 785–792.

172. Xuan, S., Wang, F., Lai, J. S. M., Sham, K. W. Y., Wang, Y.-X. J., Lee, S.-F., Yu, J. C., Cheng, C. H. K., and Leung, K. C.-F. (2011). *ACS Appl. Mater. Interfaces*, **3**, pp. 237–244.

173. Jordan, A., Scholz, R., Maier-Hauff, K., Johannsen, M., Wust, P., Nadobny, J., Schirra, H., Schmidt, H., Deger, S., Loening, S., Lanksch, W., and Felix, R. (2001). *J. Magn. Magn. Mater.*, **225**, pp. 118–126.

174. Wust, P., Gneveckow, U., Johannsen, M., Bohmer, D., Henkel, T., Kahmann, F., Sehouli, J., Felix, R., Ricke, J., and Jordan, A. (2006). *Int. J. Hyperthermia*, **22**, pp. 673–685.

175. Nie, L., Gao, L., Yan, X., and Wang, T. (2007). *Nanotechnology*, **18**, pp. 015101(1–6).

176. Lie, L., Gao, L., Feng, P., Zhang, J., Fu, X., Liu, Y., Yan, X., and Wang, T. (2006). *Small*, **2**, pp. 621–625.

177. Zhang, C., and Yang, H., eds. (2006). *Introduction to Biomaterials*, Chemical Industry Press: Beijing.

178. Liu, X., Wei, W., Yuan, Q., Zhang, X., Li, N., Du, Y., Ma, G., Yana, C., and Ma, D. (2012). *Chem. Comm.*, **48**, pp. 3155–3157.

179. Philip, J., Shima, P. D., and Raj, B. (2007). *Appl. Phys. Lett.*, **91**, pp. 203108(1–3).

180. Philip, J., Shima, P. D., and Raj, B. (2008). *Nanotechnology*, **19**, pp. 305706(1–7).

181. Das, S. K., Putra, N., Thiesen, P., and Roetzel, W. (2003). *J. Heat Transfer*, **125**, pp. 567–574.

182. Zhu, H. T., Zhang, C. Y., Tang, Y. M., and Wang, J. X. (2007). *J. Phys. Chem. C*, **111**, pp. 1646–1650.

183. Dey, K. K., Kumar, A., Shanker, R., Dhawan. A., Wan. M., Yadav, R. R., and Srivastava, A. K. (2012). *RSC Adv.*, **2**, pp. 1387–1403.

184. Djaoued, Y., Phong, V. H., and Badile, S. (1997). *Thin Solid Films*, **293**, pp. 108–112.

185. Kawachi, G., Kimura, E., Wakui, Y., Konoshi, N., Yamamoto, H., Matsukawa, Y., and Sasano, A. (1994). *IEEE Trans. Electron. Devices*, **41**, pp. 1120–1124.

186. Stulik, P., and Singh, J. (1996). *Sol. Energy Mater. Sol. Cells*, **40**, pp. 239–251.

187. Molnar, K., Mohacsy, T., Varga, P., Vazsonyi, E., and Barsony, I. (1999). *J. Lumin.*, **80**, pp. 91–97.

188. Cho, Y.-S., Yi, G.-R., Hong J.-J., Jang, S. H., and Yang, S.-M. (2006). *Thin solid Films*, **515**, pp. 1864–1871.

189. Yang, Y., Jin, S., Medvedjeva, J. E., Ireland, J. R., Metz, A. W., Ni, J., Hersam, M. C., Freeman, A. J., and Marks, T. J. (2005). *J. Am. Chem. Soc.*, **127**, pp. 8796–8804.

190. Shannon, R. D. (1976). *Acta Crystallogr.*, **A32**, pp. 751–767.

191. Metz, A. W., Ireland, J. R., Zheng, J.-G., Lobo, R. P. S. M., Yang, Y., Ni, J., Stern, C. L., Dravid, V. P., Bontemps, N., Kannewurf, C. R., Poeppelmeier, K. R., and Marks, T. J. (2004). *J. Am. Chem. Soc.*, **126**, pp. 8477–8492.

192. Yang, T. L., Zhang, D. H., Ma, J., Ma, H. L., and Chen, Y. (1998). *Thin Solid Films*, **326**, pp. 60–62.

193. Keller, J., and Kind, S. (1995). *J. Low Temp. Phys.*, **99**, pp. 293–298.

194. Yip, S. K., and Sauls, J. A. (1992). *Phys. Rev. Lett.*, **69**, pp. 2264–2267.

195. Chen, G., Langlois, J.-M., Guo, Y., and Goddard (III), W. A. (1989). *Proc. Nad. Acad. Sci. U. S. A.*, **86**, pp. 3447–3451.

196. Reich, S., and Tsabba, Y. (1999). *Eur. Phys. J. B*, **9**, pp. 1–4.

197. Reich, S., Leitus, G., Popvitz-Biro, R., Goldbourt, A., and Vega, S. (2009). *J. Supercond. Novel Magn.*, **22**, pp. 343–346.

198. Tawale, J. S., Dey, K. K., Pasricha, R., Sood, K. N., and A. K. Srivastava. (2010). *Thin Solid Films*, **519**, pp. 1244–1247.

199. Applerot, G. Lipovsky, A., Dror, R., Perkas, N., Nitzan, Y., Lubart, R., and Gedanken, A. (2009). *Adv. Funct. Mater.*, **19**, pp. 842–852.

200. Yuvraj, D., Kaushik, R., and Rao, K. N. (2010). *ACS Appl. Mater. Interfaces*, **2**, pp. 1019–1024.

201. Ren, G., Hu, D., Cheng, E. W. C., Vargas-Reusc, M. A., Reip, P., and Allaker, R. P. (2009). *Int. J. Antimicrob. Ag.*, **33**, pp. 587–590.

202. Mahapatra, O., Bhagat, M., Gopalakrishnan, C., and Arunachalam, K. D. (2008). *J. Exp. Nanosci.*, **3**, pp. 185–193.

203. Lee, M.-H. (2002). *Sol. Energy Mater. Sol. C*, **71**, pp. 537-540.

204. Kam, K. C., and Cheetham, A. K. (2006). *Mater. Res. Bull.*, **41**, pp. 1015–1021.

205. Zhang, Z., Gao, Y., Chen, Z., Du, J., Cao, C., Kang, L., and Luo, H. (2010). *Langmuir*, **26**, pp. 10738–10744.

206. Lu, Z., Li, C., and Yin, Y. (2011). *J. Mater. Chem.*, **21**, pp. 14776–14782.

207. Kang, L., Gao, G., Luo, H., Chen, Z., Du, J., and Zhang, Z. (2011). *ACS Appl. Mater. Interfaces*, **3**, pp. 135–138.

208. Durrani, S. M. A., Khawaja, E. E., Salim, M. A., Al-Kuhaili, M. F., and Al-Shukri, A. M. (2002). *Sol. Energy Mater. Sol. C*, **71**, pp. 313–325.

209. Boulova, M., and Lucazeau, G. (2002). *J. Solid State Chem.*, **167**, pp. 425–434.

210. Cazzanelli, E., Vinegoni, C., Mariotto, G., Kuzmin, A., and Purans, J. (1999). *J. Solid State Chem.*, **143**, pp. 24–32.

211. Cazzanelli, E., Vinegoni, C., Mariotto, G., Kuzmin, A., and Purans, J. (1999). *Solid State Ionics*, **123**, pp. 67–74.

212. Bach, U., Corr, D., Lupo, D., Pichot, F., and Ryan, M. (2002). *Adv. Mater.*, **14**, pp. 845–848.

213. Bonhote, P., Gogniat, E., Campus, F., Walder, L., and Grätzel, M. (1999). *Displays*, **20**, pp. 137–144.

214. Bonhote, P., Gogniat, E., Grätzel, M., and Ashrit, P. V. (1999). *Thin Solid Films*, **350**, pp. 269–275.

215. Campus, F., Bonhote, P., Grätzel, M., Heinen, S., and Walder, L. (1999). *Sol. Energy Mater. Sol. Cells*, **56**, pp. 281–297.

216. Hagfeldt, A., Vlachopoulos, N., and Grätzel, M. (1994). *J. Electrochem. Soc.*, **141**, L82–L84.

217. Ottaviani, M., Panero, S., Morzilli, S., Scrosati, B., and Lazzari, M. (1986). *Solid State Ionics*, **20**, pp. 197–202.

218. Cinnsealach, R., Boschloo, G., Nagaraja Rao, S., and Fitzmaurice, D. (1999). *Sol. Energy Mater. Sol. Cells*, **57**, pp. 107–125.

219. Cummins, D., Boschloo, G., Ryan, M., Corr, D., Rao, S. N., and Fitzmaurice, D. (2000). *J. Phys. Chem. B*, **104**, pp. 11449–11459.

220. Pichot, F., Ferrere, S., Pitts, R. J., and Gregg, B. A. (1999). *J. Electrochem. Soc.*, **146**, pp. 4324–4326.

221. Xi, Y., Song, J., Xu, S., Yang, R., Gao, Z., Hu, C., and Wang, Z. L. (2009). *J. Mater. Chem.*, **19**, pp. 9260–9264.

222. Wang, Z. L., Kong, X. Y., Ding, Y., Gao, P., Hughes, W. L., Yang, R., and Zhang, Y. (2004). *Adv. Funct. Mater.*, **14**, pp. 943–956.

223. Wang, Z. L. (2004). *J. Phys. Condens. Matter.*, **16**, pp. R829–R858.

224. Wang, X.-F., Xu, J.-J., and Chen, H.-Y. (2008). *J. Phys. Chem. C*, **112**, pp. 7151–7157.

225. Heo, Y. W., Tien, L. C., Kwon, Y., Norton, D. P., and Pearton, S. J. (2004). *Appl. Phys. Lett.*, **85**, pp. 2274 (1–3).

226. Arnold, M. S., Avouris, P., Pan, Z. W., and Wand, Z. L., (2003). *J. Phys. Chem. B*, **107**, pp. 659–663.

227. Liu, J., Zhang, Z., Zhao, Y., Su., X., Liu, S., and Wang, E. (2005). *Small*, **1**, pp. 310–313.

228. Li, Y., Bando, Y., and Goldberg, D. (2003). *Adv. Mater.*, **15**, pp. 1294–1296.

229. Yin, H., Luo, M., Yu, K., Gao, Y., Huang, R., Zhang, Z., Zeng, M., Cao, C., and Zhu, Z. (2011). *ACS Appl. Mater. Interfaces*, **3**, pp. 2057–2062.

230. Li, M., Kong, F., Li, L., Zhang, Y., Chen, L., Yan, W., and Li, G. (2011). *Dalton Trans.*, **40**, pp. 10961–10965.

231. Liao, L., Yan, B., Hao, Y. F., Xing, G. Z., Liu, J. P., Zhao, B. C., Shen, Z. X., Wu, T., Wang, L., Thong, J. T. L., Li, C. M., Huang, W., and Yu, T. (2009). *Appl. Phys. Lett.*, **94**, pp. 113106 (1–3).

232. Huang, M. H., Mao, S., Feick, H., Yan, H., Wu, Y., Kind, H., Weber, E., Russo, R., and Yang, P. (2001). *Science*, **292**, pp. 1897–1899.

233. Kind, H., Yan, H., Messer, B., Law, M., and Yang, P. (2002). *Adv. Mater.*, **14**, pp. 158–160.

234. Rini, M., Hao, Z., Schoenlein, R. W., Giannetti, C., Parmigiani, F., Fourmaux, S., Kieffer, J. C., Fujimori, A., Onoda, M., Wall, S., and Cavalleri, A. (2008). *Appl. Phys. Lett.*, **92**, pp. 181904 (1–3).

235. Chen, S., Ma, H., Yi, X., Wang, H., Tao, X., Chen, M., Li, X., and Ke, C. (2004). *Infrared Phys. Techn.*, **45**, pp. 239–242.

236. Gea, L. A., and Boatner, L. A. (1996). *Appl. Phys. Lett.*, **68**, pp. 3081 (1–3).

237. Wu, C., Wei, H., Ning, B., and Xie, Y. (2010). *Adv. Mater.*, **22**, pp. 1972–1976.

238. Driscoll, T., Kim, H.-T., Chae, B.-G., Ventra, M. D., and Basov, D. N. (2009). *Appl. Phys. Lett.*, **95**, pp. 043503 (1–3).

239. Rigney, M. P., Weber, T. P., and Carr, P. W. (1989). *J. Chromatogr.*, **484**, pp. 273–291.

240. Kawahara, M., and Nakajma, H. T. (1990). *J. Chromatogr.*, **515**, pp. 149–158.

241. Dun, H., Zhang, W., Wei, Y., Xiuqing, S., Li, Y., and Chen, L. (2004). *Anal. Chem.*, **76**, pp. 5016–5023.

242. Heuer, A. H., and Hobbs, L. W. (1981). *Science and Technology of Zirconia, Advances In Ceramics Vol. 3*, eds. Americal Ceramic Society, Columbus, OH.

243. Lu, Y. F., and Qiu, H. (2000). *J. Appl. Phys.*, **88**, pp. 1082 (1–6).

244. Aoki, T., Matsushita, T, Suzuki, K., Okuda, T. M. (2005). *J. Vac. Sci. Technol. A*, **23**, pp. 1325 (1–6).

245. Kim, D.-W., Cho, M.-W., Seo, T., and Shin, Y.-J. (2008). *Sensors*, **8**. pp. 222–235.

246. Hatanaka, T., Hirawala, T., Matsukura, M., Aloki, N., and Imai, S. United States Patent, Patent No. 5215551, Date of patent June 1, 1993, *Appl. No.*, 582, 218.

247. Kim, J. P., Yeo, J. G., Paik, U., Jung, Y. G., Park, J. G., and Hackley, V. A. (2001). *J. Korean Phys. Soc.*, **39**, pp. S197–S200.

248. Lim, D. S., Ahn, J. W., Park, H. S., and Shin, J. H. (2005). *Surf. Coat. Tech.*, **200**, pp. 1751–1754.

249. Cornel, R. M., and Schwertmann, U. *The Iron Oxides. Structure, Properties, Reactions and Uses*; VCH: Weinheim, 1996.

250. Truffault, L., Ta, M.-T., Devers, T., Konstantinov, K., Harel, V., Simmonard, C., Nevirkovets, I. P., Pineau, A., Veron, O., and Blondeau, J.-P. (2010). *Mater. Res. Bull.*, **45**, pp. 527–535.

251. Tnunekawa, S., Fukuda, T., and Kasuya, A. (2000). *J. Appl. Phys.*, **87**, pp. 1318–1321.

252. Tnunekawa, S., Wang, J.-T., Kawazoe, Y., and Kasuya, A. (2003). *J. Appl. Phys.*, **94**, pp. 3654–3656.

253. Popov, A., Lademann, J., Priezzhev, A. Myllyla, R. (2005). *J. Biomed. Opt.*, **10**, pp. 064037 (1–9).

254. Pflucker, F., Wendel, V., Hohenberg, H., Gartner, W. T., Pfeiffer, S., Wepf, R., and Gers-Barlag, H. (2001). *Skin Pharmacol. Appl. Skin Physiol.*, **14**, pp. 92–97.

255. Uchino, T., Tokunaga, H., Ando, M., and Utsumi, H. (2002). *Toxicol. in Vitro*, **16**, pp. 629–635.

256. Schulz, J., Hohenberg, H., Pflucker, F., Gartner, E., Will, T., Pfeiffer, S., Wepf, R., Wendel, V., GersBarlag, H., and Wittern, K. P. (2002). *Adv. Drug. Deliv. Rev.*, **54**, pp. 157–163.

257. Nohynek, G. J., Dudour, E. K., and Roberts, M. S. (2008). *Skin Pharmacol. Phys.*, **21**, pp. 136–149.
258. Hass, D. D., Groves, J. F., and Wadley, H. N. G. (2001). *Surf. Coat. Technol.* **146–147**, pp. 85–93.
259. Dickey, M. D., Weiss, E. A., Smythe, E. J., Chiechi, R. C., Capasso, F., and Whitesides, G. M. (2008). *ACS Nano*, **2**, pp. 800–808.
260. Lin, Z., Lee, I.-S., Choi, Y.-J., Noh, I.-S., and Chung, S.-M. (2009). *Biomed. Mater.*, **4**, pp. 015013.
261. Irwin, M. D., Servaites, J. D., Buchholz, D. B., Leever, B. J., Liu, J., Emery, J. D., Zhang, M., Song, J.-H., Durstock, M. F., Freeman, A. J., Bedzyk, M. J., Hersam, M. C., Chang, R. P. H., Ratner, M. A., and Marks, T. J. (2011). *Chem. Mater.*, **23**, pp. 2218–2226.
262. Han, S., Li, C., Liu, Z., Lei, B., Zhang, D., Jin, W., Liu, X., Tang, T., and Zhou, C. (2004). *Nano Lett.*, **4**, pp. 1241–1246.
263. Anders, S., Anders, A., Rubin, M., Wang, Z., Raoux, S., Kong, F., and Brown, I. G. (1995). *Surf. Coat. Technol.*, **76–77**, pp. 167–173.
264. Tay, B. K., Zhao, Z. W., and Chua, D. H. C. (2006). *Mater. Sci. Eng. R*, **52**, pp. 1–48.
265. Ju, H., Hwang, S., Jeong, C.-O., Park, C., Jeong, E., and Park, S.-H. (2004). *J. Korean Phys. Soc.*, **44**, pp. 956–961.
266. Fujinaka, M., and Berezin, A. A. (1983). *J. Appl. Phys.*, **54**, pp. 3582 (1–7).
267. Peng, Y., Park, C., and Laughlin, D. E. (2003). *J. Appl. Phys.*, **93**, pp. 7957 1–3
268. Aita, C. R. (2008). *J. Phys. Condens. Matter.*, **20**, pp. 264006 1–11.
269. Albertini, K. F., and Pereyra, I. (2009). *Thin Solid Films*, **517**, pp. 4548–4554.
270. Chambers, S. A. (2008). *J. Phys. Condens. Matter.*, **20**, pp. 264004 (1–8).
271. Holland, M., Stanley, C. R., Reid, W., Hill, R. J. W., Moran, D. A. J., Thayne, I., Paterson, G. W., and Long, A. R. (2007). *J. Vac. Sci. Technol. Part B.*, **25**, pp. 1706–1710.
272. Amlandrino, G., Finocchiaro, S. T., Nigro, R. F., Bongiorno, C., Spinella, C., and Fragala, I. L. (2004). *Chem. Mater.*, **16**, pp. 5559–5561.
273. Pan, J., Song, X., Zhang, J., Shen, H., and Xiong, Q. (2011). *J. Phys. Chem. C*, **115**, pp. 22225–22231.
274. Hao, Y., Meng, G., Ye, C., and Zhang, L. (2005). *Cryst. Growth Des.*, **5**, pp. 1617–1621.

275. Wang, X., Tabakman, S. M., and Dai, H. (2008). *J. Am. Chem. Soc.*, **130**, pp. 8152–8153.

276. Elam, J. W., Baker, D. A., Hryn, A. J., Martinson, A. B. F., Pellin, M. J., Hupp, J. T. (2008). *J. Vac. Sci. Technol. A*, **26**, pp. 244 (9pp).

277. Willinger, M.-G., Neri, G., Bonavita, A., Micali, G., Tobias, H., and Pinna, N. (2009). *Phys. Chem. Chem. Phys.*, **11**, pp. 3615–3622.

278. Okuyama, K., and Lenggoroz, I. W. (2003). *Chem. Eng. Sci.*, **58**, pp. 537–547.

279. Messing, G. L., Zhang, S.-M., and Jayanthi, G. V. (1993). *J. Am. Ceram. Soc.*, **76**, pp. 2707–2726.

280. Prathap, P., Subbaiah, Y. P. V., Devika, M., and Ramakrishna, K. T. (2006). *Mater. Chem. Phys.*, **100**, pp. 375–379.

281. Ulrich, G. D., and Riehl, J. W. (1982). *J. Colloid Interface Sci. Technol.*, **87**, pp. 257–265.

282. Slcanadan, G., Chen, Y.-J., Glumac, N., and Kear, B. H. (1999). *Nanostruct. Mater.*, **11**, pp. 149–158.

283. Cow, G. M., and Gonsalves, K. E., eds. (1996). *Nanotechnology, Molecularly Designed Materials*, American Chemical Society, Washington D. C., pp. 64–78.

284. Singhal, A., Slcandan, G., Wang, A., Glumac, N., Kear, B. H., and Hunt, R. D. (1999). *Nanostruct. Mater.*, **11**, pp. 545–552.

285. Wu, M. K., Windeler, R. S., Steiner, C. K. R., and Friendlander, S. K. (1993). *Aerosol Sci. Technol.*, **19**, pp. 527–548.

286. Glumac, N. G., Chen, Y. J., Slcandan, G., and Kear, B. (1998). *Mater. Lett.*, **34**, pp. 148–153.

287. Ding, J., and Kim, K.-S. (2012). *Korean J. Chem.*, **29**, pp. 54–58.

288. Albuquerque, A. S., Ardisson, J. D., and Macedo, W. A. (2000). *J. Appl. Phys.*, **87**, pp. 4352 (1–6).

289. Chen, Q., Rondinone, A. J., Chakoumakos, B. C., and Zhang, Z. J. (1999). *J. Magn. Magn. Mater.*, **194**, pp. 1–7.

290. Wang, J. F., Ponton, C. B., and Harris, I. R. (2002). *J. Magn. Magn. Mater.*, **242–245**, pp. 1464–1467.

291. Kuo, P. C., and Tsai, T. S. (1989). *J. Appl. Phys.*, **65**, pp. 4349 (1–8).

292. Chinnasamy, C. N., Jeyadevan, B., Perales-Perez, O., Shinoda, K., Tohji, K., and Kasuya, A. (2002). *IEEE Trans. Magn.*, **38**, pp. 2640–2642.

293. Mizushima, K., Jones, P. C., Wiseman, P. J., and Goodenough, J. B. (1980). *Mater. Res. Bull.*, **15**, pp. 783–789.

294. Li, J.-G., Ikegami, T., Wang, Y., and Mori, T. (2002). *J. Am. Ceram. Soc.*, **85**, pp. 2376–2378.

295. Xiang, L., Deng, X. Y., and Jin, Y. (2002). *Scr. Mater.*, **47**, pp. 219–224.

296. Borse, P. H., Kankate, L. S., Dassenoy, F., Vogel, W., Urban, J., and Kulkarni, S. K. (2002). *J. Mater. Sci., Mater. Electron.*, **13**, pp. 553–559.

297. Wang, Y., Ma, C., Sun, X., and Li, H. (2002). *Inorg. Chem. Commun.*, **5**, pp. 751–755.

298. Liu, J., Huang, X., Li, Y., Sulieman, K. M., He, X., and Sun, F. (2006). *Cryst. Growth Des.*, **6**, pp. 1690–1696.

299. Zhang, X., Wang, G., Liu, X., Wu, J., Li, M., Gu, J., Liu, H., and Fang, B. (2008). *J. Phys. Chem. C*, **112**, pp. 16845–16849.

300. Lu, C., Qi, L., Yang, J., Zhang, D., Wu, N., and Ma, M. (2004). *J. Phys. Chem. B*, **108**, pp. 17825–17831.

301. Singh, D. P., Ojha, A. K., and Srivastava, O. N. (2009). *J. Phys. Chem. C*, **113**, pp. 3409–3418.

302. Wang, W. W., Zhu, Y. J., Cheng, G. F., and Huang, H. Y. (2006). *Mater. Lett.*, **60**, pp. 609–612.

303. Manna, S., Das, K., and De, S. K. (2010). *ACS Appl. Mater. Interf.*, **2**, pp. 1536–1542.

304. Gou, X., Wang, G., Yang, J., Park, J., and Wexler, D. (2008). *J. Mater. Chem.*, **18**, pp. 965–969.

305. Li, J. Y., Xiong, S., Pan, J., and Qian, Y. (2010). *J. Phys. Chem. C*, **114**, pp. 9645–9650.

306. Zhang, Q., and Gao, L. (2003). *Langmuir*, **19**, pp. 967–971.

307. Feng, X., Zhai, J., and Jiang, L. (2005). *Angew. Chem. Int. Ed.*, **44**, pp. 5115–5118.

308. Zhang, Y. X., Li, G. H., Jin, Y. X., Zhang, Y., Zhang, J., and Zhang, L. D. (2002). *Chem. Phys. Lett.*, **365**, pp. 300–304.

309. Kasuga, T., Hiramatsu, M., Hoson, A., Sekino, T., and Niihara, K. (1999). *Adv. Mater.*, **11**, pp. 1307–1311.

310. Xi, G., and Ye, J. (2010). *Inorg. Chem.*, **49**, pp. 2302–2309.

311. Kar, A., Kundu, S., and Patra, A. (2011). *J. Phys. Chem. C*, **115**, pp. 118–124.

312. Jr, W. A., Ribeiro, C., Leite, E. R., and Mastelaro, V. R. (2009). *Cryst. Growth Des.*, **9**, pp. 3626–3631.

313. Dai, L., Cao, C., Gaon, Y., and Luo, H. (2008). *Sol. Energy Mater. Sol. Cells*, **95**, pp. 712–715.

314. Cao, C., Gao, Y., and Luo, H. (2008). *J. Phys. Chem. C*, **112**, pp. 18810–18814.

315. Zhang, S., Sahang, B., Yang, J., Yan, W., Wei, S., and Xie, Y. (2011). *Phys. Chem. Chem. Phys.*, **13**, pp. 15873–15881.

316. Liu, J., Li, Q., Wang, T., Yu, D., and Li, Y. (2004). *Angew. Chem. Int. Ed.*, **43**, pp. 5048–5052.

317. Cao, C., Gao, Y., Kangab, L., and Luo, H. (2010). *Cryst. Eng. Comm.*, **12**, pp. 4048–4051.

318. Wu, C., Zhang, X., Dai, J., Yang, J., Wu, Z., Weib, S., and Xie, Y. (2011). *J. Mater. Chem.*, **21**, pp. 4509–4517.

319. Sun, S., Zeng, H., Robinson, D. B., Raoux, S., Rice, P. M., Wang, S. X., and Li, G. (2004). *J. Am. Chem. Soc.*, **126**, pp. 273–279.

320. Zhu, L. P., Bing, N. C., Wang, L. L., Jin, H. Y., Liaob, G. H., and Wang, L. J. (2012). *Dalton Trans*, **41**, pp. 2959–2965.

321. Cao, Y., Hu, J. C., Hong, Z. S., Deng, J. F., and Fan, K. N. (2002). *Catal. Lett.*, **81**, pp. 107–112.

322. Wu, Z. G., Zhao, Y. X., Xu, L. P., and Liu, D. S. (2003). *J. Non-Cryst. Solids*, **330**, pp. 274–277.

323. Chemseddine, A., and Moritz, T. (1999). *Eur. J. Inorg. Chem.*, **1999**, pp. 235–245.

324. Oskam, G., Nellore, A., Penn, R. L., and Searson, P. C. (2003). *J. Phys. Chem. B*, **107**, pp. 1734–1738.

325. Kim, K. D., and Kim, H. T. (2001). *Powder Technol.*, **119**, pp. 164–172.

326. Moritz, T., Reiss, J., Diesner, K., Su, D., and Chemseddine, A. (1997). *J. Phys. Chem. B*, **101**, pp. 8052–8053.

327. Miao, L., Tanemura, S., Toh, S., Kaneko, K., and Tanemura, M. (2004). *J. Cryst. Growth*, **264**, pp. 246–252.

328. Miao, L., Tanemura, S., Toh, S., Kaneko, K., and Tanemura, M. (2004). *Appl. Surf. Sci.*, **238**, pp. 175–179.

329. Lee, S., Jeon, C., and Park, Y. (2004). *Chem. Mater.*, **16**, pp. 4292–4295.

330. Jung, J. H., Shimizu, T., and Shinkai, S. (2005). *J. Mater. Chem.*, **15**, pp. 3979–3986.

331. Shi, Q., Huang, W., Zhang, Y., Yan, J., Zhang, Y., Mao, M., Zhang, Y., and Tu, M. (2011). *ACS Appl. Mater. Interfaces*, **3**, pp. 3523–3527.

332. Teoh, G. L., Liew, K. Y., and Mahmood, W. A. K. (2007). *J. Sol-Gel Sci. Technol.*, **44**, pp. 177–186.

333. Pileni, M. P. (1993). *J. Phys. Chem.*, **97**, pp. 6961–6973.

334. Li, G. L., and Wang, G. H. (1999). *Nanostruct. Mater.*, **11**, pp. 663–668.

335. Zhang, D., Qi, L., Ma, J., and Cheng, H. (2002). *J. Mater. Chem.*, **12**, pp. 3677–3680.

336. Lim, K. T., Hwang, H. S., Ryoo, W., and Johnston, K. P. (2004). *Langmuir*, **20**, pp. 2466–2471.

337. Zhang, H., Zhang, X., Li, H., Qu, Z., Fan, S., and Ji, M. (2007). *Cryst. Growth Des.*, **7**, pp. 820–824.

338. Nasibulin, A. G., Ahonen, P. P., Richard, O., Kauppinen, E. I., and Altman, I. S. (2001). *J. Nanopart. Res.*, **3**, pp. 385–400.

339. Suzdalev, I. P., and Krupyianskii, Y. F. (1969). *Kinet. Katal.*, **10**, p. 1255.

340. Krupyianskii, Y. F., and Suzdalev, I. P. (1974). *Sov. Phys. JETP*, **38**, pp. 859–864.

341. Kamel, A. H., and Abdallah, A. M. (1972). *J. Appl. Chem. Biotechnol.*, **22**, pp. 599–602.

342. Au-Yeung, S. C. F., Denes, G., Greedan, J. E., Eaton, D. R., Birchall, T. (1984). *Inorg. Chem.*, **23**, pp. 1513–1517.

343. Kanungo, S. B., and Mishra, S. K. (1997). *J. Therm. Anal.*, **48**, pp. 385–401.

344. Hussein, G. A. M., Ismail, H. M., Attyia, K. M. E. (1995). *J. Anal. Appl. Pyrol.*, **31**, pp. 157–167.

345. Widmer, R., Haug, F. J., Ruffieux, P., Gröning, O., Bielmann, M., Gröning, P., and Fasel, R. (2006). *J. Am. Chem. Soc.*, **128**, pp. 14103–14108.

346. Liu, S., and Huang, K. (2004). *Sol. Energy Mater. Sol. Cell*, **85**, pp. 125–131.

347. Siegfried, M. J., and Choi, K. S. (2004). *Adv. Mater.*, **16**, pp. 1743–1746.

348. Wei, W., Cui, X., Chen, W., and Ivey, D. G. (2008). *J. Phys. Chem. C*, **112**, pp. 15075–15083.

349. Inoue, M., Kimura, M., and Inui, T. (1999). *J. Chem. Soc. Chem. Commun.*, **957**, pp. 957–958.

350. Chen, J., Huang, N. Y., Deng, S. Z., She, J. C., Xua, N. S., Zhang, W., Wen, X., and Yang, S. (2005). *Appl. Phys. Lett.*, **86**, pp. 151107–151103.

351. Zhang, Z., Sun, H., Shao, X., Li, D., Yu, H., Han, M. (2005). *Adv. Mater.*, **17**, pp. 42–47.

352. Jiang, X., Herricks, T., and Xia, Y. (2002). *Nanoletter*, **2**, pp. 1333–1338.

353. Wu, J. M., Hayakawa, S., Tsuru, K., and Osaka, A. (2002). *Scripta Mater.*, **46**, pp. 101–106.

354. Peng, X., and Chen, A. (2004). *J. Mater. Chem.*, **14**, pp. 2542–2548.

355. Varghese, O. K., Gong, D., Paulose, M., Grimes, C. A., and Dickey, E. C. (2003). *J. Mater. Res.*, **18**, pp. 156–165.

356. Sahoo, S., Husale, S., Colwill, B., Lu, T. M., Nayak, S., and Ajayan, P. M. (2009). *ACS Nano*, **3**, pp. 3935–3944.

357. Teo, J. J., Chang, Y., and Zeng, H. C. (2006). *Langmuir*, **22**, pp. 7369–7377.

358. Wang, Z., Chen, X., Liu, J., Mo, M., Yang, L., and Qian, Y. (2004). *Solid State Commun.*, **130**, pp. 585–589.

359. Chen, L., Xu, J., Tanner, D. A., Phelan, R., Meulen, M. V., Holmes, J. D., and Morris, M. A. (2009). *Chem. Eur. J.*, **15**, pp. 440–448.

360. Patil, K. C., Aruna, S. T., and Ekambaram, S. (1997). *Curr Opin Solid State Mater. Sci.*, **2**, pp. 158–165.

361. Zhai, X., Fu, Y., and Chu, G. (2006). *Nanoscience*, **11**, pp. 286–292.

362. Mimani, T. (2000). *Resonance*, **5**, pp. 50–57.

363. Kiminami, R. H. G. A., Morelli, M. R., Folz, D. C., and Clark, D. E. (2000). *Am. Cream. Sco. Bull.*, **79**, pp. 63–67.

364. Lee, S. Y., Gao, X., and Matsui, H. (2007). *J. Am. Chem. Soc.*, **129**, pp. 2954–2958.

365. Suslick, K. S., Choe, S. B., and Cichowlas, A. A., and Grinstaff, M. W. (1991). *Nature*, **353**, pp. 414–416.

366. Suslick, K. S., Cline, R. E., and Hammerton, D. A. (1986). *J. Am. Chem. Soc.*, **108**, pp. 5641–5642.

367. Mastai, Y., and Gedanken, A. (2004). Sonochemistry and other novel methods developed for the synthesis of nanoparticles, in *The Chemistry of Nanomaterials*, (Rao. C. N. R., Muller, A., and Cheetham, A. K., eds.), WILEY-VCH Verlag GmbH & Co. KGaA, Weinheim, pp. 113–169.

368. Cao, X., Prozorov, R., Koltypin, Y., Kataby, G., Felner, I., and Gedankenet, A. (1997). *J. Mater. Res.*, **12**, pp. 402–406.

369. Dhas, N. A., Koltypin, Y., and Gedanken, A. (1997). *Chem. Mater.*, **9**, pp. 3159–3163.

370. Guo, W. L., Wang, X. K., Lin, Z. M., and Zhi, S. G. (2002). *Chem. J. Chin. Univ.*, **23**, pp. 1592–1594.

371. Sukpirom, N., and Lerner, M. M. (2002). *Mater. Sci. Eng. A*, **333**, pp. 218–222.

372. Alammar, T., Birkner, A., and Mudring, A. V. (2009). *Eur. J. Inorg. Chem.*, 2009, pp. 2765–2768.

373. Wang, W. W., Zhu, Y. J., Cheng, G. F., and Huang, Y. H. (2006). *Mater. Lett.*, **60**, pp. 609–612.

374. Xi, G., He, Y., Zhang, Q., Xiao, H., Wang, X., and Wang, C. (2008). *J. Phys. Chem.*, **112**, pp. 11645–11649.

375. Liang, J., Deng, Z., Jiang, X., Li, F., and Li, Y. (2002). *Inorg. Chem.*, **41**, pp. 3602–3604.

376. Arbain, R., Othman, M., and Palaniandy, S. (2011). *Minerals Eng.*, **24**, pp. 1–9.

377. Salah, N., Habib, S. S., Khan, Z. H., Memic, A., Azam, A., Alarfaj, E., Zahed, N., and Hamedi, S. A. (2011). *Int. J. Nanomed.*, **6**, pp. 863–886.

378. Park, H. H., Zhang, X., Lee, S. W., Kim, K., Choi, D. G., Choi, J. H., Lee, J., Lee, E. S., Park, H. H., Hillb, R. H., and Jeong, J. H. (2011). *J. Mater. Chem.*, **21**, pp. 657–662.

379. Fajans, K. (1967). *Struct. Bonding Berlin*, **3**, pp. 88–105.

380. Shriver, D. F., Atkins, P. W., and Langford, C. H. (1990). *Inorg. Chem.*

381. Rao, C. N. R., and Raveau, B. (1998). Wiley-VCH: New York, **4**, pp. *Transition Metal Oxides*. By Rao, C. N. R., and Raveau, B. VCH Publishers. New York.

382. Richards, R. M. (2009). *Dekker Encyclopedia of Nanoscience and Nanotechnology*, **1–6**, pp. 2126–2139.

383. Sun, H., Zhang, Q.-F., and Wu, J.-L. (2006). *Nanotechnology*, **17**, pp. 2271–2274.

Chapter 2

Pulsed Laser Deposition of Nanostructured Oxides for Emerging Applications

Carlo S. Casari and Andrea Li Bassi

Department of Energy and NEMAS—Center for NanoEngineered Materials and Surfaces, Politecnico di Milano, via Ponzio 34/3 I-20133 Milano, Italy
Center for Nano Science and Technology @PoliMI, Istituto Italiano di Tecnologia, Via Pascoli 70/3 I-20133 Milano, Italy

carlo.casari@polimi.it

Metal oxides are presently employed in a number of advanced technological applications ranging from sensing, biology, and energy. For such applications, a good control of morphological, structural, and functional properties is required in order to improve or enhance the performance of the materials and the related devices. In this chapter, we review our work on the pulsed laser deposition of nanostructured films and surfaces showing that a fine-tuning of morphology and structure can be achieved for metal oxides such as tungsten and titanium oxide. Starting from the control of the building units constituting the material, we show that film morphology can be varied from compact to columnar and

Oxide Nanostructures: Growth, Microstructures, and Properties
Edited by Avanish Kumar Srivastava

ISBN 978-981-4411-35-6 (Hardcover), 978-981-4411-36-3 (eBook)
www.panstanford.com

nanostructured to a hierarchical assembly of nanoparticles and finally to highly porous foam-like. Film structure can be varied from amorphous to nanocrystalline with the possibility to vary the oxide phase. Finally, we discuss the possible applications in which such oxides can have good performances as functional materials.

2.1 Introduction

Metal oxides are key components in many technological applications such as heterogeneous catalysis, photocatalysis, gas sensing, energy production (e.g., third generation solar cells), biotechnology, and medicine. In all these applications, the interaction between oxide layers and a specific environment (i.e., gases, liquids, molecules/polymers and biological materials) is a key element for the required function. For this reason, oxides with high surface areas and optimized properties are required to develop enhanced or novel functions. Hence, the synthesis of nanostructured layers in a bottom-up approach, which exploits material growth mechanisms, nano-sized building blocks formation and assembling and self-organization properties, opens the possibility to synthesize nanostructured materials with tailored physical and chemical properties through a fine-tuning of the morphology and structure down to the nanoscale. Finally, the determination of structure–properties relations, together with the assessment of the role of the different deposition parameters, is fundamental to define experimental feasible guidelines for materials design and realization.

In this framework, we show that pulsed laser deposition (PLD) is particularly versatile to tailor nanostructured and nanoporous oxides for technological applications.

2.2 Pulsed Laser Deposition of Oxides with Tailored Properties

Pulsed laser deposition is a physical vapor deposition technique in which the irradiation of a solid target by means of high-energy laser pulses produces the local ablation of the target surface resulting in the ejection of material in the form of an expanding plume of plasma [1,2]. The expansion dynamics of the ablation plasma plume is strongly modified by the presence of a background gas [3]. In this case, spatial confinement and collision rate of ablated species increase with increasing the pressure, resulting in cluster nucleation and in a reduced cluster kinetic energy thus affecting the film growth processes [4,5]. In addition, the use of reactive gases such as oxygen permits to deposit oxidized species starting from metallic targets and, together with proper post-deposition annealing treatments, opens the way for the production of materials with tailored morphology, crystalline structure, and composition.

2.2.1 Deposition Parameters Affecting Film Growth

Film growth and consequently its morphology and structure are strongly influenced by the plasma plume expansion dynamics. As depicted in Fig. 2.1, when ablating in vacuum the plasma plume is characterized by an almost collisionless propagation regime of the ablated species [6]. The spatial distribution is strongly forward-directed and the observed light emission is weak [7,8]. In these conditions the deposition regime is mainly characterized by high-energy atoms or ions (tens-hundreds of eV) impinging the substrate leading to atom-by-atom deposition of compact films. Conversely, when an inert background gas at a sufficiently high pressure (i.e., 1–1000 Pa) is present in the deposition chamber the light emission from the plasma increases due to collisions producing radiative de-excitation of the ablated species, both in the body of the plume and particularly at the expansion front; the plume edge is better defined due to the presence of a travelling shock wave front, which expands reaching the so-called stopping distance (typical values are in the 10–80 mm range) [9]. The ablation plume is spatially confined [7,8] and the presence of a background gas produces high pressure, temperature and density gradients at the shock front resulting in strongly out of equilibrium conditions [10, 11], cooling of the plume and favoring cluster nucleation in the gas phase. Clusters may then diffuse in the background gas thus decreasing their kinetic energy. This condition may lead to low energy cluster deposition regime (i.e., <1 eV/atom) thus producing a cluster-assembled material in which clusters can maintain at least partially their structure and properties. If a reactive gas is used as a background atmosphere, cluster composition may also be affected by the chemical interaction with the surrounding gas. This permits to deposit clusters with controlled composition and it opens the possibility to deposit metal oxide clusters by ablating a metallic target in the presence of oxygen.

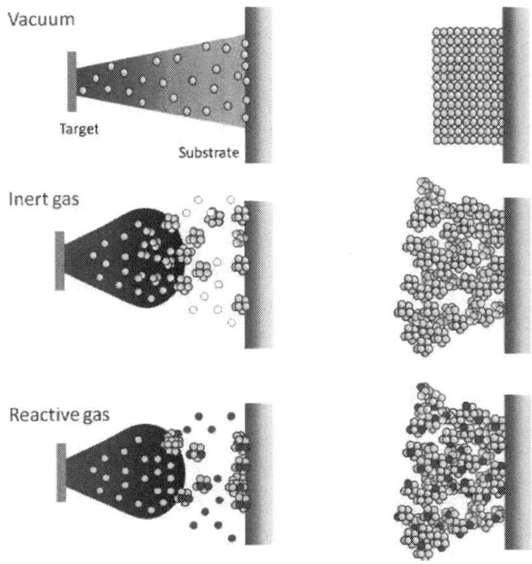

Figure 2.1 Schematic pictures of the deposition process for ablation in vacuum, in inert and in reactive atmosphere.

The main parameters affecting the plasma expansion dynamics are the type and pressure of the background gas and the laser pulse duration (i.e., nanosecond) wavelength and energy (i.e., laser fluence, which is the energy density on the target). The plume spatial confinement can be easily checked by looking the plasma plume shape and, in particular, the plume length l_p (i.e., the time-integrated visible plume length) [12]. The plume length decreases with increasing the gas pressure for a fixed gas and depends on the gas mass for different gases at the same pressure [12]. As described below, the target-to-substrate distance d_{TS} and, in particular, the relative substrate position with respect to the plume shock front is also fundamental to tune film growth.

We have shown that the morphology of the deposited films can be tuned by varying a parameter, called L, defined as the ratio of d_{TS} to l_p, thus controlling the aggregation/deposition mechanisms [12–14]. $L = d_{TS}/l_p$ is a nondimensional characteristic length indicating the relative position of the substrate with respect to the plume length (i.e., maximum distance reached by the shock front, also called stopping distance [9]). Of course, such parameter can be used when a well-defined and clearly visible shock front is formed, typically when the background gas pressure is of the order of tens Pa or less.

As shown in Fig. 2.2, in the presence of a background gas, we identify distinct deposition regimes that depend on the value of the L parameter:

(a) $L < 1$ (substrate "in plume"): Diffusion of the ablated species in the surrounding gas out of the shock front is strongly reduced. High kinetic energy of the ablated species and formation of compact and columnar films are expected.

(b) $L \sim 1$ (substrate "at the plume edge"): The substrate is placed at about the maximum position of the plume shock front. The expected effects are cluster nucleation due to the increase of interactions between ablated species and the surrounding gas, a reduced kinetic energy of the species and formation of films composed by a hierarchical assembly of nanoparticles.

(c) $L > 1$ (substrate "out of plume"): The diffusion of the formed particles in the gas increases with increasing the L value. Decrease of the kinetic energy of the species arriving on the substrate and formation of open-porous structures.

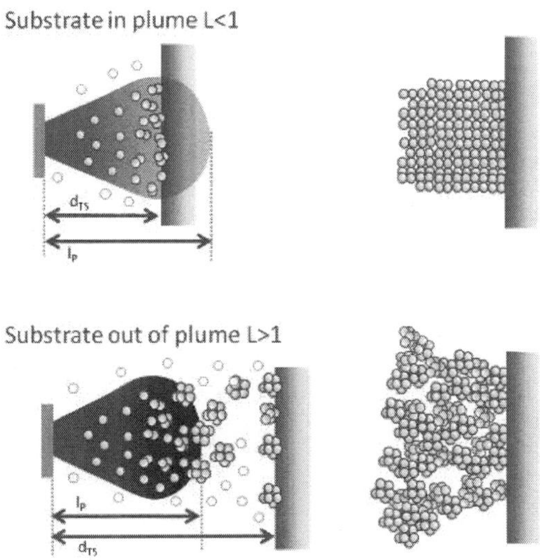

Figure 2.2 Schematic picture of the effect of the L parameter in the film growth.

By setting a relation between l_p and d_{TS}, the L parameter can be used as a control parameter to select the regime of film growth and consequently film morphology and structure [12].

2.2.2 Experimental Apparatus

The scheme of a PLD apparatus is shown in Fig. 2.3. Laser pulses are focused on the target material inside a vacuum chamber through a viewport. The target is moved to optimize the position of the laser spot on the target and to maintain a uniform ablation of the target surface.

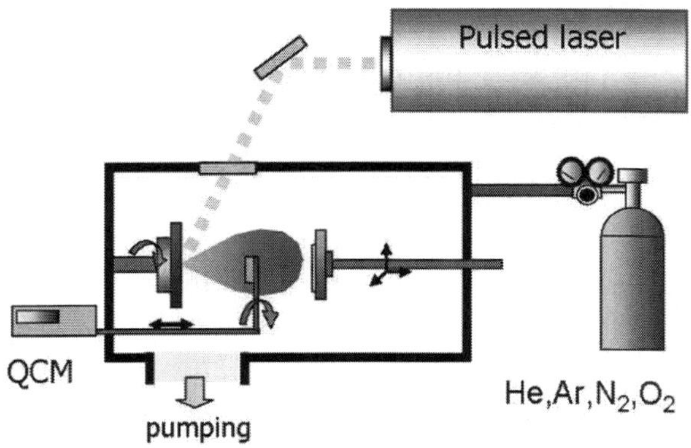

Figure 2.3 Scheme of the pulsed laser deposition apparatus.

A pumping system and a gas inlet system with mass flow controllers and pressure gauges allows pressure control from 10^{-7} Pa to atmospheric pressure. A quartz crystal microbalance (QCM) can intercept the plasma plume to measure the deposition rate at different distances from the target. Two different PLD systems have been used here, one in high vacuum (HV) with a frequency-quadrupled lamp-pumped Q-switched Nd:YAG laser (266 nm wavelength, 5–7 ns pulse duration, 180 mJ maximum pulse energy, 10 Hz repetition rate); the other is a ultra high vacuum (UHV) compatible chamber with a KrF excimer laser (248 nm wavelength, 10–15 ns pulse duration, 500 mJ maximum pulse energy, 10 Hz repetition rate). The latter apparatus is directly connected, by means of magnetically coupled transfer probes, to a UHV surface science system (base pressure < 5×10^{-11} mbar) equipped with a scanning tunneling microscope (STM).

All films were grown by PLD at room temperature. Energy density (laser fluence) was set in the 1–5 J/cm^2 range, depending on the material, to maximize deposition rates while minimizing droplet formation. Background gas pressure in the 1–1000 Pa range and target-to-substrate distance d_{TS} in the 25–100 mm range were used. The time-integrated visible plume length l_p was measured from plasma plume pictures taken with a digital camera.

2.2.3 Tuning of Morphological Properties of Oxides

The guidelines presented in Section 2.2.1 to tune film growth were exploited to deposit oxides. In the case of TiO_2, as shown in Fig. 2.4, the overall trend is in agreement with the expected general behavior: at $L < 1$ the film is compact and characterized by a columnar growth; at $L \sim 1$ a transition toward a nanoscale organized structure is observed and columns composed by fine nanoparticles open up in a ramified structure. At $L > 1$ the growth is more disordered loosing any reminiscence of vertical structures, the morphology is extremely open resembling an aerogel. For such films the surface roughness at constant thickness (100 nm) increases from 4.1 ± 0.02 nm at 10 Pa, $L < 1$ to 15.6 ± 0.7 nm at 60 Pa, $L > 1$ [15].

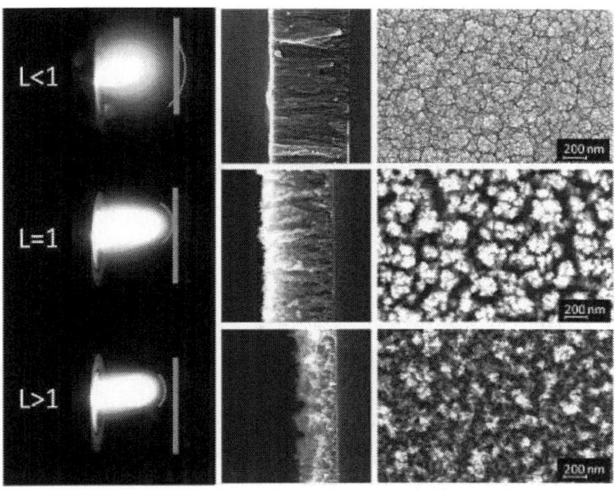

Figure 2.4 Pictures of the plasma plume produced by ablating a Ti target in Ar:O_2 (4:1) background gas. Gas pressure is varied from 10 to 60 Pa for fixed d_{TS} = 50 mm to have $L < 1$, $L \approx 1$ and $L > 1$. The laser fluence is about 3 J/cm^2. SEM images (cross-sectional and top views) of the corresponding TiO_2 films.

The variation of morphology has a direct effect on the porosity and on material density. As shown in Fig. 2.5, for TiO_2 films the density decreases from 2.6 g/cm^3 for deposition at $L < 1$ (10 Pa Ar:O_2) to about 0.3 g/cm^3 for deposition at $L > 1$ (100 Pa Ar:O_2). The general trend followed by the surface roughness and the film density is observed for other oxides such as tungsten oxide and aluminum oxide [12,16,17].

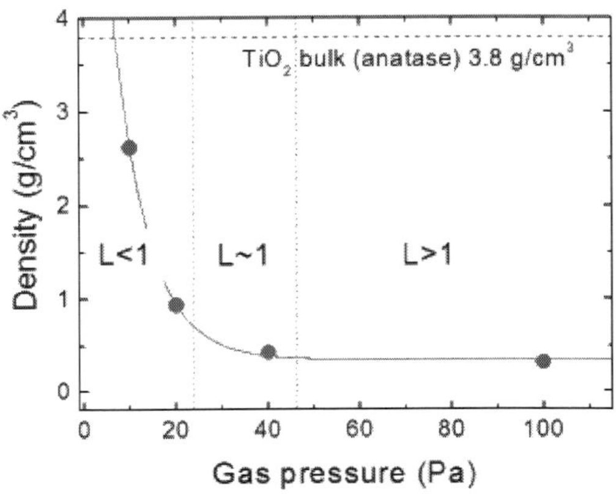

Figure 2.5 Evolution of the film density of TiO_2 produced by ablation of Ti target in the presence of Ar:O_2 (4:1) as a function of the pressure for a fixed substrate position (d_{TS} = 50 mm) and a laser fluence of about 3 J/cm^2.

We underline that the control on film morphology achievable by tuning the L value must be considered as a general guideline since L cannot uniquely define all the film properties. Anyway due to its simplicity we propose the use of this parameter to select the type of film growth, while the fine selection of film properties requires a detailed study of the specific material.

2.2.4 Tuning Structural Properties and Oxide Phase

In this section, we show an example of how it is possible to control oxide formation and the degree of structural order. In particular, the effect of increasing the pressure of an oxidizing atmosphere (dry air) when ablating a metallic W target is shown in the Raman spectra reported in Fig. 2.6. A low pressure (1 Pa) is not sufficient to promote oxide formation as suggested by the absence of any Raman signal. At 10 Pa dry air typical broad bands of amorphous tungsten oxide appear [18,19] and at about 40 Pa and beyond, a sharpening of the Raman bands toward well-defined peaks is observed. This is an indication of the increase of structural order, possibly due to the presence of nanocrystals of increasing size. Finally, at 1000 Pa dry air films show well-defined Raman peaks at 807, 718, and 270 cm^{-1}, typical of polycrystalline WO_3 in either monoclinic (γ) or triclinic (δ) phase [18].

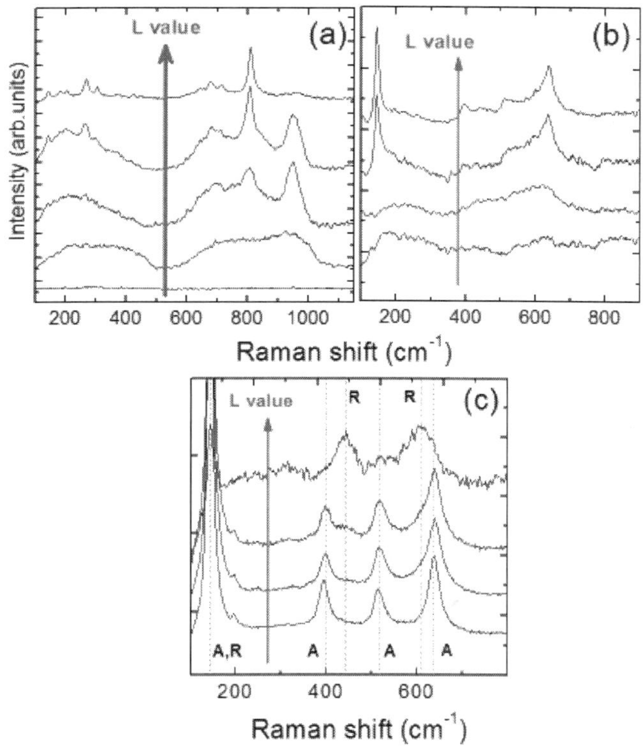

Figure 2.6 Raman spectra of (a) films produced by ablation of W target in the presence of dry air at increasing *L* values. The pressure was increased in the 1–1000 Pa range by maintaining a fixed d_{TS} = 50 mm. The excimer laser fluence is about 4.5 J/cm^2. (b) TiO_2 films produced from a Ti target at different *L* values by increasing the O_2 pressure (10–60 Pa) at a fixed d_{TS} = 50 mm and a Nd:YAG laser fluence of about 3 J/cm^2. (c) The effect of a post deposition annealing (400°C in air for 1 h) of TiO_2 films deposited at increasing pressure of dry air (10–100 Pa) at a fixed d_{TS} = 50 mm and an excimer laser fluence of about 4 J/cm^2.

A similar investigation has been conducted ablating a titanium target in an oxidizing atmosphere with the aim of controlling not only film morphology but also film structure.

Raman spectra reported in Fig. 2.6 indicate that at $L < 1$ (oxygen pressure ≤ 20 Pa) films have an amorphous titanium oxide structure, showing broad and unresolved Raman bands. Increasing L, Raman bands evolve and peaks typical of the crystalline anatase phase start to appear. In particular, a peak at 144 cm^{-1} appears, corresponding to the most intense Raman feature in anatase TiO_2 and attributed to the E_g phonon mode [20]. This peak becomes more intense with increasing L; nonetheless, films are not completely crystalline: other peaks belonging to the anatase phase at 399, 529, and 639 cm^{-1} are not well-defined and the unresolved structure in this spectral range indicates a high degree of disorder and thus the co-presence of an amorphous/disordered phase. Due to the presence of wide bands, coexistence of the rutile phase cannot be excluded a priori. Interestingly, peaks belonging to crystalline phases start to appear with $L \approx 1$ marking a transition between different deposition regimes. It is worth to underline that the spectra shown in Fig. 2.6 correspond to depositions where the L parameter was varied by changing the oxygen pressure at a fixed d_{TS}: in such case, it not possible to separate the effect of L from the effect of oxygen pressure in determining the oxidation and the film structure. Variation of the oxygen content (i.e., in a gas mixture) at a fixed L value allows to overcome this limitation, as we showed by changing the $Ar{:}O_2$ ratio at a fixed total pressure (see Ref. [21]).

In addition, we developed the capability to tune the oxide phase of TiO_2, the main phases of titanium oxide being anatase and rutile. Rutile is the most stable and usually forms at high temperature, while anatase is widely used for applications (e.g., dye-sensitized solar cells, photocatalysis). We found that annealing treatments at moderate temperature can induce the formation or ordering of different oxide phases depending on the starting disordered structure. TiO_2 films deposited at room temperature, which are characterized by a mainly disordered structure, after a thermal treatment at 400°C in air, show the increasing formation of rutile phase with respect to anatase with increasing L value as shown in Fig. 2.6. The main TiO_2 crystalline phase is anatase (Raman peaks at 143, 197, 399, 519, and 639 cm^{-1}), with an increasing rutile content (Raman peaks at 447 and 612 cm^{-1}) when increasing deposition pressure (dry air) up to 100 Pa, where a predominantly rutile film is obtained. The intensity of the crystalline rutile peaks is very low and superimposed to broad features, suggesting a highly disordered structure and very fine grains (below 10 nm). The anatase-to-rutile relative content in the film can be estimated from the analysis of Raman spectra, as reported in [20]. In PLD films we observed that the anatase-to-rutile content ratio changes from 100% to less than 22% by increasing the pressure from 10 to 100 Pa of dry air. It is interesting to observe that all the films have been treated for the same time at the same temperature of 400°C, which is well below the anatase–rutile transition temperature, suggesting the presence of small rutile nanocrystals in as deposited samples, acting as nucleation seeds for the rutile growth even at moderate temperature. Comparison of SEM images of films before and after annealing (not shown) reveals that no significant morphology modifications occur and nanoparticle size roughly coincides with crystalline domain size (measured by XRD) in the annealed samples [22]. This indicates that annealing results in structural ordering of the particles, while particle size growth and coalescence are reduced by the open morphology. Such occurrence is particularly interesting since it opens the possibility to separately control morphology and oxide phase of the deposited materials.

2.2.5 First Stages of Film Growth

In order to characterize the first stages of film growth and the size and the deposition kinetic energy of building blocks produced during the PLD process, deposits on atomically flat surfaces have been performed at low coverage before film formation to investigate isolated clusters in UHV by in situ STM [13,23]. We observed fragmentation of ablated species and damaging of the substrate surface when deposition is performed at $L < 1$. On the contrary, deposition at $L > 1$ permits to obtain isolated clusters with a substantial reduction of surface damage (see Refs. [13,14]). Figure 2.7 shows STM images of isolated W clusters obtained by depositing a few laser shots (1–30) at variable background gas pressure while maintaining the condition $L > 1$.

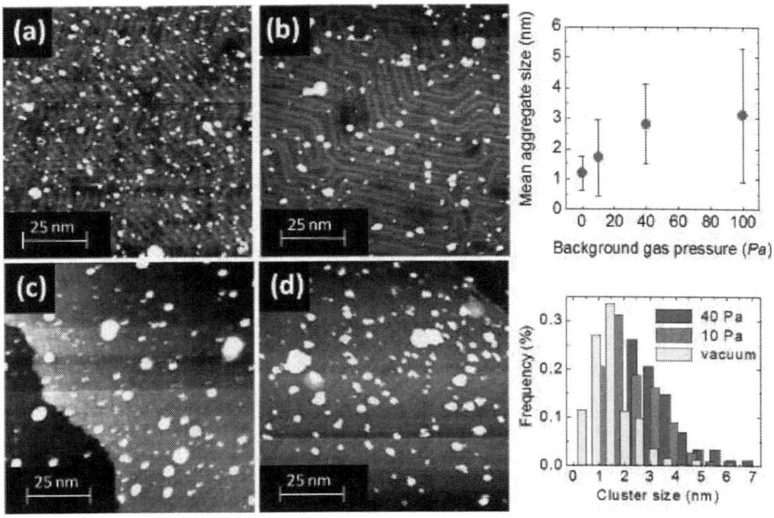

Figure 2.7 (a–d) In situ STM images of deposits produced by ablation of a W target at $L > 1$ on Au(111) at increasing Ar gas pressure (from vacuum to 100 Pa) and excimer laser fluence of 4 J/cm^2. The size (mean value and dispersion) as a function of gas pressure and the corresponding size distributions are also reported.

Mean cluster size increases from 1 to 3 nm when increasing the background gas pressure of Ar from vacuum to 100 Pa with a quite narrow size distribution, which broadens with increasing the pressure. The first stages of formation of a nanostructured film have been investigated by increasing the coverage (i.e., by increasing the number of laser shots). At variance with ablation in vacuum, the formation of a cluster-assembled system is clearly observed when ablating in the presence of a background gas [23].

2.3 Applications

In the previous sections, we presented PLD in a background atmosphere as a versatile technique to deposit nanostructured oxides such as titanium and tungsten oxide with tunable morphological and structural properties. To move toward potential employment as functional materials we developed and tested nanostructured oxides for specific applications obtaining promising results [24,25,28–30].

With the aim of exploiting tungsten oxide nanostructures for gas sensing applications, we developed WO_3 nanorods showing an increased specific sensitivity to ammonia (NH_3). By a proper thermal treatment of nanoporous films deposited by PLD at room temperature, we obtained an assembly of crystalline nanorods, as shown in Fig. 2.8 [25]. NO_2 is widely used to characterize the sensing capability of tungsten oxide, due to the strong induced signal even at low concentrations. WO_3 is an n-type oxide semiconductor, whose conductivity is determined by oxygen vacancies, which release electrons in the oxide conduction band; NO_2 oxidizes the sample through the interaction with surface oxygen vacancies, leading to a reduction in conductance [26].

With reference to ammonia sensing, the performance of PLD samples is comparable with the performance reported for other tungsten oxide nanostructures, such as nanofibers or nanoparticles and 3D nanowire networks (see [27]), but interestingly the morphology dominated by nanorods shows an increased specific sensitivity to ammonia with respect to ethanol and acetone. This selectivity is of particular interest in the development of electronic nose.

We found that in a specific range of deposition parameters, it is possible to obtain nanoparticles (about 10 nm size) hierarchically assembled in a nano- and mesostructure that resembles a "forest of trees" (see Fig. 2.8), without any use of templates or pre-patterning strategies [22].

Figure 2.8 SEM images of tungsten oxide nanorods (left) and of hierarchically structured TiO_2 (right).

Such structures are characterized by a large specific surface area (of the order of 100 m^2/g), by an anisotropic growth of columnar structures (the "trees") with respect to the substrate and by the presence of mesoscale porosity (tens of nm) between these columnar structures. All these properties make these structures interesting for applications where the surface plays a central role, and we demonstrated application of TiO_2 hierarchical structures in the fields of photocatalysis, photovoltaics, and proteomics [22,28,29].

For example, we obtained an enhanced photocatalytic activity of this TiO_2 architecture with respect to standard, commercially available, TiO_2 nanopowder (Degussa P25) [22,30]. Photocatalytic efficacy was tested by mineralization (i.e., oxidation) of stearic acid. A sharp increase in degradation kinetics is evident when using high surface area tree-like structures. It is worth noting that the as deposited amorphous tree-like structures already show an activity comparable with that of P25 powder. Upon annealing, photocatalytic activity sharply increases by almost one order of magnitude in terms of reaction rate. In particular, a complete decomposition of the organic compound is achieved after 5 h.

The same hierarchical TiO_2 nanostructures have shown interesting properties when employed as the oxide photoanode in dye-sensitized solar cells (DSSC) [28]. TiO_2 tree-like structures in fact are characterized by features that could be beneficial to DSSC performance: (a) large surface area; (b) multiscale porosity allowing effective infiltration of the electrolyte; (c) vertical preferential growth, which could result in better electron transport toward the electrode. Quite interestingly, tests in DSSC showed that hierarchical TiO_2 impairs the electron back reaction with tri-iodide electrolyte. In fact, in comparison to the standard TiO_2, electron lifetime was shown to be more than one order of magnitude longer probably due to low surface defect density or to good interparticle connection. Finally, it was shown that the mesoporous hierarchical structure does not hinder diffusive mass transport of tri-iodide when using solvent-free ionic liquid electrolytes, which are currently studied for the development of DSSC with improved performances.

References

1. Chrisey D. B., Hubler G. (eds.) *Pulsed Laser Deposition of Thin Films. Wiley & Sons: New York*, 1994.

2. Eason R. ed. *Pulsed Laser Deposition of Thin Films. Wiley: Hoboken, NJ*, 2007.

3. Willmott P. R., Huber J. R. (2000), *Rev Mod Phys*, **72**, 315.

4. Lowndes D. H., Geohegan D. B., Puretzky A. A., Norton D. P., Rouleau C. M. (1996), *Science*, **273**, 898.

5. Lowndes D. H., Rouleau C. M. (1999), *J Mater Res*, **14**, 359.

6. Itina T. E., Marine W., Autric M. (2000), *Appl Surf Sci*, **154–155**, 60–65.

7. Geohegan D. B. in *Pulsed Laser Deposition of Thin Films*, (Chrisey D. B. and Hubler G., eds.) Wiley & Sons: New York, NY, 1994, 115.

8. Geohegan D. B., Puretzky A. A., Duscher G., and Pennycook S. (1998), *J Appl Phys Lett*, **72**, 2987.

9. Amoruso S., Sambri A., Vitiello M., Wang X. (2006), *Appl Surf Sci*, **252**, 4712–4716.

10. Arnold N., Gruber J., Heitz J. (1999), *Appl Phys A*, **69**, S87-S93.

11. Chen K. R., Leboeuf J. N., Wood R. F., Geohegan D. B., Donato J. M., Liu C. L., Puretzky A. A. (1996), *Appl Surf Sci*, **45**, 96–98.

12. Bailini A., Di Fonzo F., Fusi M., Casari C. S., Li Bassi A., Russo V., Baserga A., Bottani C. E. (2007), *Appl Surf Sci*, **253**, 8130–8135.

13. Cattaneo D., Foglio S., Casari C. S., Li Bassi A., Passoni M., Bottani C. E. (2077), *Surf Sci*, **601**, 1892.

14. Casari C.S., Foglio S., Passoni M., Siviero F., Bottani C.E., Li Bassi A. (2011). *Physical Review B*, **84**, 155441.

15. Fusi M. Ph.D. Thesis Politecnico di Milano, (2009).

16. Di Fonzo F., Bailini A., Russo V., Baserga A., Cattaneo D., Beghi M. G., Ossi P. M., Casari C. S., Li Bassi A., Bottani C. E. (2006), *Catal Today*, **116**, 69–73.

17. Di Fonzo F., Tonini D., Li Bassi A., Casari C. S., Beghi M. G., Bottani C. E., Gastaldi D., Vena P., Contro R. (2008), *Appl Phys A*, **93**, 765.

18. Baserga A., Russo V., Di Fonzo F., Bailini A., Cattaneo D., Casari C. S., Li Bassi A., Bottani C. E. (2007), *Thin Solid Films*, **515**, 6465–6469.

19. Boulova M., Lucazeau G. (2002), *J Solid State Chem*, **167**, 423.

20. Li Bassi A., Cattaneo D., Russo V., Bottani C. E., Barborini E., Mazza T., Piseri P., Milani P., Ernst F. O., Wegner K., Pratsinis S. E. (2005), *J Appl Phys*, **98**, 074305.

21. Fusi M., Russo V., Casari C. S. Li Bassi A., Bottani C. E. (2009), *Appl Surf Sci*, **255**, 5334–5337.

22. Di Fonzo F., Casari C. S., Russo V., Brunella M. F., Li Bassi A., Bottani C. E. (2009), *Nanotechnology*, **20**, 015604.

23. Cattaneo D., Righetti N., Casari C. S., Li Bassi A., Bottani C. E. (2007), *Appl Surf Sci*, **253**, 7917–7921.

24. Dellasega D., Facibeni A., Di Fonzo F., Bogana M., Polissi A., Conti C., Ducati C., Casari C. S., Li Bassi A., and Bottani C. E. (2008), *Nanotechnology*, **19**, 475602.

25. Ponzoni A., Russo V., Bailini A., Casari C. S., Ferroni M., Li Bassi A., Migliori A., Morandi V., Ortolani L., Sberveglieri G., Bottani C. E. (2011), *Sens Actuators B: Chem*, **153**, 340–346.

26. Polleux J., Gurlo A., Barsan N., Weimar U., Antonietti M., Niederberger M. (2006), *Angew Chem-Int Edit*, **45**, 261–265.

27. Ponzoni A., Comini E., Sberveglieri G., Zhou J., Deng S. Z., Xu N. S., Ding Y., Wang Z. L. (2006), *Appl Phys Lett*, **88**, 203101.

28. Sauvage F., Di Fonzo F., Li Bassi A., Casari C. S., Russo V., Divitini G., Ducati C., Bottani C. E., Comte P., Graetzel M. (2010), *Nano Lett*, **10**, 2562.

29. Torta F., Fusi M., Casari C. S., Bottani C. E., and Bachi A. (2009), *J Proteome Res*, **8**, 1932–1942.

30. Brunella M. F., Diamanti M. V., Pedeferri M. P., Di Fonzo F., Casari C. S., Li Bassi A. (2007), *Thin Solid Films*, **515**, 6309.

Chapter 3

Metastable Phase Selection and Low-Temperature Plasticity in Chemically Synthesized Amorphous Al_2O_3–ZrO_2 and Al_2O_3–Y_2O_3

Ashutosh S. Gandhi,[a] Arindam Paul,[b] Shailendra Singh Shekhawat,[c] Umesh Waghmare,[c] and Vikram Jayaram[d]

[a]*Department of Metallurgical & Materials Engineering, Indian Institute of Technology Madras, Chennai 600036, India*
[b]*Crompton Greaves Ltd., Bhaskara, Kanjur Marg (East), Mumbai 400042, India*
[c]*Jawaharlal Nehru Centre for Advanced Scientific Research, Bangalore 560064, India*
[d]*Department of Materials Engineering, Indian Institute of Science, Bengaluru 560012, India*

qjayaram@materials.iisc.ernt.in

3.1 Introduction

The non-crystalline state of solids exhibits many intriguing characteristics. While silicate glasses have been used and studied for centuries, other glassy or amorphous materials are of relatively recent origin. Examples of such materials are metallic glasses,

Oxide Nanostructures: Growth, Microstructures, and Properties
Edited by Avanish Kumar Srivastava

ISBN 978-981-4411-35-6 (Hardcover), 978-981-4411-36-3 (eBook)
www.panstanford.com

chalcogenide glasses, polymeric glasses, and non-silicate amorphous oxides. The last class of amorphous materials is often difficult to obtain experimentally and therefore our understanding of their structure and properties is lagging behind those of other non-crystalline materials. Alumina forms the base for some of the interesting non-silicate amorphous materials. While multi-component glassy materials, containing small additions of silica and titania are easier to obtain and study, binary alumina–base oxide glasses (e.g., Al_2O_3–ZrO_2 and Al_2O_3–Y_2O_3) are more difficult to obtain and possess limited thermal stability. The challenges in processing and characterization of these materials have been overcome through non-conventional approaches. Interesting properties have been identified, the most exciting one being the discovery of large-scale plasticity in these materials. The amorphous phases in these alumina based binary materials crystallize into phases with excellent mechanical and optical properties. These phases are nanocrystalline with sizes below about 50 nm. Therefore a processing technique of producing bulk nanocrystalline oxides opens up as advances are made in producing bulk glasses in the first place.

Nanostructured oxides with interesting and potentially useful properties can be prepared as nanoparticles, nanorods, and thin films, as well as in bulk forms. Bulk nanostructured oxide ceramics are of particular interest for exploiting their mechanical properties. A significant amount of research effort has been directed towards consolidation of nanoparticles into bulk nanostructured oxides. All the techniques employed for this purpose involve high temperature sintering which invariably suffers from coarsening of the microstructure, some times to the extent of losing the desired nanostructure. Some of the methods adopted for achieving sintering without significant coarsening are colloidal processing, rapid heating, high-pressure cold compaction, and field-assisted sintering. Two-phase materials exhibit slower coarsening kinetics than single-phase ones. Therefore, development of nanocomposites has also been studied, using the same densification techniques as for single-phase nanomaterials. An alternative to the consolidation of crystalline nanoparticles is to produce an amorphous bulk form and crystallize it to obtain the bulk nanocrystalline material. This approach has been successfully applied to the synthesis of amorphous alumina based materials. The most widely used oxide structural ceramics are based on alumina or zirconia. The attractive mechanical properties of these materials are high hardness, fracture toughness and wear resistance. The performance of these materials may be enhanced by rendering them nanocrystalline. Therefore, obtaining a bulk amorphous form and converting it into a nanocrystalline ceramic is an attractive processing strategy for these materials.

Obviously, the key to obtaining bulk nanocrystalline oxides by crystallization is the ability to produce a bulk amorphous form. This, however, is not trivial to achieve in oxide compositions that do not contain large amounts of silica and are not easily rendered glassy. Techniques such as rapid solidification processing (RSP), vapor quenching, mechanical alloying or chemical synthesis are capable of producing amorphous phases in such compositions. These techniques produce granular forms of the amorphous materials, necessitating their consolidation into bulk forms. This step needs thermal activation for bringing about sintering. The amorphous phase is a metastable one, and is likely to crystallize upon thermal activation faster than sintering. Once crystallization occurs, the same problems that plague the sintering of nanoparticles would hinder consolidation without coarsening. The challenge, therefore, is to identify conditions under which the granular amorphous phase can be consolidated without crystallizing. Crystallization should be a controlled step that follows consolidation in the amorphous state. The identification of compositions that may be amenable to such processing requires the understanding of the thermodynamic and kinetic aspects of metastable phase selection in oxide systems.

In the following section, metastable phase selection in alumina-based binary and ternary systems is reviewed briefly. This is followed by a section on the consolidation of Al_2O_3–ZrO_2 and Al_2O_3–Y_2O_3 amorphous oxides and their crystallization behavior. Finally, it is not surprising that amorphous phases in these systems themselves possess interesting characteristics. As mentioned above, the consolidation studies have revealed large-scale plasticity in these alumina based amorphous oxides at moderate temperatures. This interesting phenomenon is discussed at the end of the chapter.

3.2 Metastable Phase Selection in Al_2O_3–ZrO_2 and Al_2O_3–Y_2O_3

The amorphous or glassy state is further away from equilibrium than metastable crystalline states in a system. Whether glass can be formed in a given composition depends on the ease of preventing crystallization. In the Al_2O_3-ZrO_2 and Al_2O_3-Y_2O_3 systems, amorphization can be achieved by rapid solidification as well as chemical synthesis. A number of crystalline metastable phases also can form in these systems. Alumina itself can be synthesized with non-equilibrium crystal structures (Levin and Brandon, 1998) as well as in the amorphous state, during chemical synthesis as well as rapid quenching. The polymorphic transformations in ZrO_2 are well known and are given below.

$$\text{Monoclinic} \xrightarrow{950\text{–}1150^\circ\text{C}} \text{Tetrgonal} \xrightarrow{2370^\circ\text{C}} \text{Cubic} \xrightarrow{2680^\circ\text{C}} \text{Liquid}$$

Yttria also exhibits polymorphism, and intermediate compounds $Y_3Al_5O_{12}$ (YAG), $YAlO_3$ (YAP) and $Y_4Al_2O_9$ (YAM) appear in the Al_2O_3-Y_2O_3 system (Fig. 3.1). These phases may form metastably in compositions away from their stoichiometry. Because of such a large number of possible phases it is expected that both the Al_2O_3-ZrO_2 and Al_2O_3-Y_2O_3 systems lend themselves to the formation of metastable phases.

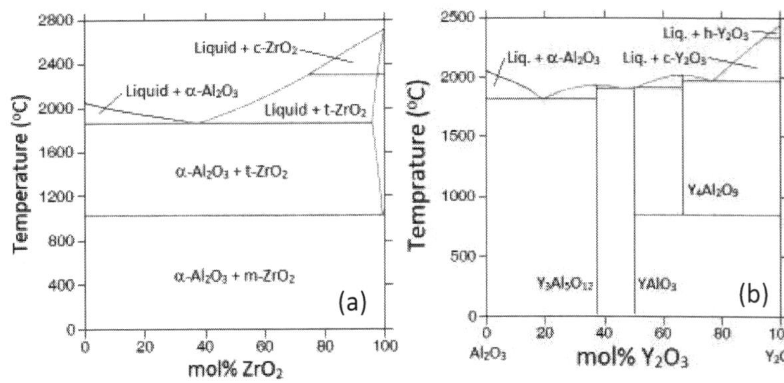

Figure 3.1 Phase diagrams for (a) the alumina–zirconia and (b) alumina–yttria systems calculated using ThermoCalc® software (h: hexagonal).

3.2.1 Phase Selection in Al_2O_3–ZrO_2 System

Initial studies on the metastable phase formation in the Al_2O_3–ZrO_2 system used RSP for accessing these phases (Jacobson and McKittrick, 1994). These reports showed that it is possible to produce an amorphous phase in alumina-rich compositions. Some recent studies have also shown amorphous phase formation in plasma sprayed materials (Vasiliev et al., 2006). Kim and Kim (1999) also showed amorphous phase forming in alumina–zirconia by RSP, although small amounts of titania were also added in their materials. Chemical synthesis routes have shown better potential to retain the amorphous phase since the temperature of synthesis is low enough to kinetically suppress crystallization. Even pure zirconia has been synthesized in the amorphous state using chemical synthesis (Molodetsky et al., 2000).

It is important that the crystallization temperature of the amorphous phase is as high as possible so that consolidation can be carried out at sufficiently high temperature. The first phase to crystallize and the subsequent phase evolution are also of interest. Initial work on chemically synthesized ZrO_2–Al_2O_3 compositions (Yamaguchi et al., 1988) was based on the alkoxide (sol-gel) route. It was reported that the crystallization of the amorphous phase led to the formation of a cubic (c) ZrO_2 solid solution up to 45 mol% Al_2O_3. Similarly, simultaneous hydrolysis of alkoxides (Inamura et al., 1994) produced amorphous ZrO_2–Al_2O_3 with up to 40 mol% Al_2O_3. Tetragonal (t) solid solution formed upon crystallization in ZrO_2–25 mol% Al_2O_3. On the other hand, it was claimed that with 40 mol% Al_2O_3, the crystallization product is cubic ZrO_2 solid solution. Subsequently, Balmer et al. (Balmer et al., 1994; Balmer et al., 1995) showed that amorphous phase in the ZrO_2–Al_2O_3 system could be obtained in almost the entire composition range by spray pyrolysis of aqueous solutions of zirconium acetate and aluminum nitrate. The main outcome of this work was that the products of crystallization in the compositions up to 40 mol% Al_2O_3 were clearly identified. Based on X-ray diffraction (XRD) results, cubic ZrO_2 appeared to crystallize first, similar to previously reported results (Yamaguchi et al., 1988). Investigation by transmission electron microscopy (TEM) revealed the presence of the 112 diffraction ring corresponding to the tetragonal ZrO_2 solid solution in the selected area diffraction patterns (SADPs). Moreover, Raman spectroscopy performed on the solid solutions established that although the tetragonality (c/a ratio) of the phase to crystallize is ~1, it is a tetragonal zirconia solid solution. These results (Balmer et al., 1994; Balmer et al., 1995) were corroborated by the studies on ZrO_2–Al_2O_3 thin films deposited using aqueous precursors (Narwankar, Speck and Lange, 1995), as well as by studies on phase selection in powders synthesized using co-precipitation (Moreau, Gervais and Douy, 1997). All the reports on chemical synthesis of ZrO_2–Al_2O_3 established that the grain sizes of the crystallized materials were in the nanometric range, with the grain size of the first product of crystallization measured to be as small as 3 nm. It is remarkable that tetragonal zirconia is the phase to crystallize when as much as 40 mol% Al_2O_3 is added to ZrO_2. This composition is equivalent to 57% Al^{3+} cation substitution for Zr^{4+}, i.e. $x = 0.57$ in $Zr_{1-x}Al_xO_{2-x/2}$. It is possible that the crystallization is not partitionless so that only a part of the Al_2O_3 present is dissolved in the tetragonal zirconia. This issue was investigated by Balmer and co-workers using ^{27}Al magic angle spinning nuclear magnetic resonance (MAS-NMR) spectroscopy (Balmer et al., 1996). The large alumina content is likely to be accommodated in disordered clusters between tetragonal zirconia crystallites. In compositions up to about 10 mol% Al_2O_3, the small tetravalent Al^{3+} ions are distributed randomly in the tetragonal solid solution and exhibit five- and six-fold coordination in the O^{2-} polyhedra. The crystallization temperature is the highest near the 1:1 molar ratio of the two oxides and decreases as the composition moves towards either pure zirconia or alumina. Both, RSP and chemically synthesized materials follow this trend.

For the compositions in which tetragonal zirconia crystallizes first, further thermal exposure leads to the formation of the metastable γ-Al_2O_3. The excess alumina dissolved in the tetragonal zirconia is rejected in this process. The formation of the equilibrium phases, monoclinic zirconia (m) and α-Al_2O_3 occurs with further thermal exposure above 1000°C. To illustrate, the typical phase evolution sequence in the ZrO_2–40 mol% Al_2O_3 composition is

$$\text{Amorphous} \xrightarrow{\sim 870^\circ\text{C}} \underset{(\text{Al}_2\text{O}_3)}{\text{t-ZrO}_2} \xrightarrow{\sim 1000^\circ\text{C}} \text{t-ZrO}_2 + \gamma\text{-Al}_2\text{O}_3 \xrightarrow{\sim 1400^\circ\text{C}} \text{m-ZrO}_2 + \alpha\text{-Al}_2\text{O}_3$$

The formation of nanocrystalline and metastable phases makes this system a good candidate for research on processing of bulk forms.

With alumina-rich compositions, it was shown that the crystallization temperature increased with the addition of zirconia (Stefanic, Music and Trojko, 2005). For small amounts of zirconia (x up to 0.2 in $Zr_{1-x}Al_xO_{2-x/2}$), the crystallization temperature was 700°C or less. It appears that x = 0.3 to 0.6 is the composition range with adequate thermal stability of the amorphous phase.

Note that the hydroxyl and nitrate radicals have a possible role in accelerating crystallization. Djuricic et al. showed that in the presence of small amount of zirconia the amorphous phase is stable up to 800°C (Djuricic et al., 1997). Hydrothermal treatment accelerates crystallization by boehmite formation. The presence of OH^- and such radicals may be detrimental to the stability of the amorphous phase.

3.2.2 Phase Selection in Al_2O_3–Y_2O_3 System

The amorphous phase is obtained with relative ease in this system by chemical synthesis routes and RSP. The crystallization temperature does not show a monotonic change with increasing yttria content. The crystallization and further phase evolution in compositions with 4, 10 and 15 mol% Y_2O_3 were studied by Ullal et al. (Ullal et al., 2001), in addition to the YAG composition (37.5% Y_2O_3). The powders were prepared by spray pyrolysis of nitrate solutions. Compositions up to 15% Y_2O_3 crystallized between 900 and 1000°C, while the 37.5% Y_2O_3 material crystallized at 800°C. Gandhi and Levi found that in compositions of 25 to 55% Y_2O_3, the amorphous phase stability generally increased with increasing yttria content (Gandhi and Levi, 2005).

Considering the phase evolution sequence, Ullal et al. showed that the amorphous phase in 4% Y_2O_3 material could be retained up to 900°C for short duration, though 1 h at this temperature was sufficient time to crystallize γ-Al_2O_3 (Table 3.1). All of the 4% yttria was retained in the γ-Al_2O_3 as an extended solid solution up to 1200°C and was rejected to form the orthorhombic YAP only when the γ-Al_2O_3 transformed to α-Al_2O_3 at ~1250°C. It is interesting that YAP formed in the place of the equilibrium YAG. Such metastable phase selection in this system makes it one of the more interesting ones. In the work by Ullal et al. the compositions of 10 and 15%

Y_2O_3 did not crystallize in a partitionless manner. The XRD evidence showed that a significant amount of amorphous phase was present after crystallization of γ-Al_2O_3 and the yttria-rich metastable hexagonal $YAlO_3$ (H-YA) crystallized from this remaining amorphous phase. The equilibrium YAG formed upon further thermal exposure, though metastable phases did not disappear completely.

The nucleation of either YAP or H-YA instead of YAG is rationalized on the basis of the complexity of the crystal structures of these phases. The structural complexity is correlated to the number of atoms per primitive cell, or the primitive cell volume. Thus, YAG with 0.866 nm^3 primitive cell volume is more complex than YAP with 0.203 nm^3. The H-YA in comparison is considered as a simpler structure as its primitive cell volume is 0.12 nm^3. This point may be used to rationalize the appearance of H-YA even when the target is to synthesize phase pure YAG, i.e. when the composition is 37.5% Y_2O_3. It is claimed in the literature that the H-YA also tolerates deviation from its stoichiometry, especially supporting excess alumina content up to the stoichiometric YAG composition (Gandhi and Levi, 2005; Ullal et al., 2001). It was argued by Ullal et al. that between YAP and H-YA the choice of the phase to nucleate depends on the parent phase. When the parent phase is γ-Al_2O_3, the similarity of cation coordination between this phase and YAP makes its nucleation rate higher due to lower interfacial energy. On the other hand, when the parent phase is amorphous, H-YA is the preferred phase for nucleation since the coordination in the H-YA is similar to that in the amorphous phase.

The work of Ullal et al. also emphasized the need for homogeneity in chemically synthesized material for obtaining single phase YAG. However, Gandhi and Levi found that YAG crystallized at 1000°C in powders of YAG composition prepared from nitrates by spray pyrolysis at 315°C which had signs of inhomogeneity in the form of yttria peaks in an otherwise amorphous XRD pattern. They also observed the H-YA upon crystallization of the YAG composition prepared by spray pyrolysis at 280°C, which showed no signs of inhomogeneity. This result showed that homogeneity is not sufficient for crystallizing YAG in 37.5% Y_2O_3 composition. Gandhi and Levi further showed that YAG was the first crystalline phase in powders of compositions from 25 to 55% Y_2O_3 when reverse co-precipitation of nitrate precursors was used for synthesis. These materials were homogeneous, and the YAG phase showed deviation from its stoichiometry as evident in its increasing lattice parameter with yttria content. Since YAG can form away from its stoichiometry, the homogeneity of the amorphous phase is not a necessary condition for YAG crystallization. It appears that the nature of the amorphous phase plays a role in the phase selected by the system as the product of crystallization. Li and co-workers carried out the synthesis of the YAG composition using inhomogeneous and homogeneous precursors and claimed that the relative diffusion coefficients of the two cations, Al^{3+} and Y^{3+} in the amorphous phase determine whether YAG can crystallize or other metastable phases form (Li et al., 2011). This is an interesting development that needs to be taken into account for establishing the reasons for the observed diversity of phase selection in the Al_2O_3–Y_2O_3 system.

The Al_2O_3–Y_2O_3 system exhibits another interesting phenomenon related to the structure of the amorphous phase. It has been reported that rapidly solidified glasses phase separate into two glasses with the same composition but different densities. This phenomenon is called polyamorphism and is not unique to this system. When an alumina–yttria liquid, in the composition range of 24 to 32% Y_2O_3, is supercooled, it undergoes density driven phase separation into two liquids. One liquid is dispersed in the other. When these liquids are further cooled, they undergo glass transition. The density difference between the liquids is due to the differences in their cation coordination and short range order. Solidification of alumina-rich compositions with large supercooling has been carried out using aero-acoustic levitation with laser heating (Greaves et al., 2008; Greaves et al., 2009), and a flame-spray technique (Rosenflanz et al., 2004). The latter produces glassy microspheres in large quantities suitable for consolidation into bulk forms.

It is clear that both, the Al_2O_3–ZrO_2 and Al_2O_3–Y_2O_3 systems are amenable to amorphization by the chemical synthesis route. The thermal stability of the amorphous phase is limited in the sense that consolidation of the amorphous powders cannot be carried out at as high temperatures as those used for the sintering of crystalline materials in these systems. Therefore consolidation studies have employed high pressures to accelerate sintering. The following section deals with the consolidation of amorphous powders in these systems.

Table 3.1 Phase evolution in Al_2O_3-Y_2O_3 powders prepared by spray pyrolysis of nitrate solutions (Ullal et al., 2001)

Temperatures of heat treatment (°C)	Composition (mol% Y_2O_3)			
	4	10	15	37.5
500	—	—	—	A
800	—	—	—	A(G)
900	A	A	A	G(A)
1000	γ(A)	γ(A)	γ(A)	G
1100	—	γ + H	γ + H	—
1150	γ	G + H + γ	G + H + γ	—
1200	γ	G + H + α + O	G + H + γ	—
1250	α + O + γ	—	G + α	—
1600	α + G + O	—	—	—

3.3 Consolidation of Amorphous Powders of Al_2O_3–ZrO_2 and Al_2O_3–Y_2O_3

The consolidation of amorphous powders into bulk forms is crucial for evaluating the structural properties of the amorphous materials. Consolidation is also a necessity if the amorphous phase is to serve as a precursor to a bulk nanostructured ceramic form. The consolidation of powders needs thermal activation for sintering processes to occur. Since crystallization temperatures of the alumina-zirconia and alumina-yttria amorphous powders are lower than 900°C for most of the compositions, consolidation temperature needs to be significantly lower than 900°C. The crystalline powders in these oxide systems are consolidated by sintering well above this temperature, in the range of 1200–1600°C. Either uniaxial or isostatic pressure is often applied to enhance the driving force for sintering. Uniaxial hot pressing, for instance, at temperatures of 1200°C and above is carried out at pressures of ~50 MPa or lower. Therefore, acceleration of sintering at temperatures below 900°C is expected to require much higher pressures. This would be in spite of the faster diffusional mass transport, and possibly more viscous flow, expected in the amorphous materials due to the presence of free volume.

With the above considerations, consolidation of amorphous $Zr_{1-x}Al_xO_{2-x/2}$ ($x = 0.57$) was carried out by Jayaram and co-workers at a quasi-hydrostatic pressure of 1 GPa at temperatures between 800 and 1000°C (Jayaram et al., 1998; Mishra et al., 1999). It was found that the amorphous alumina–zirconia powders could be consolidated to high relative densities but had crystallized at the same time. The least amount of porosity, as revealed by electron microscopy, was achieved at the lowest temperature used (700°C). Pressure consolidation at the highest temperature used (900°C) produced a material with the highest amount of porosity. This observation appears to be contrary to the established understanding of sintering, viz. the higher the temperature, the greater is the extent of sintering achieved. The explanation lay in the extent of crystallization and subsequent phase evolution that occurred under the high applied pressure. It was found that compaction at 700°C led to the crystallization of the metastable phases, t-ZrO_2 and γ-Al_2O_3. Higher temperature experiments (800 and 900°C) produced the equilibrium m-ZrO_2 and α-Al_2O_3. Thus, the higher amounts of porosity after compaction are associated with the formation of equilibrium phases. The implication is that the highly non-equilibrium phases are capable of faster sintering, at substantially lower temperatures.

At this juncture it is interesting to examine the effect of pressure on solid-solid transformations based on the theory of nucleation. Mishra et al. considered the $\gamma \rightarrow \alpha$ polymorphic transformation in alumina under high applied pressure (Mishra et al., 1999). By analogy, the results of the calculations were used for explaining the pressure driven phase evolution in amorphous alumina–zirconia. The nucleation rate in a diffusional solid-solid phase transformation is given by

$$N_{\mathrm{hom}} = \omega C_0 \exp\left(-\frac{\Delta G_{\mathrm{m}}}{kT}\right) \exp\left(-\frac{\Delta G^*}{kT}\right),$$

where ω is related to the vibrational frequency of atoms, ΔG_{m} is the free energy of migration of the atoms, and C_0 is the number of atoms per unit volume (Porter and Easterling, 1992). The free energy barrier for nucleation ΔG^* is given by

$$\Delta G^* = \frac{16\pi \Delta G_i^3}{\Delta G_t^2}, \text{ where, } \Delta G_t = \frac{\Delta G_n + \Delta G_s + \Delta G_{pv}}{\Omega_{average}},$$

where the subscripts i, n, s and p refer to the free energies of the interface, chemical driving force, strain energy of the nucleus and the additional work carried out by the applied pressure.

By using representative values of the various terms, Mishra et al. showed that the applied pressure accelerated nucleation by at least a few orders of magnitude (Table 3.2). The pressure effect, however, diminished as the temperature of transformation was decreased. It therefore emerges that the crystallization of the amorphous phase during high-pressure consolidation of powders may be suppressed by using lower temperatures. Obviously, lower pressures than used by Mishra et al. would also be useful in retaining the amorphous phase during consolidation.

Table 3.2 The estimated pressure-induced enhancement in the nucleation rate in the $\gamma \rightarrow \alpha$ transformation in alumina (R is the ratio of nucleation rate under 1 GPa pressure to that under atmospheric pressure)

Temperature (K)	900	1000	1100	1200
R (ΔG_i = 0.10 J m^{-2})	490	1200	4200	41000
R (ΔG_i = 0.15 J m^{-2})	10^9	10^{10}	10^{12}	10^{15}

Gandhi et al. used these arguments to design pressure consolidation experiments on amorphous powders of $Zr_{1-x}Al_xO_{2-x/2}$ (x = 0.57) synthesized by spray pyrolysis of nitrate precursors (Gandhi, Jayaram and Chokshi, 1997; Gandhi, Jayaram and Chokshi, 1999; Gandhi, Jayaram and Chokshi, 2001). Uniaxial hot pressing carried out in Ni-superalloy dies at 600 and 650°C, with pressures of 500 and 750 MPa brought about significant consolidation, up to relative densities in the range of 92 to 98%. Equally significant was the result that the consolidated material remained amorphous. Due to the high pressure used, the size of the compact was limited to a diameter of 5 mm, and the height was proportionately restricted to ~3 mm or less. However, these compacts are certainly bulk materials when compared to the materials obtained by rapid solidification or vapor quenching techniques. It was shown that a composition richer in alumina (Al_2O_3–20 mol% ZrO_2) could be consolidated up to 99% relative density at 650°C under a pressure of 750 MPa. Improving the powder characteristics in terms of removal of agglomerates larger than 10 μm led to a uniform distribution of residual porosity in $Zr_{1-x}Al_xO_{2-x/2}$ ($x = 0.57$).

The sintering of amorphous materials (e.g., silicate glasses) is generally brought about by viscous flow. This capillarity driven densification of a glassy porous compact is dependent on time (t), whether the viscosity of the glass is Newtonian or not. The earliest analytical models of sintering were based on a viscous mechanism. Frenkel developed a model for the early stage of sintering (Frenkel, 1945), wherein the change in the height of a sintering compact is given by

$$\frac{\Delta L}{L_0} = -\left(\frac{3\gamma}{8\eta R}\right)t.$$

Here $\Delta L = L - L_0$, the difference between the instantaneous height and the original one. The surface energy of the glass is given by γ while its viscosity is η. Sintering is directly proportional to time. Similarly, Mackenzie and Shuttleworth developed a model for the final stage of sintering by the viscous flow mechanism (Mackenzie and Shuttleworth, 1949). The densification rate is given by

$$\frac{d\rho}{dt} = \left(\frac{3\gamma}{2\eta R_0}\right)(1-\rho),$$

where ρ is the relative density and R_0 the initial particle size.

These models do not consider the presence of an externally applied pressure. Murray et al. adapted the Mackenzie Shuttleworth model to include external pressure (P_a). The relative density as a function of time is given as

$$\ln(1-\rho) = \ln(1-P_c) + \frac{3P_a}{4\eta}t,$$

where P_c is the capillarity pressure, usually $2\gamma/R$ for a sphere of radius R (Murray, Rogers and Williams, 1954).

Sintering mechanisms are also analyzed based on scaling laws (Rahaman, 1995). For viscous sintering, if the time taken to achieve densification is t_1 for the initial particle size of R_1, then the time required for the same densification for a particle size R_2 would require time t_2, with the following relationship:

$$\frac{\Delta t_2}{\Delta t_1} = \frac{R_2}{R_1}$$

In comparison, the scaling law for sintering by grain boundary diffusion is given as

$$\frac{\Delta t_2}{\Delta t_1} = \left(\frac{R_2}{R_1}\right)^4$$

The scaling law implies that smaller initial particle size should lead to faster densification in a glass powder compact.

With this background, the characteristics of the pressure consolidation of amorphous $Zr_{1-x}Al_xO_{2-x/2}$ (x = 0.57) were investigated (Gandhi et al., 2001, Gandhi and Jayaram, 2002). The most important observation was that consolidation at temperatures up to 600°C was more a function of the applied pressure than time. This behavior is shown in Fig. 3.2 for two compositions in the alumina–zirconia system. The 62% Al_2O_3 composition corresponds to x = 0.765 in $Zr_{1-x}Al_xO_{2-x/2}$, which is the eutectic composition in this system. The particle size of both the materials was below 3 μm. The initial relative density of ~45% increased to ~92 and 95% for x = 0.57 and 0.765, respectively, as the uniaxial pressure increased from 0 to 750 MPa at 600°C. Upon reaching the final pressure it was held constant for an hour and the density increased from 92 to 96% for x = 0.57 and 95 to more than 99% for 0.765. It can be argued that the time dependent densification was less because the compacts were close to the final density by the time the maximum pressure was reached. Gandhi conducted experiments wherein low pressures of 250 and 500 MPa were applied at temperatures of 450 and 600°C so that the relative density after reaching the final value of the pressure was as low as ~59% (Gandhi, 2001). The compacts were then allowed to sinter under the applied pressures for one hour. It was found that the increment in the density was insignificant compared to the increments achieved in reaching the applied pressures. It is therefore established that time-dependent densification in the amorphous $Zr_{1-x}Al_xO_{2-x/2}$ materials is far less than the pressure-sensitive densification. These observations led to the conclusion that viscous sintering is not the mechanism of sintering in this material.

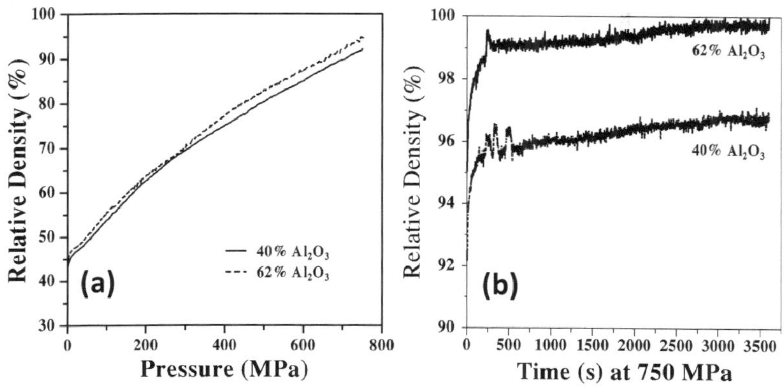

Figure 3.2 Increments in the relative density of amorphous alumina–zirconia compacts during uniaxial compaction at 600°C. (a) Consolidation as the applied pressure increases, and (b) relative density as a function of time as the pressure (750 MPa) and temperature are held constant. Consolidation occurs more due to the high pressure applied than the time given for sintering (Gandhi et al., 2001).

Further evidence for the non-viscous consolidation of amorphous $Zr_{1-x}Al_xO_{2-x/2}$ came in the form of the effect of particle size on consolidation (Gandhi and Jayaram, 2002). It was seen that the green density after compaction at room temperature, prior to high temperature pressure consolidation, was sensitive to the particle size. Fine particles of composition with $x = 0.57$ with size smaller than 3 μm could be packed to a green density as high as 43% with a cold compaction pressure of 50 MPa. Powder with a medium particle size (3 to 15 μm) could be compacted to 37% and coarse powder with particles larger than 15 μm could be compacted to ~34% green density at the same pressure. Upon applying the high pressure of 750 MPa after heating to 600°C, however, the increments in density observed in the three compacts with the different particle sizes was similar in magnitude. This observation established that the material does not obey the scaling law for viscous sintering.

Gandhi and Jayaram compared the behavior of the amorphous $Zr_{1-x}Al_xO_{2-x/2}$ with the pressure consolidation of silver powder and a borosilicate glass powder with average particle sizes of 2 and 9 µm, respectively (Gandhi and Jayaram, 2002). They showed that the amorphous alumina–zirconia powders behave more like silver powder does at room temperature (cold compaction) than borosilicate glass at high temperature (viscous sintering). Since the cold compaction of silver involves large-scale plastic deformation of the powder particles, the inevitable conclusion is that the amorphous alumina–zirconia material undergoes plastic deformation during pressure consolidation at moderate temperatures of 450 to 600°C. This hypothesis was tested by Gandhi and Jayaram (Gandhi and Jayaram, 2003) conducting uniaxial compression tests on dense compacts of $Zr_{1-x}Al_xO_{2-x/2}$ (x = 0.57). They discovered that this material is capable of plastic deformation with a yield stress of ~360 MPa. The flow stress decreases to ~240 MPa after the onset of plastic flow. This interesting phenomenon will be discussed in detail in the following section. For the moment, one can take into account the flow stress values to analyse the pressure consolidation data as if the amorphous powder of $Zr_{1-x}Al_xO_{2-x/2}$ (x = 0.57) was consolidated in exactly the same manner as a ductile metal powder, i.e. involving large-scale plastic deformation (Gandhi and Jayaram, 2002).

A simple model developed for the cold compaction of ductile metal powders is available in the literature (Akisanya, Cocks and Fleck, 1997). In this model, the macroscopic hydrostatic yield stress of the powder compact (P_y) is related to the instantaneous relative density (ρ), the initial relative density (ρ_0) and the yield stress of the material being compacted (σ_y) as follows:

$$P_y = 3\rho^2 \frac{\rho - \rho_0}{1 - \rho_0} \sigma_y$$

Gandhi et al. used their density data from the pressure consolidation experiments and the yield stress values from the compression tests to estimate P_y. The corresponding uniaxial pressure in a constrained die was calculated from the relation

$$P_y = \frac{P}{3}\left(\frac{1+\nu}{1-\nu}\right).$$

The outcome of this analysis is shown in Fig. 3.3. The agreement between the model prediction and the actual data is striking. The initial data are closer to the curve for σ_y = 360 MPa and later they are closer to the σ_y = 240 MPa curve. This is because the initial yield stress of the material is higher, followed by a yield drop as deformation proceeds. The role of plastic deformation in obtaining high relative densities by pressure consolidation of alumina–zirconia amorphous materials is therefore established beyond reasonable doubt.

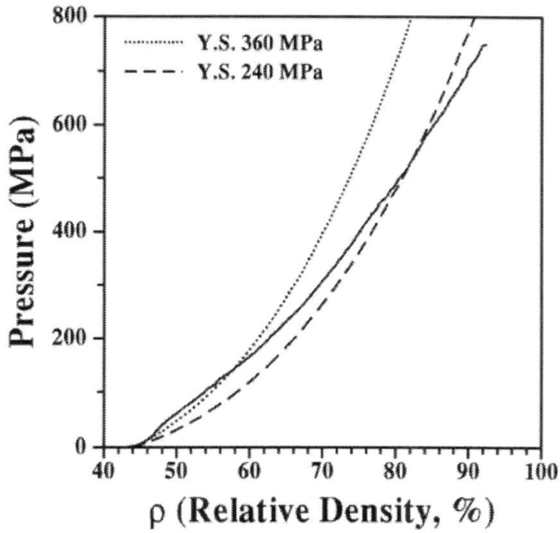

Figure 3.3 Comparison of the pressure consolidation data for amorphous $Zr_{1-x}Al_xO_{2-x/2}$ (x = 0.57) at 600°C with the prediction by the model by Akisanya et al. (Akisanya et al., 1997; Gandhi and Jayaram, 2003).

At this point, it is worth mentioning that the time-dependent changes during pressure consolidation of amorphous alumina–zirconia are not insignificant. Though the increment in density is about an order of magnitude smaller, it influences the hardness of the as-hot pressed material. The Vickers hardness of amorphous $Zr_{1-x}Al_xO_{2-x/2}$ (x = 0.57) was ~170 VHN when the pressure was removed right after reaching the maximum value of 750 MPa at 600°C. The hardness reached ~200 VHN when the pressure was held for 30 minutes at the same temperature (Gandhi and Jayaram, 2002). It is likely that the increase in the hardness is due to improved bonding at prior particle boundaries. It may also be due to structural relaxation that might occur in the amorphous material after undergoing plastic flow during consolidation.

The work by Gandhi and co-workers opened the avenue for producing bulk amorphous materials in alumina based compositions. The results on the alumina–zirconia compositions indicate the possibility that pressure consolidation at temperatures lower than the normal crystallization temperature might be carried out for amorphous materials in the alumina–yttria system. Consolidation of amorphous Al_2O_3–Y_2O_3 was first studied on the YAG composition (Choudhury, Gandhi and Jayaram, 2003) at 450 to 640°C and 750 MPa pressure. Partial crystallization could not be prevented during consolidation, even at the low temperature of 450°C. However, the consolidation behavior was qualitatively similar to alumina–zirconia. Most of the increment in relative density was achieved during the application of the pressure, rather than during holding the pressure and temperature constant for time up to 30 minutes. Interestingly, higher relative densities were achieved when the high pressure was applied at room temperature and the temperature was raised later. As the crystallization temperature of YAG is low, a composition away from this phase, with 15% yttria was synthesized (Thangamani et al., 2005). As shown in the previous section, the amorphous phase has better thermal stability in this composition range. Consequently, 98% relative density could be achieved in pressure consolidation at 750 MPa and temperature of 600°C, without crystallization. It should be noted that this high relative density was achieved in amorphous powder prepared by co-precipitation using nitrate precursors. The results of Choudhury et al. and Thangamani et al. established the generality of the processing route developed for amorphous alumina-based materials. So far we have not discussed the possible reasons for the ability of these amorphous materials to undergo pressure consolidation at moderate temperatures through large-scale plastic deformation. Since plasticity is the fundamental characteristic of these materials as apparent from their consolidation behavior, this aspect is discussed in detail in the next section.

The bulk amorphous materials serve as a precursor to nanostructured crystalline ceramics. The dense amorphous $Zr_{1-x}Al_xO_{2-x/2}$ (x = 0.57) compacts were heat treated to produce nanocrystalline ceramics, mainly consisting of m-zirconia and α-alumina. The grain size in the as-crystallized state was as small as 6 nm, which grew to ~0.3 μm after thermal exposure up to 1400°C for 1 h. The alumina–yttria materials are prone to pressure induced crystallization. It is possible, however, to carry out controlled crystallization of the bulk amorphous compacts. Sellappan et al. worked on crystallizing the consolidated bulk amorphous YAG composition into nanostructured ceramics (Sellappan et al., 2007). The most important issue here is the introduction of ultrafine porosity due to the volume shrinkage during crystallization. Application of hydrostatic pressure of 1 GPa at 1000°C was partly successful in preventing this so that the final density was 96%.

The amorphous alumina based materials discussed in this section are chemically synthesized, either by spray pyrolysis or by co-precipitation. It is possible to obtain glassy materials by rapid solidification. Rosenflanz et al. prepared micro-beads of glassy alumina–rare earth compositions, including alumina–yttria compositions, by a flame-spray technique (Rosenflanz et al., 2004). The oxide powders were fed through a hydrogen-oxygen flame and quenched in water. The estimated cooling rate was ~10^3 K s^{-1}. The glassy micro-beads were sintered by mild hot pressing, at ~905°C with a pressure of 34 MPa. The dense glassy compacts were transparent. Crystallization was carried out to develop nanostructured ceramics, which remained transparent. These materials have potential applications in optics as their refractive indices can be as high as 1.8. Interestingly, Rosenflanz et al. also reported the much higher hardness (compared to sintered alpha alumina) of the nanocrystalline materials derived by crystallizing the alumina–rare earth glasses even though they are full of metastable alumina polymorphs and other rare earth aluminates. It is not clear, however, whether ultrafine porosity is induced during crystallization due to the volume shrinkage. It is interesting that low pressures, similar to those used in conventional hot pressing, are adequate to bring about consolidation. It appears that the structure of a rapidly quenched glass is different from an amorphous material synthesized through a chemical route, so that their consolidation characteristics are significantly different. This aspect will be dealt with in the next section to analyse the plastic deformation of amorphous alumina based materials.

3.4 Plastic Deformation of Glassy Al_2O_3–ZrO_2 and Al_2O_3–Y_2O_3

The unusual consolidation characteristics of amorphous alumina-zirconia and alumina-yttria powders led to the hypothesis that these materials undergo plastic deformation at temperatures in the range of 450–700°C. Gandhi and Jayaram carried out uniaxial compression tests on dense amorphous $Zr_{1-x}Al_xO_{2-x/2}$ (x = 0.57) specimens and found large scale plasticity with strains up to 14% (Gandhi and Jayaram, 2003). Plasticity was also discovered in amorphous alumina-yttria material (Paul and Jayaram, 2011). In this section, we discuss this new and interesting phenomenon. The phenomenological description of plasticity is followed by its analysis in terms of the structure of the amorphous phase and the role of temperature and synthesis route in plastic flow, especially in the alumina-yttria system.

Gandhi and Jayaram studied the deformation behavior of amorphous $Zr_{1-x}Al_xO_{2-x/2}$ (x = 0.57) (Gandhi and Jayaram, 2003). They carried out compression tests on specimens of 2 mm × 2 mm cross section and height of ~1.8 mm. Larger size specimens could not be used because of the size restrictions on the pressure consolidation used for producing the dense material. Compressive test at 700°C (Fig. 3.4) showed elastic deformation up to a stress of ~360 MPa. At this stress, plastic deformation set in with a sharp decrease in the flow stress down to ~240 MPa. The flow stress continued to decrease, but at a much slower rate as the deformation progressed. The specimen did not fracture and when the test was stopped, a plastic strain of ~8.5% had accumulated in the specimen. It was found that the yield stress and the flow stress after the sharp decrease were not sensitive to strain rate. When the stress was removed and re-applied immediately, deformation resumed at the same stress. On the other hand, when a specimen was unloaded, cooled, reheated and then subjected to deformation, the upper yield stress was observed, followed by a sharp decrease in the flow stress. It appears that the deformed structure was restored at least partially when the material was allowed some time at the moderately high temperatures used by the investigators. The question now is what structural changes might occur during deformation. It was reported that a polished surface parallel to the compressive axis developed shear bands after the test (Fig. 3.5). Surface relief observed in a scanning electron microscope (SEM) showed that the shear bands were at an angle of ~52° to the compression axis. The true angle between the shear bands and the compressive axis could be smaller, depending on the orientation of the bands with respect to the specimen side on which they were observed.

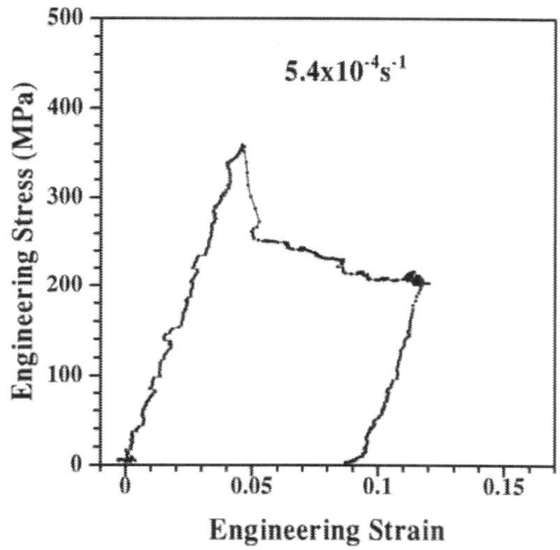

Figure 3.4 Compressive stress-strain curve for amorphous $Zr_{1-x}Al_xO_{2-x/2}$ (x = 0.57) deformed at the strain rate indicated. The temperature was 700°C (Gandhi and Jayaram, 2003).

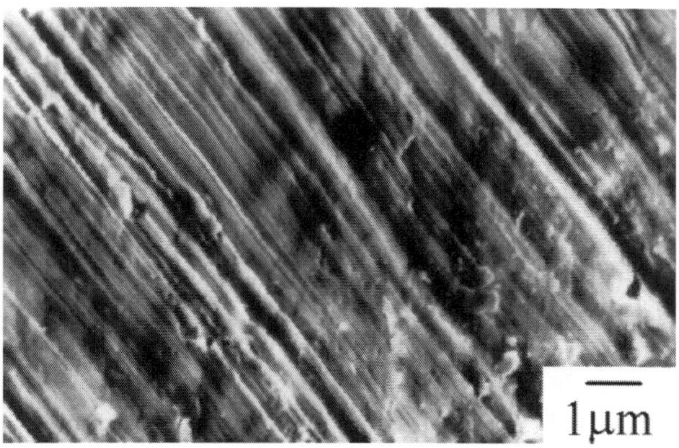

Figure 3.5 Shear bands on the surface of amorphous $Zr_{1-x}Al_xO_{2-x/2}$ (x = 0.57) deformed under compression at 700°C. The compression axis is from top to bottom (Gandhi and Jayaram, 2003).

Shear bands represent localized deformation. They are not observed in oxide glasses as they are brittle below the glass transition temperature (T_g) and undergo viscous flow above it. Only a few observations of shear bands have been made in silicate glasses during room temperature deformation during indentation. Metallic and polymeric glasses, however, exhibit shear bands more commonly when deformed below the glass transition temperature (Courtney, 1990; Meyers and Chawla, 1999). Focussing on the deformation of metallic glasses, one can see the similarity with the amorphous alumina–zirconia. At temperatures below ~$0.6T_g$, metallic glasses show elastic-plastic deformation. The yield stress is not a strong function of either the strain rate or the temperature. Large decrease in the flow stress after the onset of plastic deformation is seen in many metallic glasses. The shear bands form at angles smaller than 45° with the compressive axis during compression tests, and greater than 45°C with tensile axis when tested in tension. Annealing removes the effects of shear band formation. The exact mechanism of shear band formation in metallic glasses is not understood so far (Schuh, Hufnagel and Ramamurty, 2007). However, the similarity between the deformation of amorphous alumina–zirconia and metallic glasses raises the question of any structural similarity between the two types of materials.

The structure of metallic glasses is described as one having short range order, with a distribution of free volume over sizes equivalent to a vacancy in the corresponding crystal, down to smaller interstitial voids. The amorphous alumina–zirconia materials contain a large amount of free volume; for instance the true density of amorphous $Zr_{1-x}Al_xO_{2-x/2}$ (x = 0.57) was measured to be 3.4 g cm^{-3}, which is 68% of the density of the mixture of equilibrium phases, m-ZrO_2 and α-Al_2O_3 (Gandhi et al., 1999). Such low density may be due to a significant amount of four- and five-fold coordination of Al^{3+} (Balmer et al., 1996) as well as free volume left behind during calcination of the chemically synthesized material. The hardness of the amorphous material is lower than the crystalline phases by four to five times (Gandhi and Jayaram, 2002). This is consistent with the existence of a large amount of free volume in the amorphous phase. The structural details of the amorphous phase need to be further investigated, along with more studies on the deformation characteristics, to understand the mechanism of the plastic deformation.

Considering the deformation behavior of amorphous alumina–yttria, the similarity of the pressure consolidation characteristics with alumina–zirconia would suggest that the deformation behavior would also be similar. While plastic deformation has indeed been observed in amorphous materials in this system, there are significant differences, adding to the richness of research opportunities in this new field of plastic flow in amorphous oxides. Paul and Jayaram studied the deformation characteristics of amorphous alumina–yttria materials in the temperature range of 650 to 850°C (Paul and Jayaram, 2011). The composition with 15 mol% Y_2O_3 was selected as it was shown that the amorphous phase can be retained after pressure consolidation (Thangamani et al., 2005). Co-precipitation was used for synthesis and the calcined powders were consolidated at ~635°C and a pressure of ~730 MPa. Compressive testing at 650°C at a strain rate of 3.5×10^{-4} s^{-1} evoked an elastic-plastic response from this amorphous material. A yield stress of 500 MPa was recorded. Deformation after yielding led to a rapid decrease in the flow stress to about 300 MPa and up to 17% plastic strain could be accumulated in the specimens without failure. The large plastic deformation, however, made the specimens too fragile for further investigation and it was not possible to find out if any shear banding had occurred. Nevertheless, the similarity of the deformation characteristics with amorphous alumina–zirconia is striking. Compression testing at 725°C showed flow softening, although the decrease in the flow stress was not as rapid as described in the examples above. The specimens could be studied post deformation and shear bands were observed using atomic force microscopy (AFM) as well as SEM.

Paul and Jayaram carried out compressive tests at as high a temperature as 850°C on the amorphous alumina–yttria material, which is very close to the crystallization temperature (Paul and Jayaram, 2011). The strain rates used were of similar magnitudes as the lower temperature tests. Surprisingly, the material exhibited strain hardening at this temperature, as the flow stress increased from 500 to 625 MPa. As much as 28% plastic strain was observed without failure. The deformation appears to be uniform as no localized shear was evident in the specimen when examined by

SEM. In order to understand the reasons for strain hardening, Paul and Jayaram measured the true density of the material subjected to increasing amounts of plastic deformation. The true density of specimens deformed at 850°C increased from its initial value of 2.59 g cm^{-3} to 2.90 g cm^{-3}, which is ~12% increase, as the plastic strain increased up to 17.5%. During deformation at 725°C, the increase in density for the same strain was 8%, while at 650°C, the increase in true density was a negligible 2%, though the strain was much higher at 22%. Paul and Jayaram clearly showed that the measured increase in density was not due to a reduction in porosity but an increase in the material density. This structural densification is an intriguing phenomenon. As shown in Table 3.3, the hardness and the elastic modulus of the amorphous material also increased after plastic deformation.

Table 3.3 Changes in the properties of amorphous Al_2O_3–15 mol% Y_2O_3 due to compressive deformation

Testing temp (°C)	Bulk density, g cm^{-3} (ρ)		Hardness (H), GPa		Modulus (E), GPa	
	Undeformed	Deformed	Undeformed	Deformed	Undeformed	Deformed
850	2.59 ± 0.03	2.90 ± 0.05	3.9 ± 0.5	7.0 ± 1.0	57 ± 7.4	116.6 ± 14.4
725	2.58 ± 0.04	2.79 ± 0.04	4.4 ± 0.5	6.4 ± 0.3	52.8 ± 1.8	94.1 ± 3.2
650	2.55 ± 0.05	2.61 ± 0.05	3.4 ± 0.7	—	50.5 ± 11.5	—

The structural densification during deformation is believed to be responsible for the observed strain hardening at 850°C. Mass transport required for such densification would obviously be slower at lower temperatures and hence the structural densification was not seen at 725 and 650°C by Paul and Jayaram. It appears, however, that the low density of the amorphous materials discussed here is chiefly responsible for the unusual deformation characteristics. Data on amorphous phase density are available for the alumina–yttria system. The polyamorphism exhibited by rapidly quenched liquids in this system consists of amorphous phases with densities generally higher than those measured on chemically synthesized materials discussed so far. To give an example, McMillan and Wilding reported densities of the high-density amorphous (HDA) phase to be 3.72 g cm^{-3} and that of the low-density amorphous (LDA) phase to be 3.58 g cm^{-3} (McMillan and Wilding, 2008). This is a much smaller difference than the structural densification observed during deformation of amorphous alumina–yttria reported by Paul and Jayaram.

There are a few interesting issues concerning the existence of such high amounts of void space, or "free volume" in the pressure consolidated materials studied by Gandhi and co-workers, and Paul and Jayaram. It is possible that parts of the chemical precursors used for synthesis leaving the material during calcination would leave behind void spaces. These, however, cannot be considered as pores as their size would be commensurate with the size of the chemical species leaving, e.g. OH^- ions. Thus, a large number of voids smaller than ~1 nm may exist as a consequence of calcination. These voids add to the free volume of the material, rather than becoming fine pores. This large amount of free volume has shown remarkable kinetic stability, considering the fact the high pressures at moderately high temperatures are applied during pressure consolidation and yet the true density remains the same as that obtained after calcination. It is likely that at the higher temperatures used for deformation studies (725 and 850°C) by Paul and Jayaram the overall ionic mobility was enough to cause structural relaxation by the removal of such free volume. This concept of relaxation of free volume in glasses has been used to explain another interesting set of observations in the so-called nanoglasses (Gleiter, 2008). In metallic glasses prepared by consolidation of glassy nanoparticles, there are two different characteristic free volumes: a higher value associated with the interfaces between particles (akin to grain boundaries in crystalline metals) and a lower, equilibrium value in the interior of the particles. It has been shown that, with annealing, the excess interfacial free volume gradually relaxes into the bulk, resulting in a marked density and property evolution with time. Nevertheless, as with the rapidly cooled alumina—yttria glasses discussed earlier, it must be emphasized that the density changes seen in the evolution of the nanoglass is significantly smaller than those seen in materials made by precursor pyrolysis.

The other consideration for explaining the low true densities of the amorphous phases in the alumina-based compositions is the existing coordination structure. In alumina–yttria compositions exhibiting polyamorphism with ~4% difference of density between the polyamorphs, it has been shown that the difference in the polyamorph is not in terms of the first nearest neighbors, i.e. the coordination polyhedra, but in the linkages of these polyhedra. The change in second nearest neighbor shell accounts for the transition between the two glasses. It is intriguing to consider the structural changes responsible for the enhanced true density in amorphous alumina–yttria during deformation. The Al^{3+} ions are known to exist with four- and six-fold coordination in most of the crystalline phases. It does, however, exist with five-fold coordination in the metastable hexagonal $YAlO_3$ phase. It is possible that the amorphous phase in the alumina–yttria system has a distribution of four-, five- and six-fold coordination. Considering the alumina–zirconia system, a large fraction of five-fold coordination has been shown to exist with such multiple coordination. As regards the larger cations, Zr^{4+} and Y^{3+} (Shannon, 1976), the preferred coordination is 7 and 8, respectively. It is possible, however, that these cations occupy a six-fold site in the amorphous phases. It is therefore clear that a multitude of coordination polyhedral may exist, and their linkages are even more numerous in type. One can envisage a change in these linkages during deformation at temperatures of sufficient ionic mobility bringing about an increase in the density of the structure. Obviously, a change in the basic polyhedral structure cannot be ruled out, either. Currently the experimental data on these structural aspects of the amorphous phases are limited. Hence, theoretical estimates of amorphous phase density were made, which are discussed below.

3.5 Modelling of the Structure of Amorphous Al_2O_3–Y_2O_3

The structure of the amorphous phase of alumina–yttria, as inferred from the density and modulus measurements, is unusual. Not only are density differences between crystalline and amorphous counterparts in most materials generally much smaller, it is also unexpected to see such small elastic moduli arising purely from the lack of long range order. Because of the difficulty of making large samples and the ever present possibility of hydroxyl groups influencing the structure, it was decided to examine whether atomistic simulations could provide a plausible justification for the existence of such a metastable glass.

We have carried out simulations for minimization of energy to determine structure using GULP software (Gale, 1997). Since alumina–yttria is primarily ionic with some covalency, we used a shell model with local polarizability at oxygen sites (Gale, 1996). A very useful capability of GULP is the calculation of phonons, which essentially relate to the Hessian of total energy with respect to ionic position. Thus, it is readily possible to diagnose local stability of a structure by making sure there are no modes with imaginary frequency (all eigenvalues of the Hessian matrix are positive definite).

We used periodic boundary conditions in our simulations choosing a large enough box and systematically introducing disorder through randomness in site occupancy of the cations. The composition chosen was $(Y_2O_3)_{0.15}(Al_2O_3)_{0.85}$. The conventional unit cell of alumina with 6 formula units was used to generate a supercell of $Al_{216}O_{324}$ by repeating it thrice in the *x* and *y* and twice in the *z* directions (3-3-2 configuration). Substitution of aluminum ions by yttrium ions was made through a random position generator algorithm in the supercell to achieve the desired concentration, i.e. Y to be 15% of all cations. This algorithm was used to generate a total of 10 statistically reproducible ensembles of configurations $Y_{32}Al_{184}O_{324}$. These configurations, used as different starting configurations, were then relaxed through GULP to the equilibrium volume at zero pressure and temperature. After relaxation, these 10 configurations were then studied for the extent of segregation of yttrium ions by examining the number of Y–Y pairs within a defined radius around each Y-ion in order to eliminate those configurations in which locally distinct crystalline phases, such α-alumina could arise.

The 10 structures relaxed at constant pressure had real phonon frequencies, implying that all atoms within the supercell were at equilibrium with respect to the forces between neighboring atoms. These relaxed structures had densities in the range of 4.18–4.2 g/cc. Incidentally, this density is close to what has been reported for sputtered, amorphous alumina-yttria films of similar composition (Fukuda et al., 2002; Reyes and Drew, 2003). The configuration with the lowest lattice energy was rescaled to higher volumes so that the density is reduced to 2.8–3.0 g cm^{-3} as experimentally measured. Once again, this higher volume glassy phase configuration was relaxed under constant volume to look for local minima as shown in Fig. 3.6. It represents a scatter of local minima. The entire landscape of Fig. 3.6 was explored for local minima in constant volume GULP-relaxed structures. It revealed an interesting fact that the scatter of these minima increased at higher volumes.

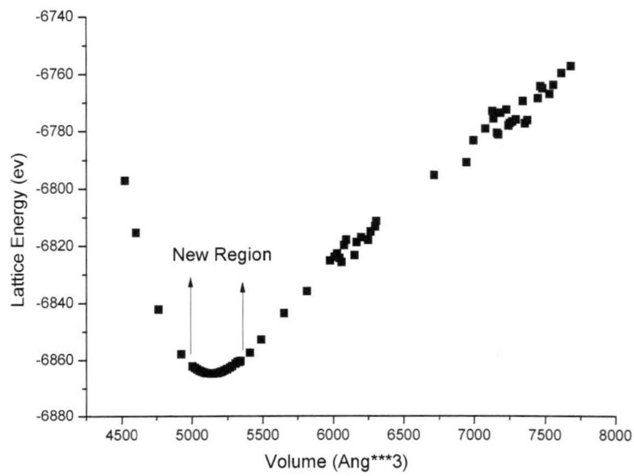

Figure 3.6 Local minima around global minimum, 5900–6400 and the glassy phase (7100–7800).

Three regions were explored, viz. (a) the region around the global minimum of 5000–5400 (Fig. 3.7), (b) 5900–6400 (Fig. 3.8), and (c) 6400–7000, (Fig. 3.9).

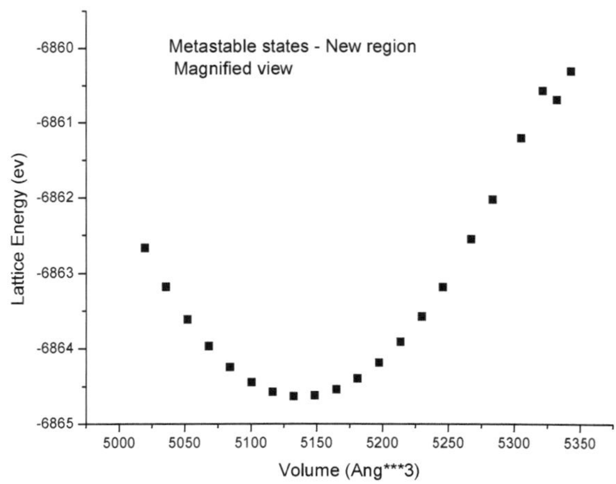

Figure 3.7 Metastable states around global minimum.

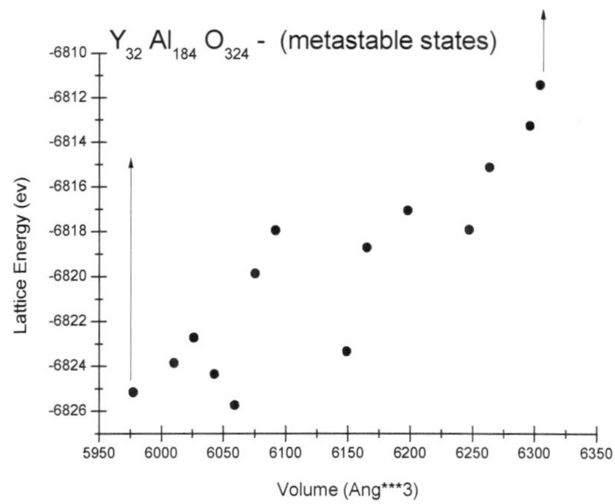

Figure 3.8 Metastable states b/w 5900 and 6400.

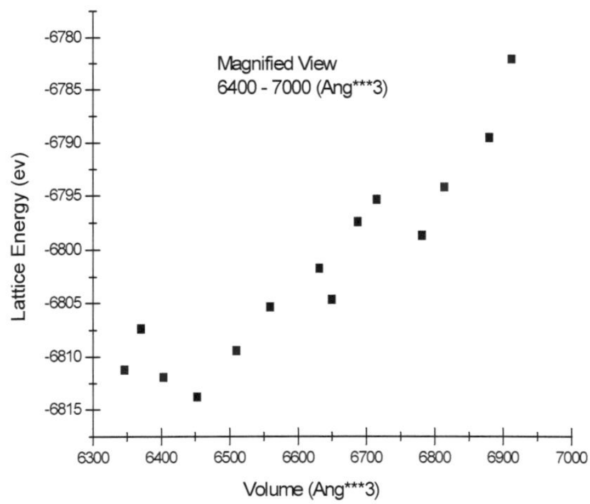

Figure 3.9 Metastable states between 6,400 and 7,000.

The increasing scatter in regions clustered around 7200 and 6000 Å^3 suggests the presence of local minima of energy, which could correspond to accessible metastable states separated by low activation barriers, one of which, near 6800 Å^3, corresponds to the experimentally measured density of the amorphous phase. To shed further light on such a structure, the distribution or density of bonds was evaluated and compared for the Al–O and Y–O bonds in the low-density amorphous phase, the most stable amorphous phase with density of 4.17 g cm^{-3} and for the stable crystalline phases for alumina and yttria. As seen in Figs. 3.10–3.15, while scatter in the bond lengths increases as the density drops, the mean Al–O distance reduces, consistent with the greater fraction of four- or five-fold coordination. The maximum in the Y–O bond length also reduces somewhat but additional smaller maxima appear at larger bond lengths, suggestive of clusters of higher than the normal six-fold coordination.

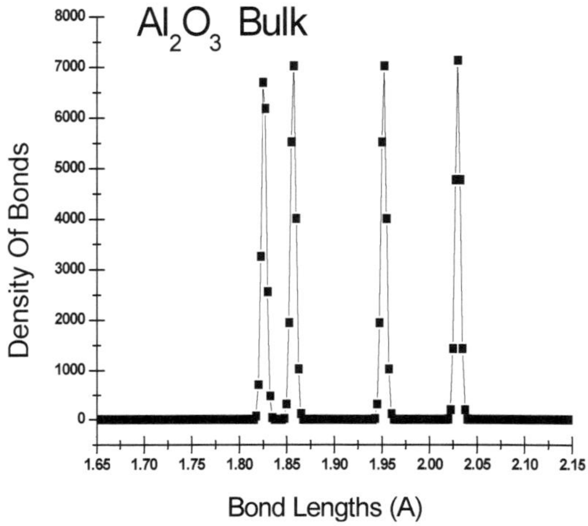

Figure 3.10 Alumina bulk structure has each Al atom bonded to four O atoms and has four kinds of bond lengths i.e., one kind of Al atoms have Al–O bond lengths of 1.826 Å and 2.030 Å (2 each) and the other kind have Al–O bond lengths of 1.856 Å and 1.952 Å (2 each). Both kinds of Al atoms are equal in number in the bulk structure.

A few observations from the results of the density of bonds calculations are as follows:

(1) In the high-density phases, Al–O or Y–O bonds span about the same range of bond-lengths as in their bulk forms.
(2) It is remarkable that in the low-density phase, while Al–O bonds are significantly shorter, Y–O bonds are actually much longer.
(3) The above finding can be traced to the coordination numbers of the cations. It is seen that 95%, 5%, 0% and 0% Al^{3+} cations have coordination of 4, 5, 6 and 7, respectively. In contrast, 20%, 57%, 20% and 3% of Y^{3+} cations have coordination of 4, 5, 6, and 7, respectively. The higher coordination of Y naturally gives longer Y–O bonds.

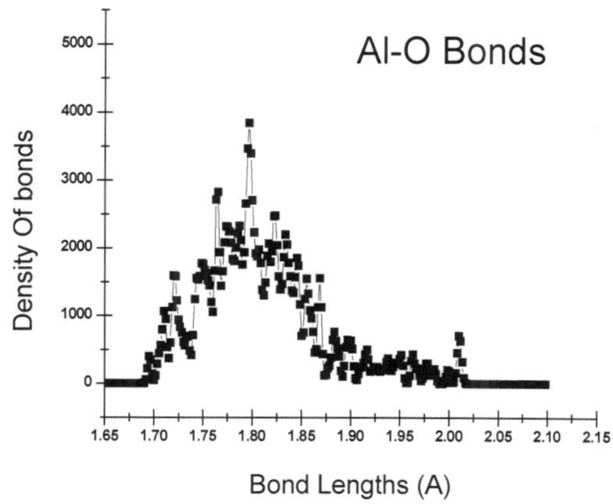

Figure 3.11 Al–O bonds in the glassy phase (low density—higher volumes).

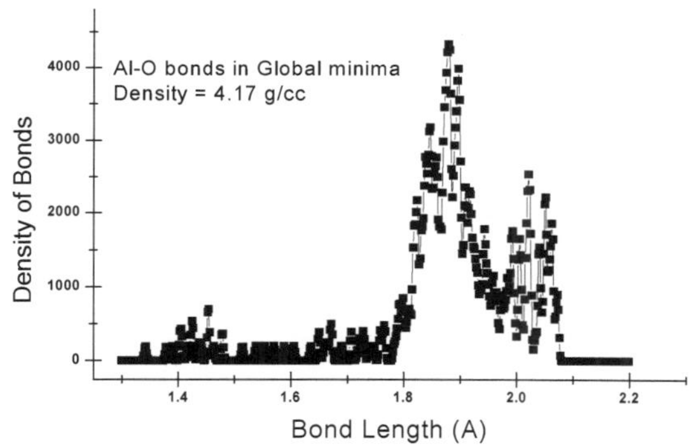

Figure 3.12 Al–O bonds in the global minimum amorphous phase.

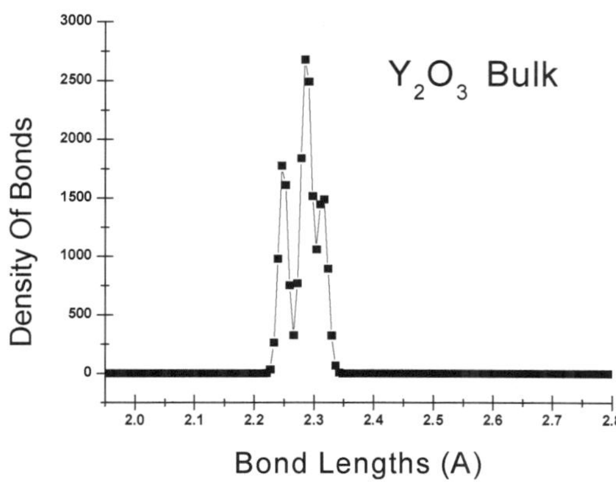

Figure 3.13 Bulk yttria (crystal structure).

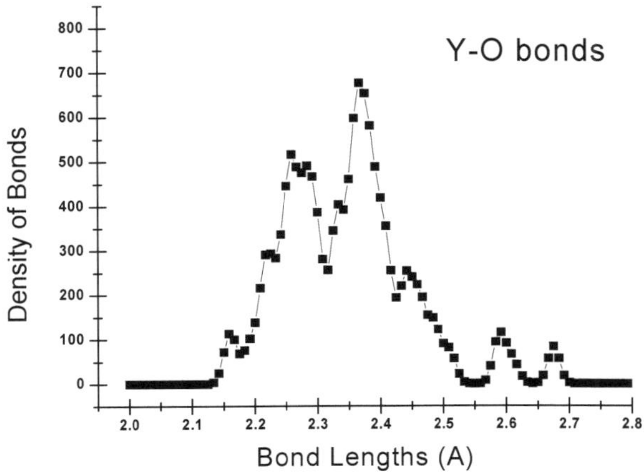

Figure 3.14 Y–O bonds in the glassy phase (low density—higher volumes).

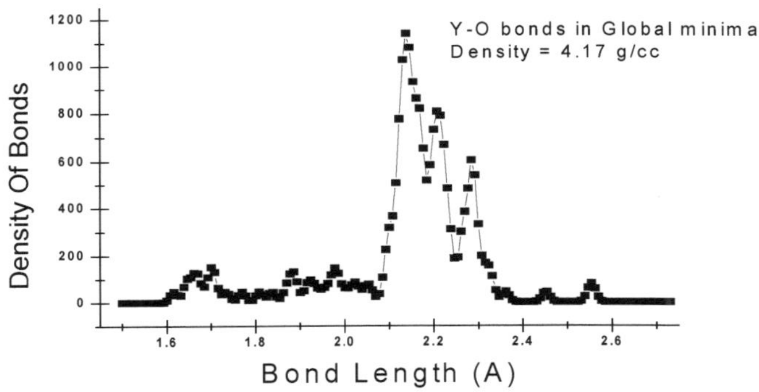

Figure 3.15 Y–O bonds in the global minimum amorphous phase.

3.6 Concluding Remarks

In this Chapter, we have shown that amorphous/glassy materials continue to provide interesting opportunities for fundamental research on their structure and properties. At the same time, they offer new avenues for developing advanced materials. Alumina-based oxides, particularly those devoid of silica and consisting of only a binary addition of either zirconia or yttria, can be synthesized through simple chemical routes such as spray pyrolysis and co-precipitation. By exploring the compositions in these systems, materials with sufficient kinetic stability of the amorphous phase can be obtained. Since the consolidation of these powders requires moderate temperatures in the range of 400 to 700°C, and pressure ~750 MPa, it is possible to prepare bulk amorphous materials. This ability to produce bulk forms facilitated studies on the mechanical behavior of these interesting materials. Both, alumina–zirconia and alumina–yttria exhibit elastic-plastic deformation at temperatures well below their crystallization temperatures. Flow localization to produce shear bands has been observed in these materials. This interesting parallel to metallic and polymeric glasses has shown the generality of the shear banding phenomenon. Since these materials are distinct in the nature of bonding that exists in them, the theories of flow localization need to consider those mechanisms that can operate irrespective of the nature of the bond.

We have made a beginning in understanding the structure of the amorphous alumina-based materials to understand how large amounts of free volume can be trapped during their synthesis. The mechanism of structural densification during plastic deformation can be understood with further experimental and theoretical research directed at describing the structure more completely and then identifying the changes occurring in it.

Future research will undoubtedly also include developing the processing route of first producing a bulk amorphous shape and crystallizing it in a controlled fashion to obtain nanostructured ceramics. One can envisage the inclusion of other oxides, such as lanthanides and titania as ternary additions to the pseudo-binary systems described in this Chapter. It is also likely that the functional properties, e.g. refractive index, phosphorescence, and laser action will be of as much interest as the mechanical properties of these materials.

References

Akisanya, A. R., A. C. F. Cocks and N. A. Fleck (1997). The yield behaviour of metal powders. *International Journal of Mechanical Sciences,* **39**, 1315–1324.

Balmer, M. L., H. Eckert, N. Das and F. F. Lange (1996). ^{27}Al nuclear magnetic resonance of glassy and crystalline $Zr_{(1-x)}Al_xO_{(2-x/2)}$ materials prepared from solution precursors. *Journal of the American Ceramic Society,* **79**, 321–326.

Balmer, M. L., F. F. Lange, V. Jayaram and C. G. Levi (1995). Development of nano-composite microstructures in ZrO_2–Al_2O_3 via the solution precursor method. *Journal of the American Ceramic Society,* **78**, 1489–1494.

Balmer, M. L., F. F. Lange and C. G. Levi (1994). Metastable phase selection and partitioning for $Zr_{(1-x)}Al_xO_{(2-x/2)}$ materials synthesized with liquid precursors. *Journal of the American Ceramic Society,* **77**, 2069–2075.

Choudhury, S., A. S. Gandhi and V. Jayaram (2003). Bulk, dense, Nanocrystalline yttrium aluminum garnet by consolidation of amorphous powders at low temperatures and high pressures. *Journal of the American Ceramic Society,* **86**, 247–251.

Courtney, T. H. (1990). *Mechanical Behavior of Materials.* Singapore: McGraw-Hill Publishing Company.

Djuricic, B., S. Pickering, P. Glaude, D. McGarry and P. Tambuyser (1997). Thermal stability of transition phases in zirconia-doped alumina. *Journal of Materials Science,* **32**, 589–601.

Frenkel, J. (1945). Viscous flow of crystalline bodies under the action of surface tension. *Journal of Physics (Moscow),* **5**, 385–391.

Fukuda, K., T. Hanada, S. Tanabe and T. Yao (2002). Physical properties and structure of rf-sputtered amorphous films in the system Al_2O_3–Y_2O_3. *Journal of the American Ceramic Society,* **85**, 915–920.

Gale, J. D. (1996). Empirical potential derivation for ionic materials. *Philosophical Magazine B-Physics of Condensed Matter Statistical Mechanics Electronic Optical and Magnetic Properties,* **73**, 3–19.

Gale, J.D. (1997). GULP: A computer program for the symmetry-adapted simulation of solids. *Journal of the Chemical Society-Faraday Transactions,* **93**, 629–637.

Gandhi, A. S., and V. Jayaram (2002). Pressure consolidation of amorphous ZrO_2–Al_2O_3 by plastic deformation of powder particles. *Acta Materialia,* **50**, 2137–2149.

Gandhi A. S. (2003). Plastically deforming amorphous ZrO_2–Al_2O_3. *Acta Materialia,* **51**, 1641–1649.

Gandhi, A. S., V. Jayaram and A. H. Chokshi (1997). Phase evolution and densification of spray pyrolysed ZrO_2–Al_2O_3 powders. In *Superplasticity in Advanced Materials—Icsam-97,* 227–232.

Gandhi, A. S., V. Jayaram and A. H. Chokshi (1999). Dense amorphous zirconia-alumina by low-temperature consolidation of spray-pyrolyzed powders. *Journal of the American Ceramic Society,* **82**, 2613–2618.

Gandhi, A. S., V. Jayaram and A. H. Chokshi (2001). Low temperature densification behaviour of metastable phases in ZrO_2–Al_2O_3 powders produced by spray pyrolysis. *Materials Science and Engineering A—Structural Materials Properties Microstructure and Processing,* **304**, 785–789.

Gandhi, A. S., and C. Levi (2005). Phase selection in precursor-derived yttrium aluminum garnet and related Al_2O_3-Y_2O_3 compositions. *Journal of Materials Research,* **20**, 1017–1025.

Gandhi, A. S. (2001). The Processing of Bulk Metastable Amorphous and Nanocrystalline ZrO_2–Al_2O_3 Ceramics by Pressure Consolidation of Amorphous Powders. Ph.D. Thesis, *Department of Metallurgy*. Indian Institute of Science, Bangalore, India.

Gleiter, H. (2008). Our thoughts are ours, their ends none of our own: Are there ways to synthesize materials beyond the limitations of today? *Acta Materialia,* **56**, 5875–5893.

Greaves, G. N., M. C. Wilding, S. Fearn, F. Kargl, L. Hennet, W. Bras, O. Majerus and C. M. Martin (2009). Liquid-liquid transitions, crystallization and long range fluctuations in supercooled yttrium oxide-aluminium oxide melts. *Journal of Non-Crystalline Solids,* **355**, 715–721.

Greaves, G. N., M. C. Wilding, S. Fearn, D. Langstaff, F. Kargl, S. Cox, Q. V. Van, O. Majerus, C. J. Benmore, R. Weber, C. M. Martin and L. Hennet (2008). Detection of first-order liquid/liquid phase transitions in yttrium oxide-aluminum oxide melts. *Science,* **322**, 566–570.

Inamura, S., H. Miyamoto, Y. Imaida, M. Takagawa, K. Hirota and O. Yamaguchi (1994). Formation and hot isostatic pressing of ZrO_2 solid-solution in the system ZrO_2–Al_2O_3. *Journal of Materials Science,* **29**, 4913–4917.

Jacobson, L. A., and J. McKittrick (1994). Rapid solidification processing. *Materials Science & Engineering R-Reports,* **11**, 355–408.

Jayaram, V., R. S. Mishra, B. Majumdar, C. Lesher and A. Mukherjee (1998). Dense nanometric ZrO_2–Al_2O_3 from spray-pyrolysed powders. *Colloids and Surfaces a-Physicochemical and Engineering Aspects,* **133**, 25–31.

Kim, H. J., and Y. J. Kim (1999). Amorphous phase formation of the pseudo-binary Al_2O_3–ZrO_2 alloy during plasma spray processing. *Journal of Materials Science,* **34**, 29–33.

Levin, I., and D. Brandon (1998). Metastable alumina polymorphs: Crystal structures and transition sequences. *Journal of the American Ceramic Society,* **81**, 1995–2012.

Li, M. J., Y. C. Wu, F. S. Yen and C. Y. Huang (2011). Influence of ionic mobility on the phase transformation route in $Y_3Al_5O_{12}$ (YAG) stoichiometry. *Journal of the European Ceramic Society,* **31**, 2099–2106.

Mackenzie, J. K., and R. Shuttleworth (1949). A phenomenological theory of sintering. *Proceedings of the Physical Society (London)*, **62**, 833–852.

McMillan, P. F., and M. C. Wilding (2008). Direct density determination of low- and high-density glassy polyamorphs following a liquid-liquid phase transition in Y_2O_3–Al_2O_3 supercooled liquids. *Journal of Non-Crystalline Solids,* **354**, 1015–1025.

Meyers, M. A., and K. K. Chawla. (1999). *Mechanical Behavior of Materials.* Upper Saddle River, NJ, USA: Prentice-Hall Inc.

Mishra, R. S., V. Jayaram, B. Majumdar, C. E. Lesher and A. K. Mukherjee (1999). Preparation of a ZrO_2–Al_2O_3 nanocomposite by high-pressure sintering of spray-pyrolyzed powders. *Journal of Materials Research,* **14**, 834–840.

Molodetsky, I., A. Navrotsky, M. J. Paskowitz, V. J. Leppert and S. H. Risbud (2000). Energetics of X-ray-amorphous zirconia and the role of surface energy in its formation. *Journal of Non-Crystalline Solids,* **262**, 106–113.

Moreau, S., M. Gervais and A. Douy (1997). Formation of metastable solid solution in the ZrO_2-rich part of the system ZrO_2–Al_2O_3. *Solid State Ionics,* **101**, 625–631.

Murray, P., E. P. Rogers and E. A. Williams (1954). *Transactions of the British Ceramic Society,* **53**.

Narwankar, P. K., J. S. Speck and F. F. Lange (1995). Phase partitioning and epitaxy of $Zr(Al)O_2$ thin-films on cubic zirconia substrates. *Journal of Materials Research,* **10**, 1756–1763.

Paul, A., and V. Jayaram (2011). Deformation and structural densification in Al_2O_3–Y_2O_3 glass. *Acta Materialia,* **59**, 82–92.

Porter, D. A., and K. E. Easterling. (1992). *Phase Transformations in Metals and Alloys.* Cheltenham (U.K.): Nelson Thornes.

Rahaman, M. N. (1995). *Ceramic Processing and Sintering.* New York, NY, USA: Marcel Dekker.

Reyes, E. A. A., and R. A. L. Drew (2003). Kinetic study of amorohous yttria-alumina fibers using differential thermal analysis. *Journal of the American Ceramic Society,* **86**, 43–48.

Rosenflanz, A., M. Frey, B. Endres, T. Anderson, E. Richards and C. Schardt (2004). Bulk glasses and ultrahard nanoceramics based on alumina and rare-earth oxides. *Nature,* **430**, 761–764.

Schuh, C. A., T. C. Hufnagel and U. Ramamurty (2007). Overview No. 144: Mechanical behavior of amorphous alloys. *Acta Materialia,* **55**, 4067–4109.

Sellappan, P., V. Jayaram, A. H. Chokshi and C. Divakar (2007). Synthesis of bulk, dense, nanocrystalline yttrium aluminum garnet from amorphous powders. *Journal of the American Ceramic Society,* **90**, 3638–3641.

Shannon, R. D. (1976). Revised effective ionic-radii and systematic studies of interatomic distances in halides and chalcogenides. *Acta Crystallographica Section A,* **32**, 751–767.

Stefanic, G., S. Music and R. Trojko (2005). The influence of thermal treatment on the phase development in HfO_2–Al_2O_3 and ZrO_2–Al_2O_3 systems. *Journal of Alloys and Compounds,* **388**, 126–137.

Thangamani, N., A. S. Gandhi, V. Jayaram and A. Chokshi (2005). Low-temperature high-pressure consolidation of amorphous Al_2O_3–15 mol% Y_2O_3. *Journal of the American Ceramic Society,* **88**, 2696–2701.

Ullal, C. K., K. R. Balasubramaniam, A. S. Gandhi and V. Jayaram (2001). Non-equilibrium phase synthesis in Al_2O_3–Y_2O_3 by spray pyrolysis of nitrate precursors. *Acta Materialia,* **49**, 2691–2699.

Vasiliev, A. L., N. P. Padture and X. Q. Ma (2006). Coatings of metastable ceramics deposited by solution-precursor plasma spray: I. Binary ZrO_2–Al_2O_3 system. *Acta Materialia,* **54**, 4913–4920.

Yamaguchi, O., M. Shirai and M. Yoshinaka (1988). Formation and transformation of cubic ZrO_2 solid-solutions in the system ZrO_2–Al_2O_3. *Journal of the American Ceramic Society,* **71**, C510–C512.

Chapter 4

Porous and Hollow Oxide Nanostructures: Synthesis, Stability and Applications

Erumpukuthickal Ashokkumar Anumol and Narayanan Ravishankar

Materials Research Centre, Indian Institute of Science, Bangalore, Karnataka 560012, India

nravi@mrc.iisc.ernet.in

Porous and hollow nanostructures of metal oxides are of great interest owing to their potential applications as catalysts, sensors and in drug delivery. In this chapter, we discuss the various synthesis strategies and general principles developed over the years to synthesize porous and hollow nanostructures of various inorganic oxides. A few important applications of these materials are also presented.

4.1 Introduction

Metal oxides are important due to the extraordinary range of properties that they exhibit in addition to being chemically

Oxide Nanostructures: Growth, Microstructures, and Properties
Edited by Avanish Kumar Srivastava

ISBN 978-981-4411-35-6 (Hardcover), 978-981-4411-36-3 (eBook)
www.panstanford.com

and thermally stable. Oxides are ideal candidates for various applications such as catalysts and catalytic supports, adsorption media, electronic components, battery materials and sensors, to name a few [1–5]. The properties of these materials are strongly dependent not only on the structure and composition/chemistry but also on their size and shape [6]. Therefore, considerable efforts have been made towards controlling the structure and morphology of these materials depending on the property in consideration. Among these porous or hollow materials are of importance due to high surface area, low density and the presence of accessible pores of various sizes, which imparts functionality/multi-functionality for many applications [2,7–9].

4.2 Porous Structure: Definition

Porous materials are those that possess ordered or random pores, which results in a high surface area. Porous metal oxides are extensively used in heterogeneous catalysis and as adsorption media and molecular sieves. These materials are generally chemically, mechanically, and thermally stable and may exhibit not only high specific surface areas but also uniform pore size distribution and regular pore arrangements. According to IUPAC nomenclature, materials with pore diameters in the range of 2–50 nm are known as mesoporous, <2 nm as microporous and >50 nm as macroporous.

4.3 Synthesis Methods for Porous Structures

Wet chemistry offers a wide variety of techniques for materials synthesis. Various wet chemical synthesis methods are developed for obtaining porous oxide materials. The synthesis protocols for porous materials are mainly focused on obtaining ordered pores, silica being the most extensively studied material. In addition to silica, porous TiO_2, ZrO_2, Al_2O_3 and SnO_2 are also research interest due to potential catalytic and sensing applications. The existing methods of synthesis can be broadly classified as template-based method and template-less method. Among this, the synthesis using a template has the advantage that it gives uniform pore sizes, which is characteristic of the template. However, it poses the problem of template removal, which results in the destruction or deformation of the porous structure. It can also result in undesirable phase transition and coarsening during the thermal treatment employed for template removal. Most of the synthesis methods make use of sol-gel technology to obtain these inorganic materials in the presence or absence of templates. Some of them use a hybrid method combining sol-gel method with other synthesis techniques such as combustion, hydrothermal or solvothermal methods.

4.3.1 Template-Assisted Methods

Template-assisted synthesis is the most common synthesis technique in the case of mesoporous oxide materials. Various methods are developed making use of surfactant liquid crystal templates or block copolymers. Microemulsion technique is also employed for the synthesis of porous oxides. The use of these templates results in various mesoporous oxide structures with controlled pore dimensions and wall thickness. Also, many of these synthesis methods results in the formation of ordered pores of various geometries, which makes them ideal candidates in many important applications such as drug/gene delivery and also as templates in the synthesis of various nanostructures.

4.3.1.1 Surfactant template

The extensively used method to synthesize porous materials includes the use of surfactant molecules as structure directing agents. Surfactants are molecules with hydrophobic and hydrophilic parts that self-organize in solution into micelles and eventually assemble into liquid crystal phases, depending on interfacial curvature and concentration. Formation of the metal oxide takes place at the surfactant/water interface and finally leads to the formation of an inorganic/organic composite material. Removal of the organic content by thermal treatment results in mesoporous structures.

In 1992, Beck et al. introduced a method for the synthesis of high surface area mesoporous molecular sieve materials M41S with uniform and controllable pore sizes, where surfactant liquid crystal phases act as the template [10,11]. M41S materials have surface areas above 700 m^2/g and hydrocarbon sorption capacities of 0.7 cc/g and greater. MCM-41, one member of this family, exhibits a hexagonal arrangement of uniform mesopores whose dimensions can be varied in the range of 15–100 Å. The pore diameter can be tuned by adjusting the alkyl chain length of the cationic surfactant used. They also investigated the role of auxiliary hydrocarbons in the reaction mixture in varying the pore diameter. The method involved sol-gel synthesis followed by hydrothermal reaction of the gel precursor. The organic surfactant is removed by either acid wash or calcination of the dried product to obtain an inorganic silica framework with a hexagonal, disordered, or cubic pore structure, depending upon the specific synthetic conditions. They proposed a liquid crystal templating mechanism for the formation of these ordered porous materials, where the silicate material forms inorganic walls between the ordered surfactant micelle. This initiated great interest among researchers considering the potential application of these mesoporous structures with the option of variable pore sizes of uniform distribution and high surface area in various fields including catalysis, adsorption. This report was followed by many attempts to synthesize such mesoporous materials of SiO_2 and other oxide materials by modifying the method of employing liquid crystal template by different surfactants and precursors [1,12,13].

Nanometre-sized mesoporous silica materials were synthesized by Fowler et al., by modifying the template-directed alkaline synthesis of MCM-41 silica [14]. The synthesis involved hydrolysis and condensation of tetraethoxysilane (TEOS) in aqueous alkaline solutions containing cetyltrimethylammonium bromide (CTAB) surfactant at room temperature. The experimental conditions were similar to those used previously to prepare hexagonally ordered channel structures for MCM-41 but with modifications. The alkaline precursor solution containing the surfactant was stirred for short duration and quenched by adding excess water followed by neutralization by HCl. This reduces the particle growth and decreased the rate of silica condensation resulting in the formation of nano-size mesoporous silica. Sadasivan et al. reported a mechanism for the formation of mesoporous silica where the nucleation of MCM-41 nanoparticles occurs heterogeneously by a disorder–order phase transition within silica–surfactant micellar aggregates less than 50 nm in size [15].

In another report, mesoporous silica nanoparticles were prepared by hydrolysis of TEOS in presence of styrene [16]. The synthesis procedure involved the use of an amino acid as the catalyst, octane as the hydrophobic-supporting reaction component, and CTAB as the surfactant. Styrene polymerization leading to the formation of polystyrene along with the formation of silica by hydrolytic condensation of TEOS took place in the CTAB surfactant micelle. Cetyltrimethylammonium bromide and polystyrene are removed by the calcination of the product resulting in the formation of mesoporous silica with high specific surface area and uniform pore size. Pore sizes in the range of 5–15 nm were obtained by varying the styrene concentration. The outer diameter was mostly controlled by the concentration of the hydrophobic molecules. The mesoporous SiO_2 thus obtained is shown in Fig. 4.1.

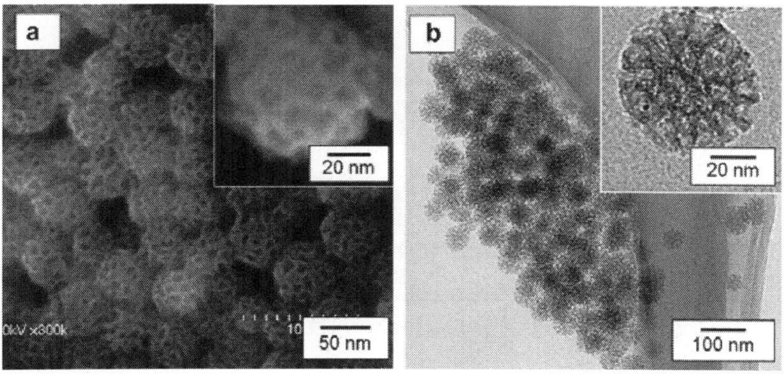

Figure 4.1 SEM (a) and TEM (b) images of mesoporous silica nanoparticles synthesized in presence of CTAB surfactant. The inset shows high-magnification image of a single silica particle [16].

Zhao et al. reported a synthesis procedure to obtain a family of highly ordered mesoporous (2–30 nm) silica structures, SBA-15, where commercially available nonionic alkyl poly(ethylene oxide) (PEO) oligomeric surfactants and poly-(alkylene oxide) block copolymers acted as templates [17]. The synthesis was carried out in acidic media. The family includes materials with cubic cage structures, the three-dimensional hexagonal cage structure, honeycomb hexagonal structure, and lamellar and continuous sponge, mesophases. Nonionic alkyl-ethylene oxide oligomeric surfactants support the formation of cubic mesoporous silica phases at room temperature, while poly(alkylene oxide) triblock copolymers favoured the formation of hexagonal mesoporous silica structure. Using aqueous acidic conditions and dilute triblock copolymer concentrations, SBA-15 has been synthesized with a highly ordered two-dimensional hexagonal mesostructure and thick uniform silica walls (31–64 Å) [18]. The thick silica walls lead to greater hydrothermal stability of SBA-15. The pore size and the thickness of the silica wall can be adjusted by varying the heating temperature (35° to 140°C) and time (11–72 h) in the reaction solution. The use of triblock copolymer in obtaining ordered pores of diverse structure and sizes initiated the use of triblock copolymers as versatile structure directing agent for the synthesis of porous inorganic materials [19–23].

Goltner et al. reported the use of novel non-ionic polymer templates with improved water solubility and a broader range of accessible molecular weights, which allows the pore diameters of ceramic nanostructures to be tuned in a wider range [20]. The method involves the sol-gel synthesis of silica in presence of a liquid crystal template of amphiphilic block copolymer. The block copolymer (30, 50 or 70% w/w with respect to water) was mixed with 1 g of distilled water followed by the addition of 2 g tetramethyl orthosilicate (TMOS) and homogenizing. The final polycondensation was achieved by exposing the siliceous mesophase to gaseous hydrochloric acid. The resulting product was calcined at 500°C (12 h under nitrogen, 16 h under oxygen) to obtain mesoporous SiO_2. The pore connectivity and wall thickness are controlled by tuning the silica concentration, whereas the morphology is determined by the structure of the block polymer solution, which acts as the template for silica cast. Dai et al. reported the use of a green template, amphiphilic dendritic polyamidoamine, to obtain mesoporous silica with pore diameters of 3–9 nm with narrow size distribution and high surface area [24]. Hierarchically assembled porous ZnO particles synthesis in aqueous medium using ZnO nanoparticles as building block and triblock copolymer P-123 as template has been reported [25].

Mesoporous silica with hexagonal pores were synthesized at room temperature by sol-gel technique in an alkaline solution using TEOS as silica precursor and cetyltrimethylammonium bromide (CTAB) as surfactant [26]. Amine functionalized mesoporous silica (NH_2-MSN) was synthesized by a modification of the above synthesis by introducing 3-aminopropyltriethoxysilane (APTES). Synthesis procedure involves the base-catalyzed sol-gel silica reactions using CTAB, TEOS, and high molar amounts of APTES in the presence of ethyl acetate. Reaction was for 24 h at room temperature. Cetyltrimethylammonium bromide was removed by either acetic acid extraction or calcinations [27]. 1,3,5-trimethylbenzene (TMB) was used as a pore expander molecule to increase the pore size from 2.7 to 5 nm. To prepare NH_2-MSNs for fluorescence microscopy applications, TRITC dye was covalently bound to the organically modified silica matrix. TEM images revealing the morphology of NH_2-MSNs with various pore sizes are given in Fig. 4.2. Various modifications of the surfactant-assisted MSN synthesis have been reported for synthesizing MSN with various pore sizes, surface areas and surface functionalization [28].

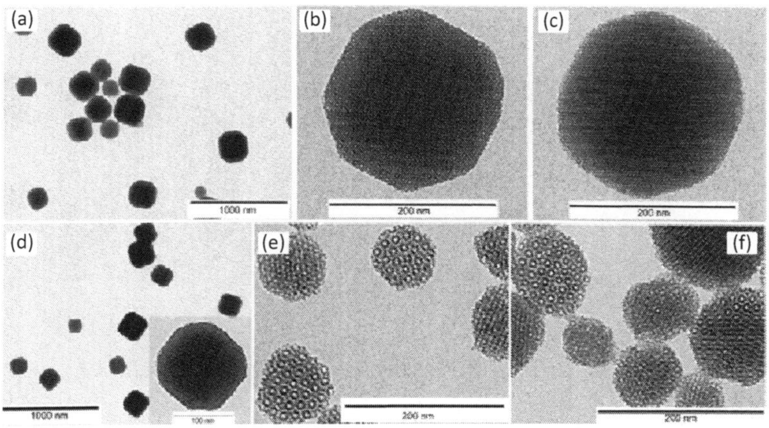

Figure 4.2 TEM images of NH_2-MSNs at low magnification (a), NH_2-MSN at [100] projection (b), NH_2-MSN at [111] projection (c), TRITC-labelled NH_2-MSNs (d), NH_2-MSNs with large pores (e) and TRITC-labelled large-pore NH_2-MSNs (f) [27].

4.3.1.2 Emulsion templating

Emulsions come under the class of colloids, which result from the mixing of two or more immiscible liquids. Emulsions can be used as a soft template. Oil-in-water-type emulsion is used as template for the synthesis of macroporous oxide materials. An emulsion templating procedure was reported by Imhof and Pine for the synthesis of macroporous TiO_2, ZrO_2 and SiO_2. The synthesis made use of a stable non-aqueous emulsions obtained with formamide as the polar liquid and isooctane as the oil [29]. Metal oxide sol has been made by mixing a chemically modified metal alkoxide with formamide, which contains a little water, such that the resulting metal/water ratio is in the range 3–10. This partially hydrolyses the alkoxide so that it becomes soluble in formamide but does not react further and eliminates the need of alcohol. Previously prepared emulsion was dispersed in this sol. The droplet volume fraction was adjusted to the desired porosity by centrifugation. Gelation is induced by adding a small amount of ammonia to increase the pH and takes place over several hours. Before gelation, monodisperse droplets self-assemble into a colloidal crystal when their volume fraction of droplets exceeds, 50%. The gel is aged to complete the reaction, and then washed in alcohol, dried, and heat treated to remove residual organics. SEM images of the porous TiO_2 synthesized are given in Fig. 4.3.

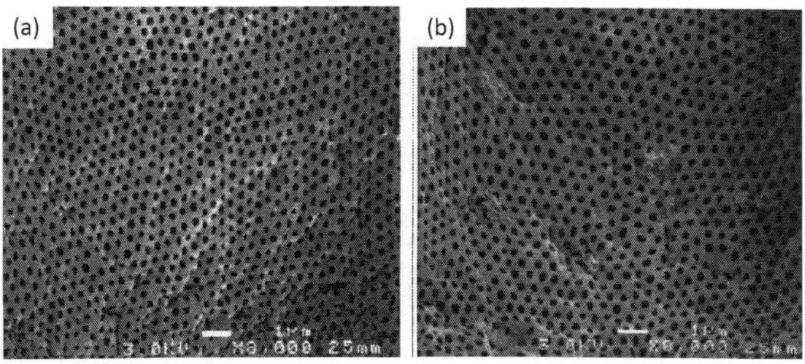

Figure 4.3 (a) and (b) shows SEM images of porous titania calcined at 1000°C, prepared from two slightly different emulsion fractions. Scale bars, 1 μm [29].

Mesoporous CeO_2 nanoparticle aggregates were synthesized by combining sol-gel and inverse mini-emulsion technique with cerium nitrate hexahydrate as precursor [30]. With the addition of triblock copolymers (PEO–PPO–PEO) as template in the miniemulsion droplets, aggregates of CeO_2 were obtained with crystallite sizes as small as 5 nm, exhibiting mesopores of 28.5 nm in diameter and a specific surface area of 158 m^2/g after calcination at 400°C. Combining PE/B-*b*-PEO as stabilizing surfactant with a template such as CTAB, Pluronic P123 or Pluronic PE 6800, the specific surface area is increased up to 255 m^2/g. Formation of porous TiO_2 is also reported by a mini-emulsion technique [31].

Mostly, soft templates are being employed in the synthesis of porous inorganic materials. Few hard template methods are reported, for example, anodic alumina membranes are used as templates in the synthesis of mesoporous structures [32]. Mesoporous carbon is also used in nanocasting of mesoporous materials such as ZnO [33].

4.3.2 Template-Less Methods

Template-less methods are investigated in the synthesis of porous materials considering the fact that template removal is a critical problem in the template based methods, which makes it complicated. A few notable template-less synthesis methods reported are discussed below.

4.3.2.1 Hydrothermal/solvothermal synthesis

Hydrothermal synthesis refers to the various techniques of crystallizing substances from high-temperature aqueous solutions where the solvent is heated above the normal boiling point in a sealed reaction vessel. The method is called solvothermal synthesis when any solvent other than water is being used. Hydrothermal/solvothermal synthesis has been employed in obtaining porous metal oxides such as TiO_2 and SnO_2. Nanoporous TiO_2 spheres were synthesized by a two step method involving the controlled hydrolysis of TTIP followed by solvothermal treatment in ethanol at 240°C [34]. By varying the concentration of TTIP and water, well-defined spheres (sphere 1) and deformed spherical structures (sphere 2) were synthesized. The spheres consisted of TiO_2 nanoparticles of 10–15 nm sizes with high surface area having internal pore sizes of ~10 nm. The TiO_2 thus synthesized were used for the fabrication of thick film TiO_2 electrode for dye-sensitized solar cells. The voids between the TiO_2 spheres in the electrode served as the channels for electrolyte diffusion. Figure 4.4 shows the morphologies of the amorphous TiO_2 precursor obtained by hydrolysis of TTIP, the crystalline mesoporous TiO_2 obtained by solvothermal treatment and the electrode fabricated. Another report presents a method involving solvothermal and sol-gel combination to obtain mesoporous TiO_2 beads with controllable pore sizes in the range of 14–22.6 nm [35]. Spindle-shaped porous TiO_2 mesocrystals with a single crystal-like structure was synthesized by an additive-free solvothermal technique [36]. Tetrabutyl titanate is used as the precursor and acetic acid as solvent. Fluorinated mesoporous TiO_2 was synthesized by combining sol-gel technique with solvothermal synthesis [37]. The surface area obtained was 132 m^2/g and pore diameter was 8.5 nm. A microwave-assisted solvothermal synthesis was reported for the synthesis of high surface area mesoporous TiO_2 using titanium butoxide precursor in

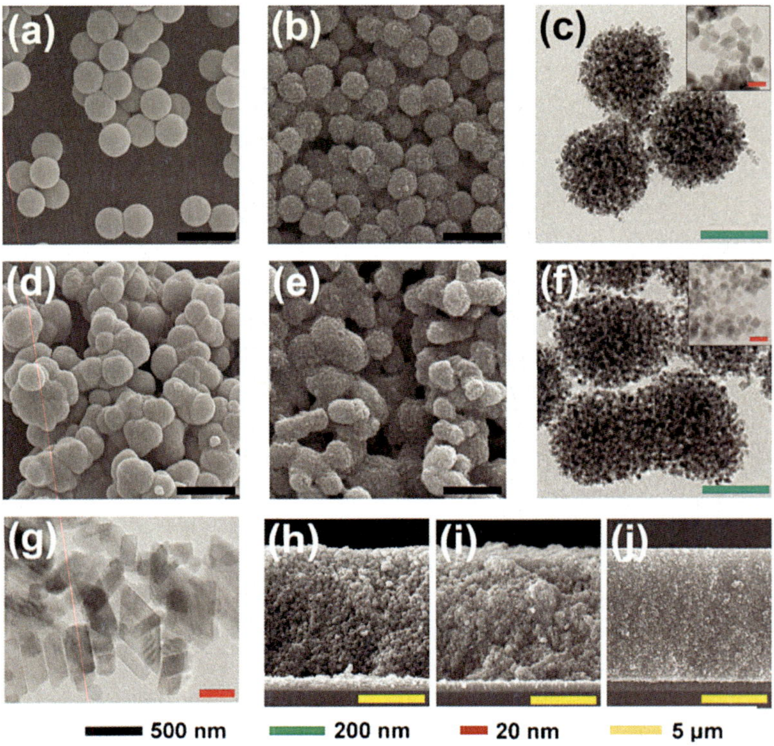

Figure 4.4 SEM image of TiO_2 sample (sphere 1) before (a) and after (b) hydrothermal treatment at 240°C. (c) is the TEM image of the sample after hydrothermal treatment showing the presence of fine particles in the aggregate. (d) and (e) are the SEM images of sphere 2 before and after hydrothermal treatment. TEM image of the sample after hydrothermal treatment is shown in (f). (g) is the TEM image of TiO_2 NPs. Cross-sectional SEM images for the nanoporous TiO_2 layers prepared from sphere 1 (h), sphere 2 (i), and TiO_2 NPs (j) [34].

an alcohol-water mixed solvent [38]. Similarly microporous SnO_2 also has been synthesized by a microwave-assisted hydrothermal method [39].

Porous SnO_2 spheres were synthesized by a solvothermal method without the use of any surfactant in ethanol medium using $SnCl_4$ precursor [40]. The porous oxide made up of interconnecting SnO_2 crystallites of 5–10 nm has a high surface area of 76.45 m^2/g. Etching induced inside-out dissolution of the inner region is proposed as the mechanism for the formation of these porous nanostructures. Porous SnO_2 obtained by this is shown to exhibit excellent gas sensing property towards 2-chloroethanol and formaldehyde vapour due to the high surface area and porous nanostructures. Porous ZnO also has been synthesized by hydrothermal method involving CTAB as surfactant [41]. A combined sol-gel/hydrothermal method was employed for the synthesis of monodisperse mesoporous anatase TiO_2 beads with high surface area and tunable pore sizes by using hexadecylamine as structure directing agent [42].

4.3.2.2 Combustion/annealing synthesis

Solution combustion is used for the synthesis of various nanocrystalline metal oxides. Mukherjee et al. employed a synthesis route combining sol-gel and combustion methods to obtain nanoporous anatase TiO_2 with a thin interconnected film-like morphology [43]. The method combines the advantages of a conventional sol-gel method for the formation of porous structures and combustion methods to yield highly crystalline, phase pure nanoporous anatase. The synthesis procedure involves the use of titanium (IV) isopropoxide with oleic acid and oleylamine with different ratios to form a viscous yellowish-brown mixture. The glass beaker containing the reaction mixture was placed in the furnace at 400°C, which undergoes combustion generating nanoporous anatase TiO_2. The surfactants aid the formation of the pores in the initial stage. The high temperature during combustion results in an increase in the pore size due to the foaming action of the organics at high temperatures and also results in the formation of highly crystalline phases as shown in Fig. 4.5.

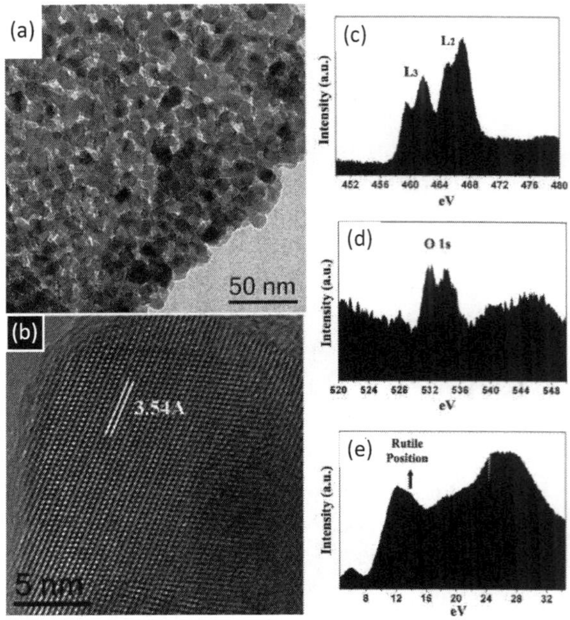

Figure 4.5 (a) TEM image of TiO_2 showing the nanoporous morphology, (b) High-resolution image of a TiO_2 crystallite with the lattice fringes corresponding to the (101) plane of anatase, (c) Near-edge fine structure of the Ti L-core loss region showing the characteristic splitting, corresponding to the anatase phase and (d) O 1s core electron excitation EELS spectrum. The presence of two broad peaks 8–15 eV away from the O 1s edge is a characteristic feature of the anatase phase. (e) Low-energy loss spectra of TiO_2 confirm the phase purity of anatase [43].

Annealing a loose compact of oxide nanorods is shown to produce interconnected porous structures [44]. Partial sintering of the compact leads to spherodization of the nanorods and their fusion at the contact regions resulting in an interconnected porous microstructure. The synthesis is applicable to various systems including ZnO and TiO_2.

4.3.2.3 Aggregation

Aggregation of particles can also lead to the formation of porous structures. Aggregation is classified as diffusion-limited and reaction-limited depending on the aggregation kinetics. In diffusion-limited aggregation, irreversible fast aggregation results in the formation of rigid clusters. However, in most aggregation processes many encounters between pairs of clusters are required before two clusters combine to form a larger cluster. Under these conditions the aggregation process is limited not by cluster diffusion but by chemical details, which determine how many collisions is required for cluster-cluster bonding. This is known as reaction-limited aggregation [45]. We have reported the formation of Pt aggregates in aqueous medium by reaction limited aggregation of Pt nanoparticles [46]. Pd aggregates and bimetallic aggregates were synthesized by a similar aggregation method but in an organic medium by tuning the dielectric constant of the medium [47,48]. The method is reported for the synthesis oxide aggregates as well. For example mesoporous aggregates of TiO_2 consisting of 2–3 nm size crystalline particles have been synthesized by reaction limited aggregation of TiO_2 nanoparticles formed during the reaction in room temperature ionic liquid using $TiCl_4$ [49]. The sponge-like TiO_2 obtained, which is shown in Fig. 4.6, exhibits a very high surface area of 554 m^2/g. In another case, reaction-limited aggregation was reported to result in mesoporous TiO_2 in boric acid solution without any organic structure directing agents [50].

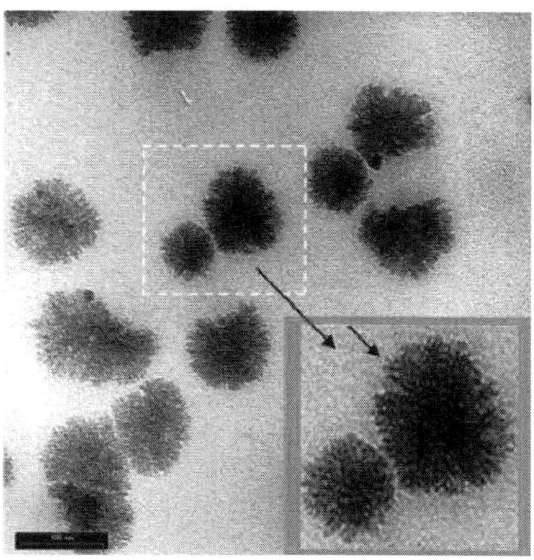

Figure 4.6 TEM image of spherical sponge-like superstructures of TiO_2, formed by reaction limited aggregation of TiO_2 nanoparticles. The inset is a 2× magnification of the image. The scale bar is 100 nm [49].

4.3.2.4 Anodization

Anodization refers to the electrochemical oxidation of a metal which acts as the anode. Anodization is widely used for the synthesis of porous and tubular Al_2O_3, TiO_2 and ZrO_2 [51–56]. The morphology of the anodization product, pore size and ordering depends on various parameters such as anodization voltage, electrolyte and solution temperature. Various mechanisms are proposed for the formation of such porous structures [57–59]. Porous anodic alumina is utilized in a number of applications, including filtration membranes, protective coating and as template material for the synthesis of 1D nanostructure of various materials.

4.4 Applications of Porous Structures

Porous materials are investigated for various potential applications of such structures due to its unique properties such as high surface area, low density and accessible pores. This includes drug delivery, catalysis and sensing, Li-ion battery and solar cell applications.

4.4.1 Drug Delivery

Bio-medical applications of inorganic materials especially SiO_2 has been studied in-depth in the recent past. Mesoporous SiO_2 nanoparticle (MSN) is suggested as drug carrier as SiO_2 is biocompatible and non-toxic. The functionalization of MSNs with organic moieties or other nanostructures directs controlled release and molecular recognition capabilities to these mesoporous materials for drug/gene delivery and biosensing applications [7,60]. High drug loading capacity and fast release rate of the drug is required for the practical use of these materials as drug carriers. High surface area, pore volume and tunable pore sizes qualifies MSN as a potential drug carrier. Ease of functionalization is an added advantage. The release of drugs from different mesoporous silica matrices has been found to be mainly diffusion controlled. The pore architecture, the host–guest chemical interaction, pH of the medium and the dissolution properties of the loaded drug also affects the release rate of drugs. Feasibility of SBA-16 microspheres as drug carrier for poorly soluble Carvedilol (CAR), an antihypertensive agent, was studied and compared to 2D hexagonal MCM-41. In comparison with 2D cylindrical channels of MCM-41, 3D continuous and interconnected pores of SBA-16 could reduce the diffusion resistance and facilitate the facile transport of drugs from the inner pores to the outer dissolution media and this result in faster releasing of the drug [61]. Mellaerts et al. reported an enhanced release of the hydrophobic drug itraconazole from ordered mesoporous SBA-15 [62]. Their findings indicate the existence of a critical pore size for enhancing the drug release. Mesoporous silica TUD-1 was studied for drug delivery considering the model drug ibuprofen. TUD-1 carrier was found to increase the dissolution rate of ibuprofen in comparison to the crystalline drug. Mesoporous silica offers both interior pore and exterior particle surfaces for loading different guest molecules, which is useful for controlling the sequence of release for different drug molecules, which is important in co-delivery applications. Zhao et al. reported that phenylboronic acid-functionalized mesoporous silica nanoparticles (MSN) can act as an efficient co-delivery system for glucose-responsive controlled release of insulin and cyclic adenosine monophosphate (cAMP) [63]. Lu et al. reported their study on a hydrophobic anticancer drug, camptothecin (CPT), which was loaded into the pores of fluorescent mesoporous silica nanoparticles (FMSNs) and delivered the drug into a variety of human cancer cells to induce cell death. FMSN used was synthesized by high-temperature base-catalyzed sol-gel technique. The results suggest that the mesoporous silica nanoparticles might be used as a vehicle to overcome the insolubility problem of anticancer drugs.

A honeycomb mesoporous silica nanoparticle (MSN) system with 3 nm pores is shown to be able to transport DNA and chemicals into isolated plant cells and intact leaves [64]. MSN was loaded with the gene and its chemical inducer and capped the ends with gold nanoparticles to keep the molecules from leaching out. Uncapping the gold nanoparticles released the chemicals and triggered gene expression in the plants under controlled release conditions. This suggests that MSNs may offer new possibilities in target-specific delivery of proteins, nucleotides and chemicals in plant biotechnology.

4.4.2 Catalysis and Sensing

Mesoporous silica based catalyst is reported for the heterogeneous catalysis of liquid-phase organic reactions [65]. Knoevenagel and

Michael reactions are examples. Mesoporous silica supported $AlCl_3$ was described several years ago as a novel solid acid for Friedel–Crafts alkylation Reactions. Ordered mesoporous silica as catalytic support in heterogeneous asymmetric catalysis has been reported by Sahoo et al. [66]. Chiral transition metal complexes anchored on functionalized mesoporous silica was shown to catalyse organic asymmetric synthesis. Porous Mn_2O_3 with high surface area is shown to be a good catalyst for CO oxidation [67]. Microporous SnO_2 synthesized by a microwave-assisted hydrothermal synthesis is shown to posses high sensitivity towards ethanol gas [39].

4.4.3 Li-Ion Batteries

Li-ion batteries have highest energy density among batteries and therefore being used in portable devices. New materials are investigated for the active components viz., cathode, anode and electrolyte. Anode materials with high theoretical capacity such as SnO_2, Co_3O_4 and Si are facing the disadvantage of pulverization during repeated cycling. Porous materials are shown to help in preventing this as it provides voids which can accommodate the volume expansion due to lithiation. Spindle shaped, nanoporous anatase mesocrystals with single crystal-like structure are reported to show improved cycling performance as anode material in Li-ion battery, which is attributed to the porosity and single crystalline nature [36]. Porous SnO_2 and Co_3O_4 are also investigated in this context [68–70].

4.4.4 Solar Cells

Dye-sensitized solar cells are looked upon as one of the most promising future energy sources. Low cost and simplicity in fabrication adds to the advantage of using solar energy. A dye is used as the sensitizer along with a wide band semiconductor such as TiO_2, SnO_2 and ZnO. Mesoporous forms of these oxides are of interest due to its high surface area facilitating to increased dye loading and availability of pores which assist electrolyte diffusion. Mesoporous TiO_2 synthesized by reaction limited aggregation has achieved high loading of the dye, increased electron transfer efficiency, and the photocurrent of the cell, contributing to the significant improvement of the energy conversion efficiency of the dye-sensitized solar cells in comparison to commercially available

P25 [50]. Hydrothermally synthesized mesoporous TiO_2 with interconnected spheres were tested as electrode for dye-sensitized solar cells. This provided pores of two different dimensions and it is to increase the efficiency [34]. Due to the submicrometre-sized particle size and high specific surface areas resulting in high dye loading, the mesoporous TiO_2 beads prepared by a combined solvothermal and sol-gel method can enhance the light harvesting and thereby increase the photon-to-current conversion efficiency compared to P25 nanoparticles [35]. A light conversion efficiency of 7.20% has been achieved by using these mesoporous TiO_2 beads as electrodes in DSSCs, which is significantly higher than that of standard Degussa P25 TiO_2 electrodes of similar thickness (5.66%). Mesoporous F–TiO_2 spheres synthesized by sol-gel combined solvothermal method showed superior photocatalytic activity due to the synergistic effect of large accessible surface area and pore size, fully crystallized anatase nanocrystallites, and surface fluorination [37]. Enhanced solar energy conversion efficiency is achieved by using a coral-like porous SnO_2 structure compared to spherical SnO_2 [71]. Porous ZnO was investigated for photocatalytic phenol degradation and the performance is shown to be better than PC-500 TiO_2 [25].

4.4.5 Templates

Other applications include the use of the mesoporous oxide materials as templates for synthesizing various nanostructures [72]. Nanocasting is extensively used in the synthesis of nanomaterials. Anodic alumina is the most used template in this case. However, the pore size is >20 nm and therefore mesoporous silica is used for obtaining nanostructures, for example nanowire arrays of smaller diameters. Mesoporous carbon with various pore structures have been synthesized using mesoporous silica such as MCM-48 and SBA-15 as template [73].

Mesoporous MgO synthesized by thermal decomposition of anhydrous magnesium acetate is shown to be more active to CO_2 adsorption in comparison with commercial MgO [74].

4.5 Hollow Structures: Definition

Hollow structures are defined as structures having a cavity present, which results in interesting properties such as low density, high surface area, improved mechanical strength and tunable optical properties compared to the solid particles [75–77]. The unique properties of hollow structures lead to its viability in various applications in the field of catalysis, drug delivery and electrochemical cells such as Li-ion battery and fuel cells [78–80].

4.6 Synthesis Methods for Hollow Structures

Syntheses of a large number of hollow structures of various oxides have been reported in the recent past. Hollow structures of different morphology such as spheres, tubes and cubes have been synthesized and studied for various applications. The synthesis methods include template-assisted methods and template-less methods.

4.6.1 Template-Assisted Methods

The most explored synthesis protocol for hollow oxide materials involve the use of templates to obtain a core–shell structure where the metal-oxide preferentially forms on the template. The removal of the template gives rise to hollow oxide structures. The size and shape of the hollow structure depend on the template. Hard templates such as polystyrene spheres and colloidal silica are extensively used to get monodisperse hollow structures of metals, semiconductors and ceramics. Metal oxides are also used as hard templates. Soft template methods where surfactant micelle, emulsion droplets or gas bubbles formed during the reaction acts as the template is also adopted in the synthesis of hollow nanostructures.

Hard templating using polymeric or silica templates results in monodisperse hollow particles with well-defined shapes. The ease of synthesis and commercial availability of these templates in various sizes make them the ideal template to synthesize hollow nanostructures. However, the chemical incompatibility between the template and the oxide formed in many cases results in incomplete coating or no coating on the template, which results in including an additional step involving the surface functionalization of the template to favour the growth of the oxide on them resulting in a uniform shell. Another disadvantage lies in the post-processing for the removal of the template. Chemical treatment for selective etching of the template and calcinations at high temperature for removing organic templates are the most common method for template removal. Chemical treatment requires the need of a chemical which will exclusively dissolve off the template leaving behind the shell intact resulting in hollow structures. The challenge is to select an appropriate reagent as most of them can react with both the template as well as the oxide shell present on that. The calcination step to remove the template can cause undesirable grain growth of the oxide formed.

Soft templates can overcome some of these disadvantages of hard templates but the control of size, shape and uniformity of the hollow product is difficult in comparison to hard templating.

4.6.1.1 Polymeric template

Polymeric templates such as polystyrene spheres are common template of choice in hollow structure synthesis. Zhong et al. reported a method for synthesizing mesoporous TiO_2 and SnO_2 by utilizing crystalline arrays of monodisperse polystyrene beads as template [81]. The template was infiltrated with the sol-gel precursor solution in a glove box which hydrolyzed into the oxide ceramics on exposure to the moisture in air. Removal of the template by dissolution in toluene resulted in the formation of hollow structures of the ceramic. The void size of these hollow spheres was determined by the diameter of the polystyrene template, and the thickness of the ceramic wall was tuned in the range of 30–100 nm by changing the concentration of the precursor. Jiang et al. reported a lost-wax approach to synthesize monodisperse colloids including solid colloids, core–shell as well as hollow structures [82]. The method involved three steps; first step being the formation of a silica colloidal crystal by convective assembly method. Then these thin-film arrays of SiO_2 are used as templates to generate macroporous polymers containing spherical and ordered voids. The macroporous polymers can be filled with different liquid precursor solutions which by then solidify to fill the cavities. Removal of the host macroporous polymer by dissolution in an appropriate solvent, or by heating, leaves behind an array of colloids that retain the initial silica template geometry. For the synthesis of ceramic colloids, the macroporous polystyrene films are first immersed in an alcoholic solution of metal alkoxide. When it is removed, the alkoxides which adhere to the inner polymer surfaces condense as they react with the moisture in air, forming hollow spherical shells. As the colloids are formed through successive deposition of thin layers of ceramics, shell thickness can be precisely controlled. The adhesion between the polymeric template and the ceramic formed, which depends on the interfacial energy, decides whether a solid product or hollow product is formed. The use of polystyrene resulted in the formation of hollow structure, whereas the use of PMMA resulted in a solid structure. The morphology of the hollow structure is shown in Fig. 4.7.

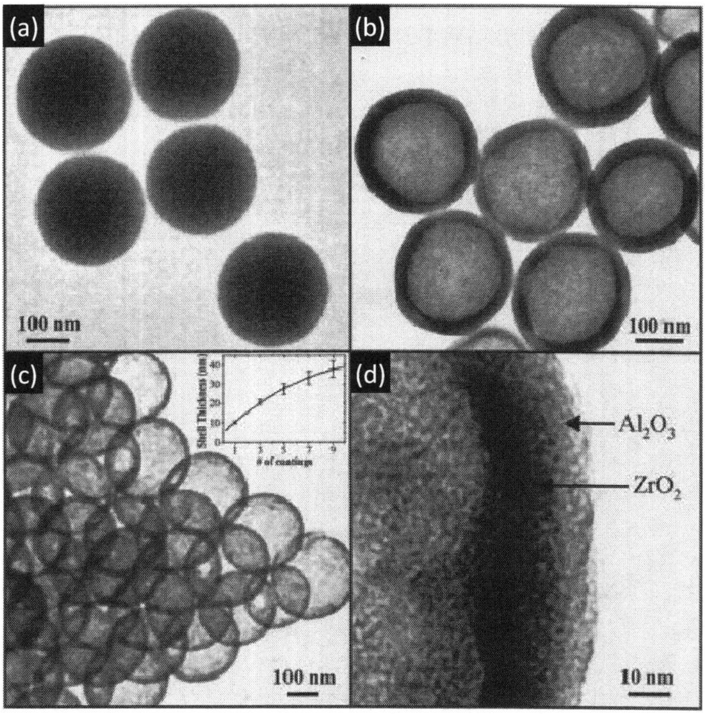

Figure 4.7 TEM images of (a) Solid titania colloids formed using PMMA template after coating seven times, (b) Hollow titania colloids obtained by using polystyrene templates after coating seven times, (c) Thinner shell hollow titania colloids replicated from the same template as (b) after coating only twice, (d) Core–shell structure of a hollow zirconia–alumina colloid formed from successive deposition of alumina and then zirconia on the surface of a polystyrene template [82].

Polystyrene (PS) beads are employed as the template for the synthesis of various hollow oxide nanostructures. ZnO–TiO_2 composite hollow spheres less than 500 nm size where synthesized by PS beads as templates [83]. The void size of these hollow are determined by the diameter of the PS template particles. ZnO : TiO_2 ratio and shell thickness of these hollow structures can be varied by tuning the concentration of precursors, $Zn(Ac)_2$:$2H_2O$ and $Ti(OEt)_4$ salts in reaction media. The SEM micrographs of the samples at various stages of synthesis are shown in Fig. 4.8. Deng et al. used sulphonated PS spheres as templates to synthesize ZnO hollow spheres [84]. In their method ZnO is formed by reacting adsorbed Zn ions on the template with NaOH in solution. The base present dissolved the template and consequently no template removal was needed. Liu and Nakashima reported a method for the fabrication of hollow nanospheres of Nb_2O_5, CeO_2, and V_2O_5 by templating the polymeric micelle of PS-PVMP-PEO, which has a PS core, a PVMP shell, and a PEO corona in aqueous solutions [85].

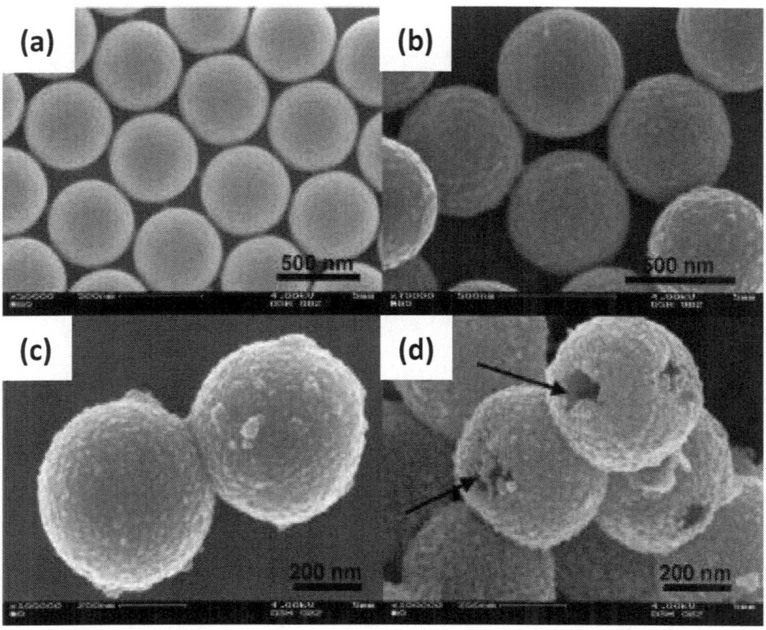

Figure 4.8 SEM images of (a) PS template particles (b) PS/ZnO core–shell composite particles (c) PS/ZnO–TiO_2 core–shell composite particles and (d) ZnO–TiO_2 hollow spheres obtained after template removal [83].

A one-step method for the synthesis of double-shelled composite hollow spheres by using sulphonated polystyrene hollow spheres as templates was demonstrated by Yang et al. [86]. Controlled sulphonation of the hollow PS template resulted in the formation of sulphonated outer layer of tunable thickness, sulphonated inner layer, and unsulphonated middle layer. The sulphonation process gives layers with sulphonic acid groups that allow the formation of a wide variety of functional composites with double shelled structure by sol-gel method. Template removal by calcination leads to hollow structure formation. TiO_2 and Fe_3O_4 hollow structures were synthesized by this method. The amine functionalized polystyrene beads can also be used as a template to prepare TiO_2 hollow nanospheres [87]. Titanium *t*-butoxide was used as the precursor. Monodispersed core–shell spheres were produced by this reaction and after calcination porous, uniform-sized TiO_2 hollow spheres were obtained.

4.6.1.2 Silica template

The ease of synthesis of SiO_2 spheres of various sizes and their commercial availability make it one of the most common templates. Park et al. have described a method for the preparation of uniform hollow metal oxide particles through calcination and etching processes on silica@coordination polymer core/shell structures. Silica@coordination polymer core–shell microspheres were synthesized by solvothermal method [88]. The product obtained is calcined to transform the coordination polymers into metal oxides resulting in silica@M_2O_3 followed by an etching process for the removal of silica template to obtain hollow metal oxides. The method is demonstrated for different oxides of composition M_2O_3 where M = Gd^{3+}, Eu^{3+}, Y^{3+}. Joo et al. employed a synthesis method where colloidal silica spheres act as the hard template in obtaining mesoporous hollow silica [89]. Colloidal silica spheres synthesized from TEOS were deposited with TiO_2 and SiO_2 layers through sol-gel processes. In order to induce crystallization of the amorphous TiO_2 layer, the sample was calcined and finally etching of the inner and outer silica by NaOH resulted in mesoporous anatase TiO_2

shells. Mesoporous hollow CeO_2 is also reported to be synthesized by a hard template route involving the use of SiO_2 spheres as the template [90]. The constituent CeO_2 crystallites were of the dimension of 2–5 nm. A high value of BET surface area was observed, which is credited to the internal surface and surface roughness. Chen et al. reported a method to synthesize hollow mesoporous silica using Stober silica as template [91]. Mesoporous silica coating on Stober silica was achieved by co-condensation of TEOS and octadecyltrimethoxysilane in H_2O/ethanol/ammonia solution resulting in core–shell structure with Stober silica core and mesoporous silica shell. Hydrothermal treatment of these core–shell structures in ammonia solution leads to the preferential etching of the Stober silica core leading to rattle type structure by partial dissolution or hollow mesoporous silica particle by complete dissolution of the core depending on the concentration of the ammonia solution. The open structure of the shell facilitates the penetration of the hydroxyl ions to selectively etch the core part, which has much more Si-OH groups than the mesoporous shell.

4.6.1.3 Other oxide materials as template

Oxides are also used as templates in synthesizing hollow structures. The structural or chemical difference between the template oxide and the oxide shell formed helps in preferential removal of the oxide core which acted as the template resulting in hollow structure of the shell material. For example α-Fe_2O_3 spindles of different aspect ratios were used as a template in obtaining non-spherical hollow TiO_2 nanostructure [92]. For a typical TiO_2 coating, 35 mg α-Fe_2O_3 spindles was dispersed in a mixture of 138 mL of ethanol, 7 mL of deionized water, and 13 mL of PVP solution (15 mg/mL) in water, to which 25 mL of TiF_4 (0.04M) solution was added. After ultrasonication for less than 1 min, 30 mL of the suspension was poured into a 40 mL Teflon-lined stainless-steel autoclave and heated in an air-flow electric oven at 180°C for 3 h. Hollow TiO_2 was obtained when Fe_2O_3 was preferably dissolved from the product by dilute HCl treatment. The morphology transition with different reaction time is shown in Fig. 4.9.

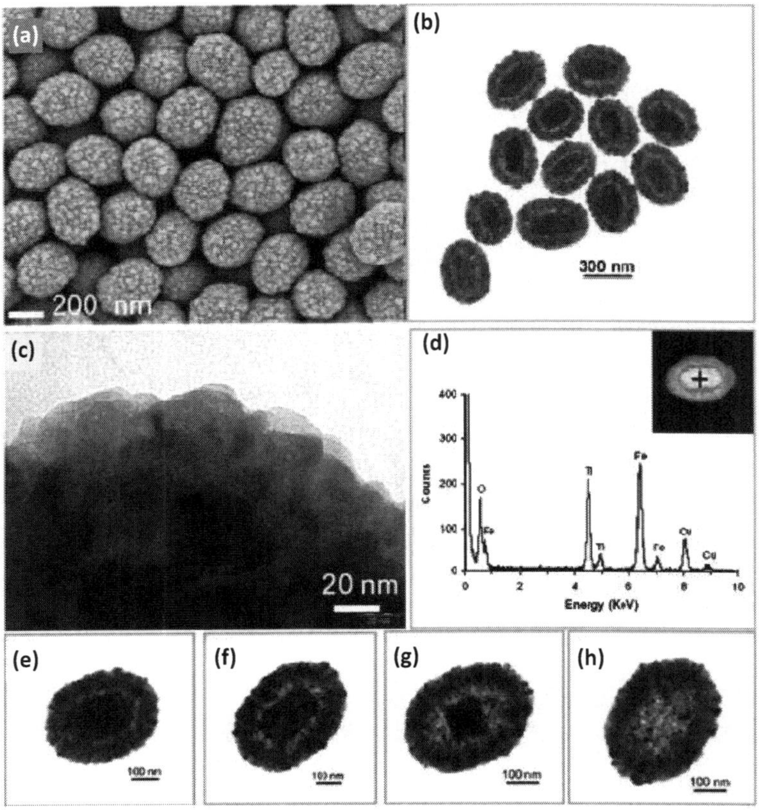

Figure 4.9 (a) is the SEM image and (b) and (c) are the TEM images of as-synthesized hematite/TiO_2 core/shell particles, (d) is energy-dispersive X-ray spectrum taken from the centre of a single core/shell particle shown in the inset; the Cu signals are generated from the grid and (e–h) are TEM images after treating hematite/TiO_2 core/shell particles with HCl (0.2 M) at 100°C for 0, 6, 14 and 24 h, respectively, [92].

Piao et al. reported a wrap–bake–peel process, where spindle shaped β-FeOOH is converted to hollow iron oxide particle [93]. The procedure involved the synthesis of β-FeOOH by hydrolysis of $FeCl_3 \cdot 6H_2O$ followed by silica coating to obtain a core–shell structure. Calcination of the β-FeOOH-Silica core–shell in air resulted in the formation of hollow magnetite particle coated with silica. In order to obtain haematite particles, the air calcined samples were further heated under H_2. Removal of silica by NaOH etching gives hollow magnetite/haematite particles. Wu et al. presented a similar β-FeOOH templating method to obtain hollow core-double shelled nanostructures [94]. In this synthesis β-FeOOH core/organosilicate shell structure is first prepared, followed by a simple decomposition/reduction process to create mesoporous structure in the outer silica shell, convert β-FeOOH into a magnetic Fe_3O_4 nanocapsule as the inner shell, and generate a large hollow core. The large pore inside the hollow iron oxide is suggested to incorporate various drugs and biomolecules with a high loading capacity and therefore the hollow structure can be used for drug delivery.

Carbon nanoparticles are also being used as hard templates in the synthesis of hollow structures. Bang et al. employed a sonochemical method to deposit Fe on carbon spheres by the decomposition of $Fe(CO)_5$ and Fe in the composite transformed to Fe_2O_3 upon drying [95]. The heat released during this oxidation causes the carbon core to ignite and the heat generated by the combustion of carbon causes the crystallization of iron oxide shell to form α-Fe_2O_3. Carbon spheres are used as template in the synthesis of In_2O_3 hollow particles as well [96].

4.6.1.4 Soft template

Soft templates such as surfactant micelle and gas bubbles formed during the reaction are reported to act as template in the synthesis of hollow nanostructures. ZnO_2 hollow structures were synthesized by a soft template route were in-situ generated O_2 gas bubbles, formed by decomposition of H_2O_2 during the hydrothermal treatment of ZnO powders in aqueous solution containing H_2O_2, served as the template. ZnO hollow spheres were obtained by annealing the ZnO_2 hollow spheres at 180°C in air [97]. This method has the advantage that there is no template removal step required in obtaining hollow structure. In another report 1H, 1H, 2H, 2H-perfluorodecylpyridinium chloride was used as the surfactant template in the synthesis of hollow silica particles with vesicle-like structure [98]. The co-assembly of fluorinated surfactant micelle with silica is proposed as the formation mechanism and a high stirring rate is shown to result in elongated multi-chambered hollow structure. Hollow mesoporous silica particles were synthesized by modifying a method for mesoporous silica synthesis [99]. 6.4 mL of deionized water, 1.05 mL of ethanol, and 1.04 mL of 25 wt% cetyltrimethylammonium chloride were mixed and stirred at room temperature for 30 min to which 1.85 g of triethanolamine was added and further stirred for 30 min until dissolved to control the pH. A mixture of 0.692 mL of TEOS and an appropriate amount of DMF was added to this with stirring. The product was precipitated by adding ethanol and washed by hydrochloric acid in ethanol solvent. These mesoporous silica spheres are reported to be formed through an interfacial reaction at the surface of liquid droplets and are different from the surfactant bilayer vesicle structures. The effect of DMF concentration on the morphology of the product is shown in Fig. 4.10.

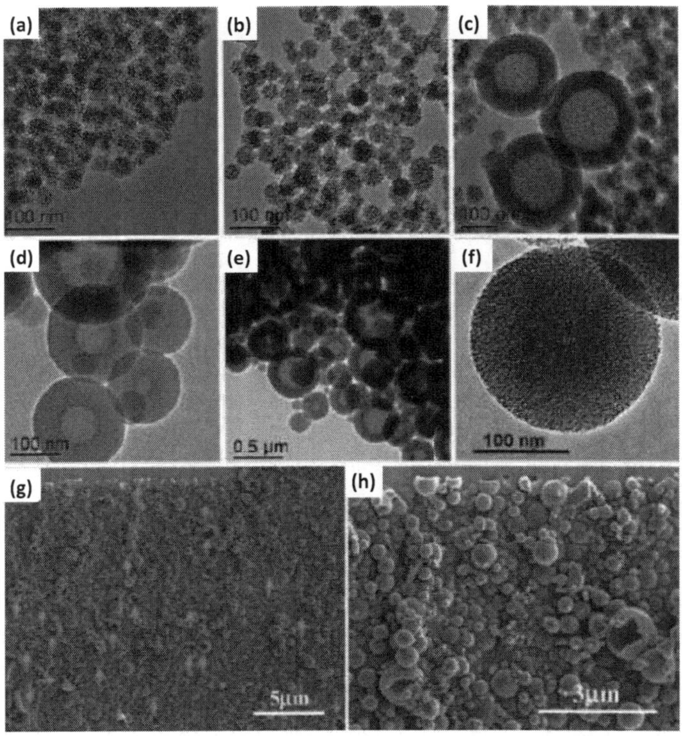

Figure 4.10 TEM images of SiO_2 particles obtained with different amounts of DMF; (a) No DMF, (b) 0.2 mL of DMF, (c) 0.976 mL of DMF, (d) 2 mL of DMF and (e) 4 mL of DMF. (f) A single sphere with almost no cavity obtained by using 4 mL of DMF. (g, h) SEM images of hollow spheres synthesized with 2 mL of DMF. Broken spheres are indicated with arrows [99].

Other reports based on soft template methods include a hollow Co_3O_4 nanocubes synthesis by the use of a surfactant template [100], a microemulsion method for the synthesis of SnO_2 particles of 15–25 nm sizes, which shows a good response to CO in a concentration range of 50 to 300 ppm [101] and hollow silica particles synthesis by templating ABC triblock polymeric micelle [79,85,102,103].

4.6.2 Template-Less Methods

Template-less methods for synthesis of hollow structures are of advantage considering the fact that there is no problem of template removal. The methods can be classified based on the mechanism of hollow structure formation. Kirkendall effect and Ostwald ripening are the most explored mechanisms. Other than this, hydrothermal synthesis is also reported in the synthesis of hollow structures.

4.6.2.1 Kirkendall effect

Kirkendall effect refers to the mutual diffusion process through an interface of two metals with different diffusivity where vacancy diffusion occurs to compensate for the difference of the material flow, which results in the movement of the initial interface. Formation of hollow nanocrystals through Kirkendall effect was observed by Yin et al. in 2004 in which pores form because of the difference in diffusion rates between two components in a diffusion couple [104]. In their study cobalt nanocrystals reacted with oxygen and either sulphur or selenium resulting in the formation of hollow oxide or chalcogenides. Jin fan et al. exploited Kirkendall effect in the fabrication of $ZnAl_2O_4$ spinel [105]. In their experimental study ZnO nanowires were coated with Al_2O_3 resulting in core-shell structure. During the annealing of the core–shell structure at 700°C, interfacial reactions resulted in the formation of spinel $ZnAl_2O_4$ with a hollow core due to Kirkendall effect. The morphology of the ZnO nanowires coated with Al_2O_3 and the $ZnAl_2O_4$ formed is given in Fig. 4.11.

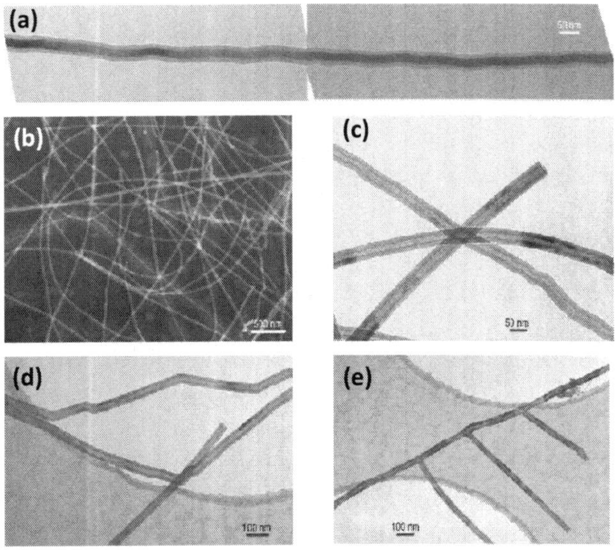

Figure 4.11 Core–shell nanowires to nanotubes by means Kirkendall diffusion (a) TEM image of ZnO–Al_2O_3 core–shell nanowire. SEM (b) and TEM (c–e) images of $ZnAl_2O_4$ spinel nanotubes [105].

A synthetic route, based on the nanoscale Kirkendall effect, to produce colloidal solutions of monodisperse hollow maghemite nanoparticles with controllable particle size and shell thickness was reported by Cabot et al. [106]. They show that there exists a limitation on the size of the hollow nanostructures that can be synthesized in solution, which is dependent on the iron diffusivities and the growth of the crystal size domains in the polycrystalline oxide shells. Khurshid et al. reported the chemical synthesis of iron oxide hollow nanoparticles [107]. Core/shell Fe/Fe-oxide nanoparticles were synthesized by thermal decomposition of $Fe(CO)_5$ at high temperature. Particles above 8 nm had core/shell morphology, whereas smaller particles exhibit hollow morphology. Annealing the core/shell particles at high temperatures in air also resulted in the formation of hollow spheres. In the case of thermally activated Kirkendall process, the particles do not fully transform into hollow structures but many irregular shaped voids exist inside each particle. Synthesis of one-dimensional chainlike arrays of Fe_3O_4 hollow nanospheres by aging iron nanoparticles in aqueous solution was also reported to be due to Kirkendall effect [108].

Peng et al. reported the synthesis of hollow Fe_3O_4 nanoparticles by controlled oxidation of amorphous core–shell Fe–Fe_3O_4 nanoparticles [109]. The core–shell particles were obtained by high-temperature solution-phase decomposition of [$Fe(CO)_5$] resulting in the formation of amorphous Fe and air oxidation of the amorphous Fe nanoparticles at room temperature. Kirkendall effect directs the formation of hollow structure in the controlled oxidation of Fe–Fe_3O_4, in which Fe metal diffuses faster outward than oxygen does inward, and Fe_3O_4 forms at the metal–oxide interface rather than in the interior of the core as shown in Scheme 4.1. The TEM images of the Fe–Fe_3O_4 and hollow Fe_3O_4 are shown in Fig. 4.12.

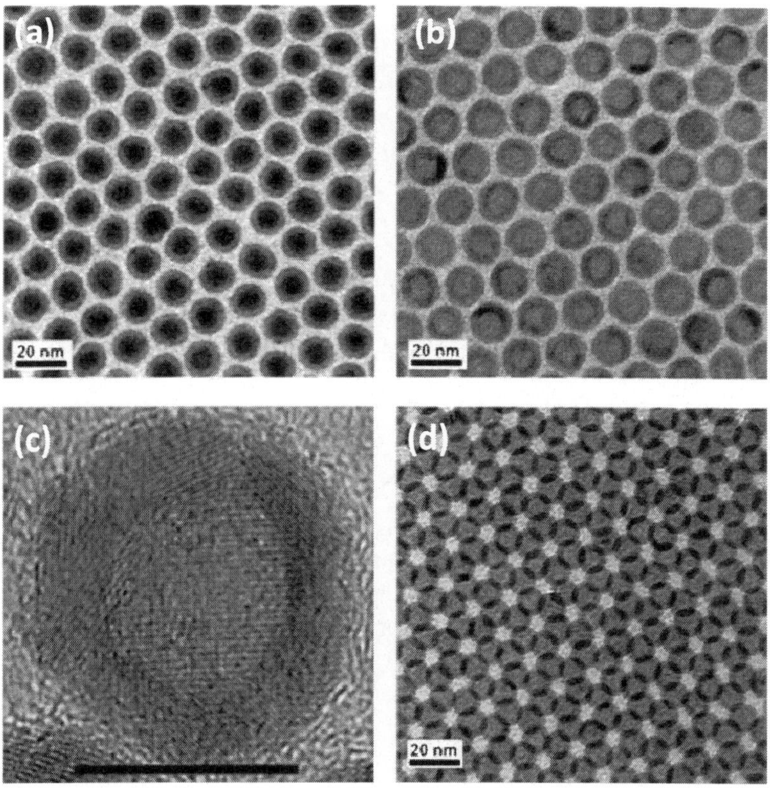

Figure 4.12 TEM images of (a) 13 nm Fe–Fe_3O_4 nanoparticle seeds, (b) 16 nm hollow Fe_3O_4 nanoparticles obtained by the oxidation of Fe–Fe_3O_4 nanoparticles, (c) a single hollow Fe_3O_4 nanoparticle (scale bar 10 nm), and (d) a superlattice array of the hollow Fe_3O_4 nanoparticles [109].

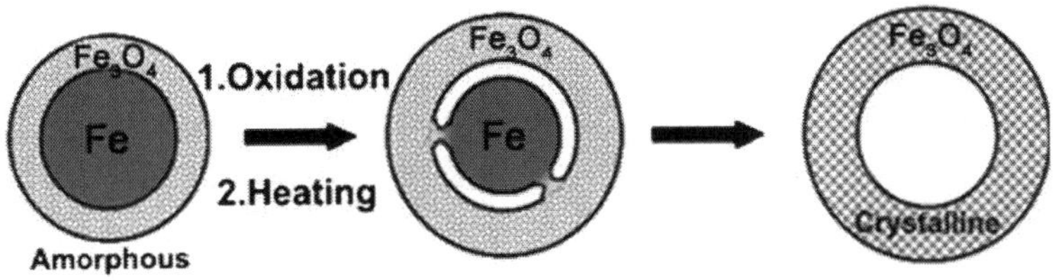

Scheme 4.1 Synthesis of core–shell–void $Fe–Fe_3O_4$ and hollow Fe_3O_4 nanoparticles from $Fe–Fe_3O_4$ nanoparticle seeds [109].

MnO and iron oxide particles are shown to undergo a transformation to hollow structure in TOPO on thermal treatment [110]. The hollow nanoparticles synthesized were amorphous and composed of oxygen, phosphorus, and metal. The alkylphosphonic acid impurity in technical grade TOPO is shown to be responsible for the etching process. The opposite diffusion of phosphorus and metal cations within metal oxide nanocrystals accompanied by Kirkendall effect was suggested as the possible formation mechanism of the hollow nanoparticles. Hung et al. investigated the oxidation of Cu nanoparticles of ~8 nm sizes protected with tetradecylphosphonic acid (TDPA) in hexane and chloroform [111]. In chloroform, where TDPA solubility is high, hollow Cu nanoparticles transferred to Cu_2O by Kirkendall effect on reaction of Cu with dissolved oxygen. Size-dependent behaviour of nanoscale Kirkendall effect has been investigated by Railsback et al. [112]. Their investigation on the oxidation of Ni nanoparticles of varied sizes concluded that the fast self-diffusion of Ni compared with its rate of diffusion through the NiO shell results in single voids for the smaller sizes (9 and 26 nm), but multiple voids form in the case of 96 nm particles, because self-diffusion is not fast enough to cause the voids to combine into a single void. It is also shown that oxidation occurs quickly at elevated temperatures but the final structures were independent of the oxidation temperature.

4.6.2.2 Ostwald ripening

Ostwald ripening refers to the growth mechanism of crystals where larger particles grow at the expense of smaller particles in an inhomogeneous system. Ostwald ripening is proposed as the mechanism of hollow structure formation in many different system, most of the synthesis route being the hydrothermal/solvothermal synthesis. For example, hollow SnO_2 aggregates were synthesized by a simple hydrothermal method using potassium stannate trihydrate as a precursor in an ethanol-water mixed solvent [113,114]. The method involved no surfactant or template. During the first stage of reaction, solid nanospheres were formed by hydrolysis of stannate. The formation of the mesoporous aggregate is by to oriented attachment of the nanocrystallites. With prolonged hydrothermal treatment, these solid nanospheres converted to hollow spheres as a result of Ostwald ripening. The mechanism is represented as a schematic in Fig. 4.13a and the morphology transition is as shown in the TEM images (Figs. 4.13b–d).

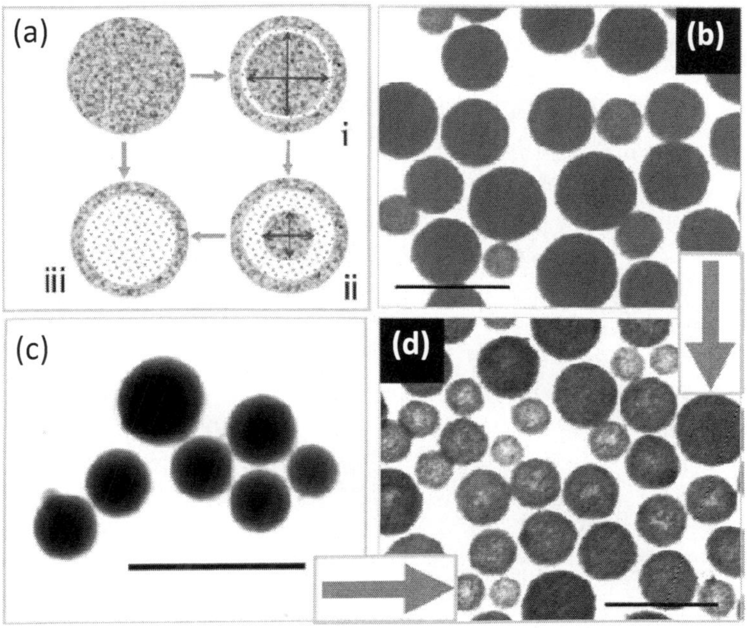

Figure 4.13 (a) Schematic illustration of the proposed inside-out ripening mechanism, TEM images of (b) SnO_2 nanospheres obtained with 6 h reaction at 150°C, (c) amorphous SnO_2 nanospheres obtained by aging the reaction mixture for several days at room temperature and (d) SnO_2 hollow nanospheres synthesized at 150°C for 24 h with the reaction mixture aged for several days. All three scale bars = 500 nm [113].

Similar mechanism of Ostwald ripening was proposed in the formation of hierarchical hollow SnO_2 structure made up of SnO_2 sheets during a hydrothermal synthesis using $SnSO_4$ [115]. Cao et al.

reported a synthesis method for ceria hollow nanospheres involving a template-free microwave-assisted hydrothermal method [116]. The formation is explained as an Ostwald ripening coupled self-templated, self-assembly process. In another report monodisperse CeO_2 hollow spheres composed of nano-octahedra were synthesized by a one-step solution approach involving hydrothermal reaction at 180°C [117]. $CeCl_3$ was used as the precursor and PVP as surfactant. Based on both the formation process of hollow interiors and the size evolution of the octahedral building blocks, Ostwald ripening is proposed as the mechanism involved. Aggregation followed by Ostwald ripening is reported for the synthesis of hollow structures of many oxide systems such as Sn-doped TiO_2, TiO_2, $ZnWO_4$, Cu_2O and $BaTiO_3$ [118–123].

4.6.2.3 Other methods

Hydrothermal syntheses of hollow structures are also reported. Hollow SnO_2 octahedra synthesis by a template-free hydrothermal route is reported [124]. The formation of SnO_2 octahedron is based on the two-dimensional attachment of nanocrystallites giving an overall {111}-like termination for each octahedral surface in the presence of ethylenediamine and 2-propanol/water. A hydrothermal synthesis to obtain $Ni(OH)_2$, which on annealing results in hollow NiO single crystal, has been reported [125]. Acid etching of Stober silica by a hydrothermal treatment is shown to result in hollow silica [126]. A surfactant-assisted solvothermal method is shown to produce hollow CeO_2 [127].

MnO_2 hollow structures are reported to be formed by self assembly of nanosheets or nanoplatelets [67]. Yang et al. reported a synthesis method where high yields of hollow Cu_2O nanocubes of 28 ± 5 nm in edge length were prepared by a one-pot synthesis in aqueous solution at pH 12.20 in the absence of any surfactant at room temperature [77]. Porous oxide nanoparticles of Fe_2O_3 were synthesized by controlled oxidation of Fe nanoparticles at 250°C followed by acid etching [128]. ZnO hollow particles were synthesized by a selective etching of Zn–ZnO core–shell structure [129].

4.6.2.4 Hollow nanostructures from nanoparticle aggregates

We have developed a general method for the synthesis of hollow nanostructures by partial sintering of mesoporous nanoparticle aggregates [130]. The experiment involves the air annealing of mesoporous aggregates of metal oxide nanoparticles. Metal oxides including CeO_2, SnO_2 and TiO_2 were selected for the study. The morphological evolution of CeO_2 aggregates on heat treatment is shown in Fig. 4.14. As-synthesized mesoporous CeO_2 consisted of ~3 nm particles aggregated randomly in spherical structure. Sintering of the aggregate by heat treatment at 700°C resulted in the formation of multiple faceted holes in each cluster. At 900°C the clusters transform to single crystalline hollow spheres as shown in Fig. (4.14c), confirmed by the high resolution TEM. The presence of the void is revealed also by the Fresnel contrast seen in Fig. 4.14c. A similar morphology transformation to hollow structure was observed for SnO_2 and TiO_2 aggregates. However, in these cases the resultant hollow structures were polycrystalline with larger grain size compared to the starting mesoporous aggregate. The hollow structures obtained have a similar dimension to that of the starting aggregates.

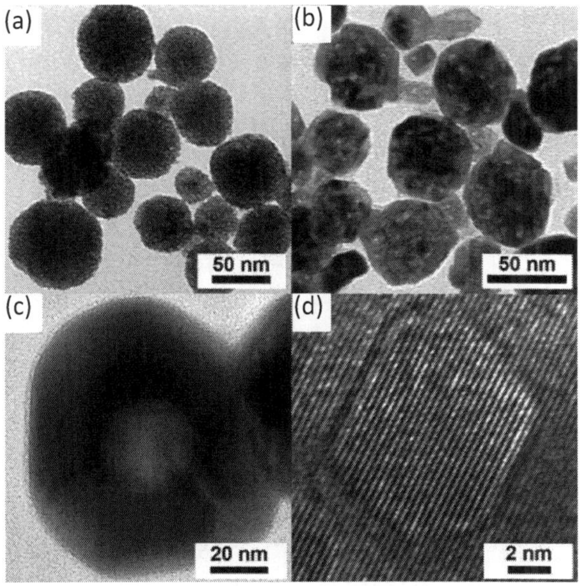

Figure 4.14 Stages of formation of hollow single crystals of CeO_2 from porous aggregate on annealing (a) mesoporous CeO_2 nanoparticle aggregate at 250°C, (b) faceted structure with multiple faceted pores at 700°C, (c) hollow single crystal obtained at 900°C and (d) high resolution image from a hollow particle showing the single crystal nature [130].

We also investigated the stability of spherical nanoporous aggregates using phase-field simulations that explore systematically the effect of various kinetic parameters, such as grain boundary diffusion (M_{GB}), surface diffusion (M_S) and grain boundary mobility (L) on the pathways for microstructural evolution [131]. Surface diffusion dominates the early stage of sintering resulting in the formation of a shell. Grain boundary mobility determines the grain growth; when L is large the resulting microstructure shows few numbers of grains. The microstructure evolution with different values of M_{GB} and L are summarized in the form of "sintering map", as shown in Fig. 4.15 for $\tau = 2000$, and $\tau = 32000$. The early stage evolution shows the formation of a surface shell in all the cases; while a central core is also formed in systems with high GB mobility ($L = 10$), which is absent in systems with moderate GB mobility ($L = 1$). The late stage microstructures show that, for a given M_{GB}, the surface shell has fewer grains in systems with higher GB mobility, and for a given L, the central pore is smaller in systems with higher GB diffusivity. In the system with $L = 10$ and $M_{GB} = 1000$ the aggregate is fully dense at $\tau = 32000$. The simulation results were consistent with the experimental observation and different combinations of surface and GB diffusivity and GB mobility show four distinct microstructural pathways leading to 100% density: multiple closed pores, hollow shells, hollow shells with a core, and multiple interconnected pores.

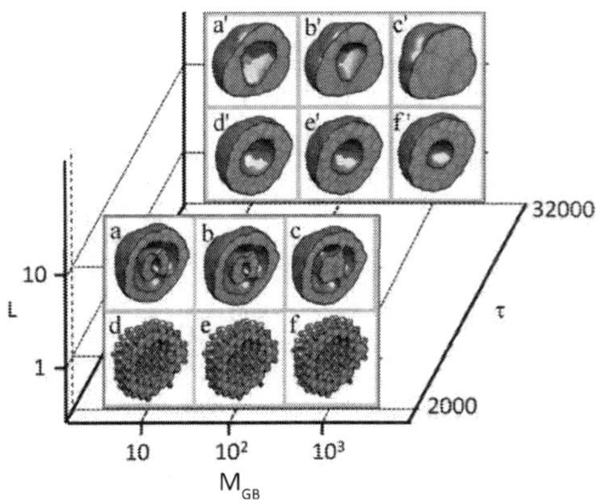

Figure 4.15 Sintering map of microstructures obtained from 3D phase field simulations in systems with varying grain boundary mobilities and grain boundary diffusivities represented by L and M_{GB}, respectively, at different time ($\tau = 2000$) and ($\tau = 32000$) [131].

4.7 Applications of Hollow Nanostructures

4.7.1 Drug Delivery

Hollow nanostructures are potential drug delivery candidates as they have a cavity which can be used to store and deliver drug molecules. Biocompatible and nontoxic oxide materials are investigated for this purpose. Cheng et al. reported an investigation on cisplatin storage and release using porous hollow nanoparticles (PHNPs) of Fe_3O_4 [128]. Cisplatin is an anticancer drug but with a drawback of targeting both tumour and healthy tissue and poor water solubility. The PHNPs were synthesized by controlled oxidation of Fe NPs at 250°C followed by acid etching. The open pores in the polycrystalline shell facilitated the cisplatin diffusion into the cavity of the hollow structure. The porous shell was stable in neutral or basic physiological conditions, and cisplatin escape from the cavity through the same pores was a diffusion-controlled slow process. In low pH (<6) conditions, the pores were subjected to etching, resulting in wider pore gaps and faster release of cisplatin. On coupling with herceptin to the surface, the cisplatin-loaded hollow NPs could target to breast cancer SK-BR-3 cells. This indicates that the low pH-responsive PHNPs of Fe_3O_4 can be exploited as a cisplatin delivery vehicle for target-specific therapeutic applications.

Hollow core, magnetic, and mesoporous double-shell nanostructures consisting of silica and Fe_2O_3 (HMMNS-R/P) are tested for biomedical applications [94]. The as-synthesized nanocomposites showed good blood compatibility and very low cytotoxicity towards MCF-7, HeLa, and *L*929 cells. Investigations on the bioimaging and drug delivery properties indicate that, the composite is a potential drug carrier and can be used for simultaneous bioimaging and therapeutic applications. Hollow mesoporous silica spheres with different pore sizes synthesized by selective etching strategy were explored to store molecules with various sizes such as traditional anticancer drugs and biomolecules and it shows a high loading capacity for anticancer drug, doxorubicin [91].

4.7.2 Li-Ion Battery Anode

Oxide materials such a SnO_2 are investigated for Li-ion battery applications as they have as much higher theoretical specific lithium storage capacity than the currently used graphite. However, the large volume change during charging/discharging processes leads to poor cyclability, which hinders the application of these materials as anode. One possible strategy to diminish the pulverization and thus enhance the structural stability is to design hollow nanostructures. Hollow nanostructures of metal oxides such as SnO_2, TiO_2 including hollow spheres, nanotubes, etc., are therefore investigated as potential anode material for Li-ion battery [132]. Lou et al. reported that the TiO_2 hollow particles have excellent capacity retention over extended cycling, and they are able to deliver a reversible Li storage capacity of 172 mAh/g after 35 cycles [92]. Hollow nanospheres of SnO_2 synthesized by a solvothermal technique are shown to exhibit better cyclability and superior Li storage capacity where a capacity comparable to its theoretical density is obtained even after 30 cycles [113]. Hierarchical hollow structures of SnO_2 exhibited high capacities and excellent cycle performances as anode materials for lithium ion batteries, which has a reversible capacity of 545 mAh/g up to 50 cycles [115].

4.7.3 Catalysis and Sensing

Photocatalytic degradation of dye is investigated with many oxide semiconductors in the recent past as it is of great concern to convert the dyes into environmental friendly chemicals. A comparison of photocatalytic degradation of RB and BR K-2BP by commercial ZnO powders and ZnO hollow spheres showed that the hollow structures had considerably higher photocatalytic activity for the degradation of dyes than the commercial ZnO powders [84]. $ZnO–TiO_2$ hollow spheres synthesized by template method has been tested for photo catalytic degradation of rhodamine 6 G dye and is shown to have superior activity compared to other structures reported in literature [83]. Hollow CeO_2 spheres were used as catalytic support for gold and the resulting supported catalyst showed CO oxidation at room temperature [116]. Hollow SnO_2 spheres of ~15 nm sizes show a good response to CO in a concentration range of 50–300 ppm [101].

4.8 Conclusions

Synthesis and applications of porous and hollow nanostructures of oxide materials are discussed. Template-assisted synthesis and template-less synthesis are the major classification. In template-assisted methods, the hard templating results in the controllable morphology and uniform size of the hollow spheres, which is determined by the template shape and size but it poses the problem of template removal, which is usually carried out by high-temperature annealing or chemical treatment to preferentially remove the template. Soft templating has the advantage of easy removal of the templates but obtaining monodispersity in the size and shape of the product is difficult. The shape and size of the soft template is sensitive to various factors such as reaction pH, polarity of the medium, etc. Thus template-less methods are being looked upon as the alternative as it can overcome the demerits of template based synthesis methods. Ostwald ripening and Kirkendall effect are the important techniques among them. Though the mechanism of formation is well investigated and applied in the synthesis of many materials, it has the disadvantage that it is not general and can be applied to only selective systems involving a diffusion couple with specific diffusivity difference. Methods based on Ostwald ripening are also not general and the mechanism involving the formation of the nanoparticle aggregate and the following ripening is not well understood and therefore does not provide the conditions necessary for synthesizing a new system with the background of existing reports. Thus different synthesis techniques are being employed in obtaining porous and hollow nanostructures of metal oxides. These materials are being investigated for applications including catalysis, sensing, drug delivery, Li-ion battery and solar cells.

Acknowledgement

Funding for various aspects of our work presented here was from the Department of Science and Technology, Government of India. The microscopy facilities were a part of the Advanced Facility for Microscopy and Microanalysis at IISc.

References

1. Attard, G.S., M. Edgar and C.G. Goltner (1998). Inorganic nanostructures from lyotropic liquid crystal phases, *Acta Mater.*, **46**(3), pp. 751–758.
2. Pietron, J.J., R.M. Stroud and D.R. Rolison. (2002). Using three dimensions in catalytic mesoporous nanoarchitectures, *Nano Lett.*, **2**(5), pp. 545–549.
3. Jiang, J., et al. (2011). Building one-dimensional oxide nanostructure arrays on conductive metal substrates for lithium-ion battery anodes, *Nanoscale*, **3**(1), pp. 45–58.
4. Zhai, T., et al. (2009). A comprehensive review of one-dimensional metal-oxide nanostructure photodetectors, *Sensors*, **9**, pp. 6504–6526.
5. Royer, S., and D. Duprez. (2011). Catalytic oxidation of carbon monoxide over transition metal oxides, *Chem. Cat. Chem.*, **3**(1), pp. 24–65.
6. Nair, H., et al. (2008). Effects of metal oxide domain size, dispersion, and interaction in mixed WO_x/MoO_x catalysts supported on Al_2O_3 for the partial oxidation of ethanol to acetaldehyde, *J. Phys. Chem. C.*, **112**(5), pp. 1612–1620.
7. Rosenholm, J.M., C. Sahlgren and M. Linden (2010). Towards multifunctional, targeted drug delivery systems using mesoporous silica nanoparticles: opportunities & challenges, *Nanoscale*, **2**(10), pp. 1870–1883.
8. Gao, C., and S. Che (2010). Organically functionalized mesoporous silica by co-structure-directing route, *Adv. Funct. Mater.*, **20**(17), pp. 2750–2768.
9. Trewyn, B.G., et al. (2008). Biocompatible mesoporous silica nanoparticles with different morphologies for animal cell membrane penetration, *Chem. Eng. J.*, **137**(1), pp. 23–29.
10. Beck, J.S., et al. (1992). A new family of mesoporous molecular sieves prepared with liquid crystal templates, *J. Am. Chem. Soc.*, **114**(27), pp. 10834–10843.
11. Kresge, C.T., et al. (1992). Ordered mesoporous molecular sieves synthesized by a liquid-crystal template mechanism, *Nature*, **359**(6397), pp. 710–712.
12. Goltner, C.G. and M. Antonietti (1997). Mesoporous materials by templating of liquid crystalline phases, *Adv. Mater.*, **9**(5), pp. 431–436.
13. Goltner, C.G., et al. (1998). Nanoporous silica from amphiphilic block copolymer (ABC) aggregates: control over correlation and architecture of cylindrical pores, *Chem. Commun.*, (21), pp. 2287–2288.

14. Fowler, C.E., et al. (2001). Nanoscale materials with mesostructured interiors, *Adv. Mater.*, **13**(9), pp. 649–652.

15. Sadasivan, S., et al. (2002). Nucleation of MCM-41 nanoparticles by internal reorganization of disordered and nematic-like silica–surfactant clusters, *Angew. Chem. Int. Ed.*, **41**(12), pp. 2151–2153.

16. Nandiyanto, A.B.D., et al. (2009). Synthesis of spherical mesoporous silica nanoparticles with nanometer-size controllable pores and outer diameters, *Microporous Mesoporous Mater.*, **120**(3), pp. 447–453.

17. Zhao, D., et al. (1998). Nonionic triblock and star diblock copolymer and oligomeric surfactant syntheses of highly ordered, hydrothermally stable, mesoporous silica structures, *J. Am. Chem. Soc.*, **120**(24), pp. 6024–6036.

18. Zhao, D., et al. (1998). Triblock copolymer syntheses of mesoporous silica with periodic 50 to 300 angstrom pores, *Science*, **279**(5350), pp. 548–552.

19. Yang, P., et al. (1998). Triblock-copolymer-directed syntheses of large-pore mesoporous silica fibers, *Chem. Mater.*, **10**(8), pp. 2033–2036.

20. Göltner, C.G., et al. (1999). Nanoporous silicas by casting the aggregates of amphiphilic block copolymers: the transition from cylinders to lamellae and vesicles, *Adv. Mater.*, **11**(5), pp. 395–398.

21. Groenewolt, M., et al. (2005). Polyisobutylene-block-poly(ethylene oxide) for robust templating of highly ordered mesoporous materials, *Adv. Mater.*, **17**(9), pp. 1158–1162.

22. Hayward, R.C., B.F. Chmelka and E.J. Kramer (2005). Crosslinked poly(styrene)-block-poly(2-vinylpyridine) thin films as swellable templates for mesostructured silica and titania, *Adv. Mater.*, **17**(21), pp. 2591–2595.

23. Kim, J.K., et al. (2010). Functional nanomaterials based on block copolymer self-assembly, *Prog. Polym. Sci.*, **35**(11), pp. 1325–1349.

24. Dai, H., et al. (2012). A green process for the synthesis of controllable mesoporous silica materials, *Microporous Mesoporous Mater.*, **147**, pp. 281–285.

25. Xu, F., et al. (2007). Hierarchically assembled porous ZnO nano-particles: synthesis, surface energy, and photocatalytic activity, *Chem. Mater.*, **19**(23), pp. 5680–5686.

26. Suteewong, T., et al. (2010). Ordered mesoporous silica nanoparticles with and without embedded iron oxide nanoparticles: structure evolution during synthesis, *J. Mater. Chem.*, **20**(36), pp. 7807–7814.

27. Suteewong, T., et al. (2011). Highly aminated mesoporous silica nanoparticles with cubic pore structure, *J. Am. Chem. Soc.*, **133**(2), pp. 172–175.

28. Hoshikawa, Y., et al. (2009). Mesoporous silica nanoparticles with remarkable stability and dispersibility for antireflective coatings, *Chem. Mater.*, **22**(1), pp. 12–14.

29. Imhof, A., and D.J. Pine (1997). Ordered macroporous materials by emulsion templating, *Nature*, **389**(6654), pp. 948–951.

30. Nermeen, N., et al. (2011). Mesoporous CeO_2 nanoparticles synthesized by an inverse miniemulsion technique and their catalytic properties in methane oxidation, *Nanotechnology*, **22**(13), pp. 135606.

31. Rossmanith, R., et al. (2008). Porous anatase nanoparticles with high specific surface area prepared by miniemulsion technique, *Chem. Mater.*, **20**(18), pp. 5768–5780.

32. Platschek, B., A. Keilbach and T. Bein (2011). Mesoporous structures confined in anodic alumina membranes, *Adv. Mater.*, **23**(21), pp. 2395–2412.

33. Chernikov, A., et al. (2011). Photoluminescence properties of ordered mesoporous ZnO, *J. Phys. Chem. C.*, **115**(5), pp. 1375–1379.

34. Kim, Y.J., et al. (2009). Formation of highly efficient dye-sensitized solar cells by hierarchical pore generation with nanoporous TiO_2 spheres, *Adv. Mater.*, **21**(36), pp. 3668–3673.

35. Chen, D., et al. (2009). Mesoporous anatase TiO_2 beads with high surface areas and controllable pore sizes: a superior candidate for high-performance dye-sensitized solar cells, *Adv. Mater.*, **21**(21), pp. 2206–2210.

36. Ye, J., et al. (2011). Nanoporous anatase TiO_2 mesocrystals: additive-free synthesis, remarkable crystalline-phase stability, and improved lithium insertion behavior, *J. Am. Chem. Soc.*, **133**(4), pp. 933–940.

37. Pan, J.H., et al. (2011). Controllable synthesis of mesoporous F-TiO_2 spheres for effective photocatalysis, *J. Mater. Chem.*, **21**(30), pp. 11430–11438.

38. Periyat, P., et al. (2010). Rapid microwave synthesis of mesoporous TiO_2 for electrochromic displays, *J. Mater. Chem.*, **20**(18), pp. 3650–3655.

39. Xi, G., et al. (2008). Synthesis of crystalline microporous SnO_2 via a surfactant-assisted microwave heating method: a general and rapid method for the synthesis of metal oxide nanostructures, *J. Phys. Chem. C.*, **112**(31), pp. 11645–11649.

40. Li, Z., et al. (2011). Porous SnO_2 nanospheres as sensitive gas sensors for volatile organic compounds detection, *Nanoscale*, **3**(4), pp. 1646–1652.

41. Yadav, R., A. Pandey, and S. Sanjay (2007). ZnO porous structures synthesized by CTAB-assisted hydrothermal process, *Struct. Chem.*, **18**(6), pp. 1001–1004.

42. Chen, D., et al. (2010). Synthesis of monodisperse mesoporous titania beads with controllable diameter, high surface areas, and variable pore diameters (14–23 nm), *J. Am. Chem. Soc.*, **132**(12), pp. 4438–4444.

43. Mukherjee, B., C. Karthik, and N. Ravishankar (2009). Hybrid sol-gel combustion synthesis of nanoporous anatase, *J. Phys. Chem. C.*, **113**(42), pp. 18204–18211.

44. Mukherjee, B., B. Viswanath, and N. Ravishankar (2010). Functional nanoporous structures by partial sintering of nanorod assemblies, *J. Phys. D: Appl. Phys.*, **43**(45), pp. 455301.

45. Meakin, P., and F. Family (1988). Structure and kinetics of reaction-limited aggregation, *Phys. Rev. A*, **38**(4), pp. 2110–2123.

46. Viswanath, B., et al. (2009). Nanoporous Pt with high surface area by reaction-limited aggregation of nanoparticles, *Langmuir*, **25**(5), pp. 3115–3121.

47. Halder, A., et al. (2011). Porous, catalytically active palladium nanostructures by tuning nanoparticle interactions in an organic medium, *Nanoscale*, **3**(2), pp. 725–730.

48. Anumol, E.A., et al. (2011). Nanoporous alloy aggregates: synthesis and electrocatalytic activity, *J. Mater. Chem.*, **21**(24), pp. 8721–8726.

49. Zhou, Y., and M. Antonietti (2003). Synthesis of very small TiO_2 nanocrystals in a room-temperature ionic liquid and their self-assembly toward mesoporous spherical aggregates, *J. Am. Chem. Soc.*, **125**(49), pp. 14960–14961.

50. Shao, W., et al. (2010). Highly efficient dye-sensitized solar cells by using a mesostructured anatase TiO_2 electrode with high dye loading capacity, *Ind. Eng. Chem. Res.*, **49**(19), pp. 9111–9116.

51. Roy, P., S. Berger and P. Schmuki (2011). TiO_2 nanotubes: synthesis and applications, *Angew. Chem. Int. Ed.*, **50**(13), pp. 2904–2939.

52. Yeonmi, S., and L. Seonghoon (2009). A freestanding membrane of highly ordered anodic ZrO_2 nanotube arrays, *Nanotechnology*, **20**(10), pp. 105301.

53. Li, A.P., et al. (1998). Hexagonal pore arrays with a 50–420 nm interpore distance formed by self-organization in anodic alumina, *J. Appl. Phys.*, **84**(11), pp. 6023–6026.
54. Guo, L., et al. (2009). Synthesis and growth mechanism of zirconia nanotubes by anodization in electrolyte containing Cl^{-}, *J. Solid State Electrochem.*, **13**(9), pp. 1321–1326.
55. Lu-Ning Wang, and J.-L. Luo (2011). Fabrication and mechanical properties of anodized zirconium dioxide nanotubular arrays, *J. Phys. D: Appl. Phys.*, **44**, pp. 075301(8pp).
56. Yu, X., et al. (2006). Formation of nanoporous titanium oxide films on silicon substrates using an anodization process, *Nanotechnology*, **17**, pp. 808–814.
57. Thamida, S.K. and H.-C. Chang (2002). Nanoscale pore formation dynamics during aluminum anodization, *Chaos*, **12**(1), pp. 240–251.
58. O'Sullivan, J.P., and G.C. Wood (1970). The morphology and mechanism of formation of porous anodic films on aluminium, Proceedings of the Royal Society of London. Series A, *Mathemat. Phys. Sci.*, **317**(1531), pp. 511–543.
59. Hebert, K.R., et al. (2012). Morphological instability leading to formation of porous anodic oxide films, *Nat. Mater.*, **11**, pp. 162-166.
60. Slowing, I.I., et al. (2007). Mesoporous silica nanoparticles for drug delivery and biosensing applications, *Adv. Funct. Mater.*, **17**(8), pp. 1225–1236.
61. Hu, Y., et al. (2012). 3D cubic mesoporous silica microsphere as a carrier for poorly soluble drug carvedilol, *Microporous Mesoporous Mater.*, **147**(1), pp. 94–101.
62. Mellaerts, R., et al. (2007). Enhanced release of itraconazole from ordered mesoporous SBA-15 silica materials, *Chem. Commun.*, (13), pp. 1375–1377.
63. Zhao, Y., et al. (2009). Mesoporous silica nanoparticle-based double drug delivery system for glucose-responsive controlled release of insulin and cyclic AMP, *J. Am. Chem. Soc.*, **131**(24), pp. 8398–8400.
64. Torney, F., et al. (2007). Mesoporous silica nanoparticles deliver DNA and chemicals into plants, *Nat. Nano*, **2**(5), pp. 295–300.
65. Clark, J.H., D.J. Macquarrie and S.J. Tavener (2006). The application of modified mesoporous silicas in liquid phase catalysis, *Dalton Trans.*, (36), pp. 4297–4309.

66. Sahoo, S., A. Bordoloi, and S. Halligudi (2011). Ordered mesoporous silica as supports in the heterogeneous asymmetric catalysis, *Catal. Surv. Asia*, **15**(3), pp. 200–214.

67. Ching, S., et al. (2011). Self-assembly of manganese oxide nanoparticles and hollow spheres. Catalytic activity in carbon monoxide oxidation, *Chem. Commun.*, **47**(29), pp. 8286–8288.

68. Han, Y., et al. (2011). Porous SnO_2 nanowire bundles for photocatalyst and Li ion battery applications, *Cryst. Eng. Comm.*, **13**, pp. 3506–3510.

69. Liu, J., et al. (2010). Anisotropic Co_3O_4 porous nanocapsules toward high-capacity Li-ion batteries, *J. Mater. Chem.*, **20**, pp. 1506–1510.

70. Shim, H.-W., et al. (2011). Highly reversible lithium storage in bacillus subtilis-directed porous Co_3O_4 nanostructures, *ACS Nano*, **5**(1), pp. 443–449.

71. Liu, J., et al. (2010). A novel coral-like porous SnO_2 hollow architecture: biomimetic swallowing growth mechanism and enhanced photovoltaic property for dye-sensitized solar cell application, *Chem. Commun.*, **46**(3), pp. 472–474.

72. Yang, H., and D. Zhao (2005). Synthesis of replica mesostructures by the nanocasting strategy, *J. Mater. Chem.*, **15**(12), pp. 1217–1231.

73. Lee, J., S. Han and T. Hyeon (2004). Synthesis of new nanoporous carbon materials using nanostructured silica materials as templates, *J. Mater. Chem.*, **14**(4), pp. 478–486.

74. Bian, S.-W., et al. (2010). A template-free, thermal decomposition method to synthesize mesoporous MgO with a nanocrystalline framework and its application in carbon dioxide adsorption, *J. Mater. Chem.*, **20**(39), pp. 8705–8710.

75. Shan, Z.W., et al. (2008). Ultrahigh stress and strain in hierarchically structured hollow nanoparticles, *Nat. Mater.*, **7**(12), pp. 947–952.

76. Wang, H., et al. (2011). Single-crystalline rutile TiO_2 hollow spheres: room-temperature synthesis, tailored visible-light-extinction, and effective scattering layer for quantum dot-sensitized solar cells, *J. Am. Chem. Soc.*, **133**(47), pp. 19102–19109.

77. Yang, Z., C.-K. Chiang and H.-T. Chang (2008). Synthesis of fluorescent and photovoltaic Cu_2O nanocubes, *Nanotechnology*, **19**, pp. 025604 (7pp).

78. Cabot, A., et al. (2009). Magnetic domains and surface effects in hollow maghemite nanoparticles, *Phys. Rev. B*, **79**(9), pp. 094419.

79. Sasidharan, M., et al. (2011). Synthesis, characterization and application for lithium-ion rechargeable batteries of hollow silica nanospheres, *J. Mater. Chem.*, **21**(36), pp. 13881–13888.

80. Hu, J., et al. (2011). Fabrication and application of inorganic hollow spheres, *Chem. Soc. Rev.*, **40**(11), pp. 5472–5491.

81. Zhong, Z., et al. (2000). Preparation of mesoscale hollow spheres of TiO_2 and SnO_2 by templating against crystalline arrays of Polystyrene Beads, *Adv. Mater.*, **12**(3), pp. 206–209.

82. Jiang, P., J.F. Bertone and V.L. Colvin (2001). A lost-wax approach to monodisperse colloids and their crystals, *Science*, **291**(5503), pp. 453–457.

83. Agrawal, M., et al. (2009). A facile approach to fabrication of ZnO–TiO_2 hollow spheres, *Chem. Mater.*, **21**(21), pp. 5343–5348.

84. Deng, Z., et al. (2008). A facile method to fabricate ZnO hollow spheres and their photocatalytic property, *J. Phys. Chem. B.*, **112**(1), pp. 16–22.

85. Liu, D., and K. Nakashima (2009). Synthesis of hollow metal oxide nanospheres by templating polymeric micelles with core–shell-corona architecture, *Inorg. Chem.*, **48**(9), pp. 3898–3900.

86. Yang, M., J. Ma, C. Zhang, Z. Yang and Y. Lu (2005). General synthetic route toward functional hollow spheres with double-shelled structures 13, *Angew. Chem. Int. Ed.*, **44**(41), pp. 6727–6730.

87. Yang, J., J.U. Lind and W.C. Trogler (2008). Synthesis of hollow silica and titania nanospheres, *Chem. Mater.*, **20**(9), pp. 2875–2877.

88. Park, J.-U., et al. (2011). Facile synthetic route for thickness and composition tunable hollow metal oxide spheres from silica-templated coordination polymers, *Adv. Mater.*, **23**(28), pp. 3161–3164.

89. Joo, J.B., et al. (2011). Mesoporous anatase titania hollow nanostructures though silica-protected calcination, *Adv. Funct. Mater.*, Volume 22, Issue 1, pages 166–174.

90. Strandwitz, N.C., and G.D. Stucky (2009). Hollow microporous cerium oxide spheres templated by colloidal silica, *Chem. Mater.*, **21**(19), pp. 4577–4582.

91. Chen, Y., et al. (2009). Hollow/rattle-type mesoporous nanostructures by a structural difference-based selective etching strategy, *ACS Nano*, **4**(1), pp. 529–539.

92. Lou, X.W., and L.A. Archer (2008). A general route to nonspherical anatase TiO_2 hollow colloids and magnetic multifunctional particles, *Adv. Mater.*, **20**(10), pp. 1853–1858.

93. Piao, Y., et al. (2008). Wrap-bake-peel process for nanostructural transformation from [beta]-FeOOH nanorods to biocompatible iron oxide nanocapsules, *Nat. Mater.*, **7**(3), pp. 242–247.

94. Wu, H., et al. (2011). A hollow-core, magnetic, and mesoporous double-shell nanostructure: in situ decomposition/reduction synthesis, bioimaging, and drug-delivery properties, *Adv. Funct. Mater.*, **21**(10), pp. 1850–1862.

95. Bang, J.H., and K.S. Suslick (2007). Sonochemical synthesis of nanosized hollow hematite, *J. Am. Chem. Soc.*, **129**(8), pp. 2242–2243.

96. Guo, Z., et al. (2008). Template synthesis, organic gas-sensing and optical properties of hollow and porous In_2O_3 nanospheres, *Nanotechnology*, **19**, pp. 345704 (9pp).

97. Cheng, S., et al. (2009). Soft-template synthesis and characterization of ZnO_2 and ZnO hollow spheres, *J. Phys. Chem. C.*, **113**(31), pp. 13630–13635.

98. Tan, B., et al. (2005). Fluorinated-surfactant-templated synthesis of hollow silica particles with a single layer of mesopores in their shells, *Adv. Mater.*, **17**(19), pp. 2368–2371.

99. Li, J., et al. (2010). Interfacially controlled synthesis of hollow mesoporous silica spheres with radially oriented pore structures, *Langmuir*, **26**(14), pp. 12267–12272.

100. He, T., et al. (2006). Co_3O_4 nanoboxes: surfactant-templated fabrication and microstructure characterization, *Adv. Mater.*, **18**(8), pp. 1078–1082.

101. Gyger, F., et al. (2010). Nanoscale SnO_2 hollow spheres and their application as a gas-sensing material, *Chem. Mater.*, **22**(16), pp. 4821–4827.

102. Khanal, A., et al. (2007). Synthesis of silica hollow nanoparticles templated by polymeric micelle with core–shell–corona structure, *J. Am. Chem. Soc.*, **129**(6), pp. 1534–1535.

103. Sun, Q., et al. (2003). The formation of well-defined hollow silica spheres with multilamellar shell structure, *Adv. Mater.*, **15**(13), pp. 1097–1100.

104. Yin, Y., et al. (2004). Formation of hollow nanocrystals through the nanoscale Kirkendall effect, *Science*, **304**(5671), pp. 711–714.

105. Jin fan, H., et al. (2006). Monocrystalline spinel nanotube fabrication based on the Kirkendall effect, *Nat. Mater.*, **5**(8), pp. 627–631.

106. Cabot, A., et al. (2007). Vacancy coalescence during oxidation of iron nanoparticles, *J. Am. Chem. Soc.*, **129**(34), pp. 10358–10360.

107. Hafsa, K., et al. (2011). Chemically synthesized hollow nanostructures in iron oxides, *Nanotechnology*, **22**(26), pp. 265605.

108. Huang, J., et al. (2009). One-dimensional chainlike arrays of Fe_3O_4 hollow nanospheres synthesized by aging iron nanoparticles in aqueous solution, *J. Phys. Chem. C.*, **113**(28), pp. 12067–12071.

109. Peng, S., and S. Sun (2007). Synthesis and characterization of monodisperse hollow Fe_3O_4 nanoparticles, *Angew. Chem. Int. Ed.*, **46**(22), pp. 4155–4158.

110. An, K., et al. (2008). Synthesis of uniform hollow oxide nanoparticles through nanoscale acid etching, *Nano Lett.*, **8**(12), pp. 4252–4258.

111. Hung, L.-I., et al. (2010). Room-temperature formation of hollow Cu_2O nanoparticles, *Adv. Mater.*, **22**(17), pp. 1910–1914.

112. Railsback, J.G., et al. (2010). Size-dependent nanoscale Kirkendall effect during the oxidation of nickel nanoparticles, *ACS Nano*, **4**(4), pp. 1913–1920.

113. Lou, X.W., et al. (2006). Template-free synthesis of SnO_2 hollow nanostructures with high lithium storage capacity, *Adv. Mater.*, **18**(17), pp. 2325–2329.

114. Ho, S.Y., A.S.W. Wong and G.W. Ho (2008). Controllable porosity of monodispersed tin oxide nanospheres via an additive-free chemical route, *Cryst. Growth Des.*, **9**(2), pp. 732–736.

115. Yin, X.M., et al. (2010). One-step synthesis of hierarchical SnO_2 hollow nanostructures via self-assembly for high power lithium ion batteries, *J. Phys. Chem. C.*, **114**(17), pp. 8084–8088.

116. Cao, C.-Y., et al. (2010). Ceria hollow nanospheres produced by a template-free microwave-assisted hydrothermal method for heavy metal Ion removal and catalysis, *J. Phys. Chem. C.*, **114**(21), pp. 9865–9870.

117. Yang, Z., et al. (2009). Fabrication of monodisperse CeO_2 hollow spheres assembled by nano-octahedra, *Cryst. Growth Des.*, **10**(1), pp. 291–295.

118. Li, J., and H.C. Zeng (2007). Hollowing Sn-doped TiO_2 nanospheres via Ostwald ripening, *J. Am. Chem. Soc.*, **129**(51), pp. 15839–15847.

119. Yang, H.G., and H.C. Zeng (2004). Preparation of hollow anatase TiO_2 nanospheres via Ostwald ripening, *J. Phys. Chem. B.*, **108**(11), pp. 3492–3495.

120. Zhao, W., et al. (2009). One-step template-free synthesis of $ZnWO_4$ hollow clusters, *J. Mater. Sci.*, **44**(12), pp. 3082–3087.

121. Li, X., et al. (2008). Polymer-induced generation of anatase TiO_2 hollow nanostructures, *Microporous Mesoporous Mater.*, **112**(1–3), pp. 641–646.

122. Chang, Y., J.J. Teo and H.C. Zeng (2005). Formation of colloidal CuO nanocrystallites and their spherical aggregation and reductive transformation to hollow Cu_2O nanospheres, *Langmuir*, **21**(3), pp. 1074–1079.

123. Tian, X., et al. (2010). Nearly monodisperse ferroelectric $BaTiO_3$ hollow nanoparticles: size-related solid evacuation in Ostwald-ripening-induced hollowing process, *Cryst. Growth Des.*, **10**(9), pp. 3990–3995.

124. Yang, H.G., and H.C. Zeng (2004). Self-construction of hollow SnO_2 octahedra based on two-dimensional aggregation of nanocrystallites, *Angew. Chem. Int. Ed.*, **43**(44), pp. 5930–5933.

125. Teng, F., et al. (2008). Formation of hollow NiO single crystals and Ag/NiO flowers, *Mater. Res. Bull.*, **43**(12), pp. 3562–3569.

126. Yu, Q., et al. (2011). Hydrothermal synthesis of hollow silica spheres under acidic conditions, *Langmuir*, **27**(11), pp. 7185–7191.

127. Song, Y., et al. (2010). Preparation of CeO_2 hollow spheres via a surfactant-assisted solvothermal route, *J. Mater. Sci.*, **45**(15), pp. 4158–4162.

128. Cheng, K., et al. (2009). Porous hollow Fe_3O_4 nanoparticles for targeted delivery and controlled release of cisplatin, *J. Am. Chem. Soc.*, **131**(30), pp. 10637–10644.

129. Zeng, H., et al. (2008). ZnO-based hollow nanoparticles by selective etching: elimination and reconstruction of metal-semiconductor interface, improvement of blue emission and Photocatalysis, *ACS Nano*, **2**(8), pp. 1661–1670.

130. Anumol, E.A., et al. (2010). Surface diffusion driven nanoshell formation by controlled sintering of mesoporous nanoparticle aggregates, *Nanoscale*, **2**, pp. 1423–1425.

131. Mukherjee, R., et al. (2011). Thermal stability of spherical nanoporous aggregates and formation of hollow structures by sintering-A phase-field study, *ACS Nano*, **5**(4), pp. 2700–2706.

132. Yoon, S., and A. Manthiram (2011). Hollow core–shell mesoporous TiO_2 spheres for lithium ion storage, *J. Phys. Chem. C.*, **115**(19), pp. 9410–9416.

Chapter 5

Doped Tin Oxide Nanomaterials for Chlorine and Hydrogen Gas Detection

Allen Chaparadza,[a] Hoang Tran,[b] and Shankar B. Rananavare[b]

[a]*Department of Chemistry, Marquette University, Milwaukee, Wisconsin 53201, USA*
[b]*Department of Chemistry, Portland State University, SRTC Building, Portland, Oregon 97206, USA*

ranavas@pdx.edu

5.1 Introduction

Sensors have a great influence in many areas such as environmental monitoring, domestic safety, public security, and industrial quality control. Gas sensors are most commonly deployed to detect not only innocuous gases such as water and oxygen in automobiles but also life-threatening gases such as carbon monoxide (at home), chlorine (a low-cost weapon of mass destruction (WMD)), etc. The modern semiconductor industry relies very heavily on maintaining a safe work environment during manufacturing by employing a vast array of sensors capable of detecting gases such as phosphine,

Oxide Nanostructures: Growth, Microstructures, and Properties
Edited by Avanish Kumar Srivastava

ISBN 978-981-4411-35-6 (Hardcover), 978-981-4411-36-3 (eBook)
www.panstanford.com

hydrogen fluorides, and hydrogen sulfide. In addition, the need for detecting air-borne pathogens and potentially monitoring spread of contagious diseases remains at the forefront of modern gas sensor research. Due to this huge application range, the need for cheap, small, low-power-consuming and reliable solid-state gas sensors has grown over the years and has triggered an intense research activity worldwide to overcome metal oxide sensors drawbacks, summed up in improving the well-known "3S": Sensitivity, Selectivity, and Stability [1].

A commonly exploited sensing mechanism of a gas sensor is the modulation of electrical conductivity of semiconductor due to surface reactions such as oxidation or reduction caused by exposure to different gases. These surface reactions depend on the availability active centers and the defects on the surface layer of materials; the sensor response is usually determined by the surface-to-volume ratio of materials [2]. Consequently, compared to conventional materials such as bulk or thin films used in gas sensors applications, zero- and one-dimensional nanomaterials such as nanoparticles, nanowires, nanorods, nanoribbons, and nanotubes have a high gas sensor response due to their high surface-to-volume ratios. Higher surface-to-volume ratio implies that large surface activities, reactions at grain boundaries [3], which can lead to complete depletion of carriers in grains changing electrical transport characteristics, i.e., conductance [4]. Currently gas sensors of nanostructured metal oxide semiconductors have been widely favored because of their small dimension, low power consumption, low cost, on-line operation, environmental stability, corrosion resistance and high compatibility with microelectronic processing [5]. Among the gas sensors, tin oxide (SnO_2)-based sensors are most common, although highly specific and sensitive SnO_2 sensors are not yet available. Attempts have been made to fine-tune the sensor selectivity, sensitivity, and stability by varying the SnO_2 crystal structure and morphology, dopants, contact geometries, operation temperature, or mode of operation, etc. [6]. Traditionally, these sensors have employed a relatively high operational temperature of about 200°C. This high-temperature operation improves the response time, yet it has a drawback of degrading their energy efficiency [1]. Oxidizing species increase the resistance of n-type semiconductors while reducing species cause an opposite effect. A complementary response is expected for a p-type SnO_2.

The preparation of tin oxide films can be classified into two categories, namely vapor phase deposition and solution-based crystal growth. Examples of vapor phase deposition techniques are laser ablation [7], thermal evaporation [8], VLS growth [9], carbo-thermal reduction [4,10], and rapid oxidation [11]. These techniques produce higher-quality nanostructures but are relatively expensive when compared to solution-based techniques. Solution-based growth techniques, such as sol-gel synthesis, allow control of the crystal growth process through control of nucleation and seed ripening processes [12]. This method of synthesis permits control over particle size and doping and produce high volume of materials at lower temperatures. Crystal morphology and microscopic phases can be controlled by simple calcinations processes. Such solution processed materials can be spin-cast, dip-coated, silk printed [13] or compressed to produce functioning sensors inexpensively [14].

In this chapter, we offer a limited review of our work on gas sensing using doped tin oxide nanostructures to enhance chemical selectivity by specific dopant–analyte binding interactions. Section 5.2 provides the synthesis and characterization details of the doped nanostructures. For device fabrication we have mostly used thick films prepared by simple powder compression (i.e., pellets) techniques. Section 5.3 provides studies that contrast the carrier transport mechanisms in n-versus p-doped SnO_2. In Section 5.4, we highlight the applications, that is, detection of Cl_2 and H_2 gases. The conclusions and exciting prospects for extending these sensors to the next generation of analytes appear in Section 5.5.

5.2 Synthesis and Characterization of Nanomaterial-Based Devices for Chlorine and Hydrogen Sensing

5.2.1 Preparation of Li(p-Type) and Sb (n-Type)-Doped SnO_2 Nanoparticles

For synthesis of p-type nanoparticles, $SnCl_4\cdot 5H_2O$ was used as the source of tin and LiCl (>95%, Fischer Scientific) as the source of p-type dopant. For synthesis of undoped tin oxide nanoparticles,

$SnCl_4 \cdot 5H_2O$ (98% Fischer Scientific) and H_2O were mixed in a 1:1 ratio and few drops of hydrochloric acid added until pH was 0.4. The pH of the resulting solution was adjusted using 30% NH_4OH under stirring and then resulting precipitate was filtered. The precipitate was washed with water until Cl^- free ($AgNO_3$ was used to test for the presence of Cl^- ions). It was initially dried at 100°C for 2 h followed by calcination at 480°C for 3 h. For Li-doped SnO_2 varying amounts of lithium chloride were added to the initial solution ranging from 5 wt% to 20 wt% and then preceded as above. For synthesis of n-type nanoparticles LiCl was replaced by $SbCl_3$ (98% pure from Fischer Scientific). Note that the actual level of doping in nanoparticles/nanowires could be different than the concentration of dopant atoms used during synthesis.

5.2.2 n-Doped Tin Oxide Nanowires

The precursors tin nanoparticles were synthesized by reducing a thoroughly mixed solution of 50 mL of 0.05 M $SnCl_4 \cdot 5H_2O$ and 0.5 g of 1,10-phenanthroline with 100 mL of 0.1 M $NaBH_4$ aqueous solution. The reducing solution was introduced drop wise to synthesize phenanthroline capped Sn nanoparticles [15]. Nanoparticles were separated from the reaction mixture by centrifuging (10,000 rpm for 15 min) after 2 h of reaction time. The precipitate was dried at 50°C for 2 h on a hotplate. 0.2 g of 1,10-phenanthroline capped Sn nanoparticles powder was mixed with a mixture of 0.4 g of NaCl and 0.6 g of KCl, grounded into a fine powder, and heated at 750°C for 2 h in a furnace. The molten mixture was then slowly cooled to room temperature. The solidified product was washed several times with water, to remove KCl and NaCl and dried in air at 100°C overnight. For n-doping, varying amount of antimony chloride (1–5% atomic weight with respect to Sn) was added to the initial solution of $SnCl_4 \cdot 5H_2O$ and 1,10-phenanthroline. The rest of the synthetic procedure was identical to the synthesis of undoped nanowires as described above.

5.2.3 p-Doped Tin Oxide Nanowires

The amounts of lithium chloride added to the initial reaction mixture ranged from 1 to 5 wt%. The precursors Sn nanoparticles coated with a thin layer of tin oxide were synthesized without capping agents by adding 0.35 g $SnCl_4 \cdot 5H_2O$ with 0.1 M $NaBH_4$ aqueous solution containing LiCl. The mixture turned black, indicating the formation of oxide coated Sn nanoparticles. The precipitate was collected by filtering and was left to dry in air over night. 0.2 g of the above precursor was mixed with 0.2 g of 1,10-phenanthroline, 0.31 g of LiCl and 0.6 g of KCl, grounded into a fine powder, and was kept for 2 h at temperatures above the melting point of the two salt mixtures. The molten mixture was then slowly cooled to room temperature with cooling rate of 4°C/minute. The products were collected and purified as in the case of undoped nanowires.

5.2.4 Characterization of Li- and Sb-Doped SnO_2

A battery of analytical techniques such as TEM, SEM, nuclear reaction analysis (NRA), nuclear magnetic resonance (NMR), X-ray diffraction (XRD) and conductivity measurements were used to characterize the nanomaterials. Precipitation pH was found to play an important role when it came to lithium doping. ^{7}Li nuclear reaction analysis showed that Li is not detected (i.e., <0.005 atomic % level) when the solution pH during synthesis is less than seven while above pH 7 relatively small amounts of Li is incorporated [14] (see Fig. 5.1 for a sample of NRA and NMR results confirming Li doping).

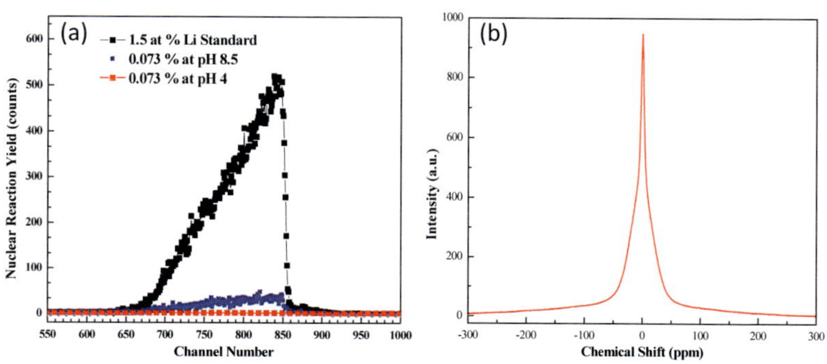

Figure 5.1 (a) NRA results for 0.073% Li-doped SnO_2 prepared at two different pH. Channel number is proportional to energy. The standard sample was made 1.5% LiCl in NaCl. (b) ^{7}Li NMR plot for Li-doped SnO_2 nanoparticles.

It is well known that Sn tends to hydrolyze in aqueous solutions forming charged and uncharged species $SnO(OH)_3^-$ at pH ≥ 8 and $SnO(OH)_2^0$ at pH < 7, respectively. Therefore lithium insertion into SnO_2 takes place possibly via the charge neutralization through the following scheme [16]:

$$SnO(OH)_3^- + Li^+ \rightarrow SnO(OH)_3Li \rightarrow \text{Li-doped } SnO_2$$

In acidic conditions, however, the absence of negatively charged tin complexes hinders lithium insertion. Figure 5.2a shows XRD results for Sb-doped SnO_2. The material showed sharp and well-defined peaks corresponding to the cassiterite structure of SnO_2 (JCPDS-XRD card No-211250) [17]. The presence of small amounts of Sb in the material did not affect the crystalline nature the of the SnO_2 particles. The TEM image shown in Fig. 5.2b reveals that the particles had a size around 15 nm with some agglomerated particles of varying sizes ranging between 20 and 30 nm as seen on the SEM image (Fig. 5.2c).

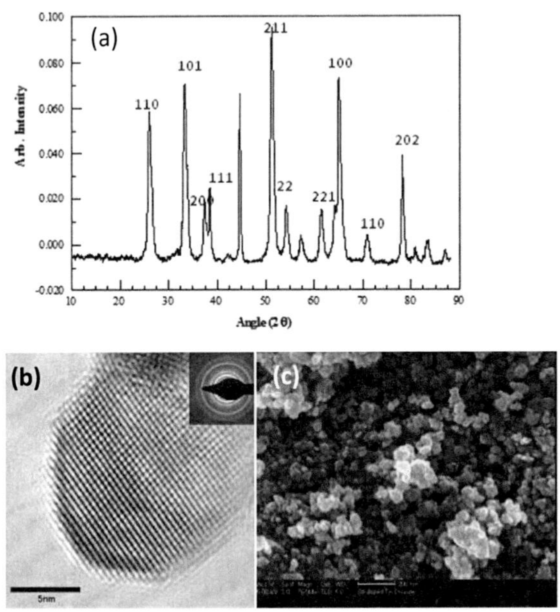

Figure 5.2 (a) XRD pattern of 0.1% Sb-doped SnO_2. (b) HRTEM of Sb-doped SnO_2 (insert shows the electron diffraction pattern). (c) SEM image of 0.1% Sb-doped SnO_2. Reprinted with permission from A. Chaparadza, S. B. Rananavare, *Nanotechnology*, **19**, 245501 (2008).

Figure 5.3a shows a typical SEM image of Sb-doped SnO_2 nanowires grown via the molten salt method. The doped nanowires have a broader size distribution than undoped nanowires, perhaps due to incorporation and morphological effects of dopant atoms. Also, note that the smaller diameter nanowires tended to fuse with one other. Since such structures are rarely observed in the nucleation and growth of nanowires mediated by the vapor–liquid–solid (VLS) mechanism, the mechanism of growth of these nanowires is likely to be governed by the surface energy of nanowires walls as controlled by the capping agent phenanthroline. Figure 5.3b is a high-resolution TEM image of the Sb-doped nanowire along with electron diffraction pattern (inset), showing single crystallinity of the nanowires. It can also be seen that the nanowires growth mechanism is neither VLS (no spherical catalyst droplet on the tip) nor solution–liquid–solid (SLS). The nanowires showed a very straight and smooth surface morphology suggesting a 1-D Ostwald-ripening mechanism [18] in which nanoparticles deposit onto, dissolve into each other, and extrude to form a 1D nanostructures.

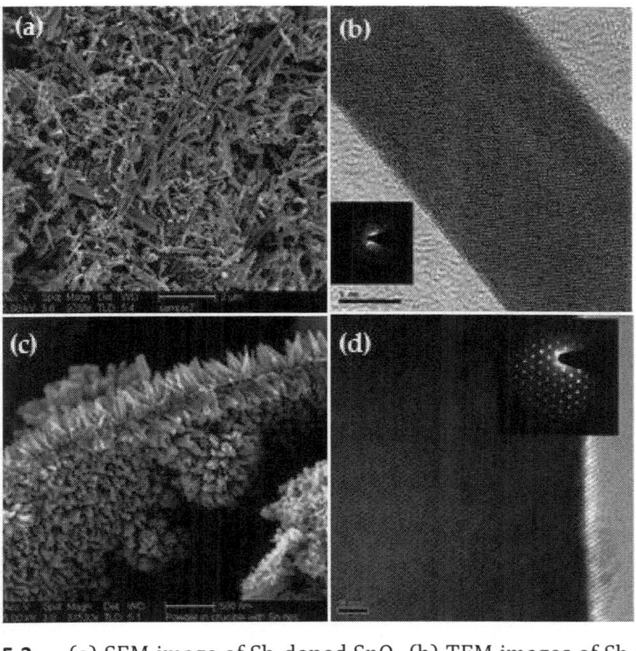

Figure 5.3 (a) SEM image of Sb-doped SnO_2 (b) TEM images of Sb-doped SnO_2 synthesized at 750°C (c) SEM image of Li-doped SnO_2 (d) TEM image of Li-doped SnO_2 synthesized at 480°C.

Figure 5.3c,d shows SEM and HR-TEM images of Li-doped nanowires synthesized at 480°C. The length of wires varied from 200 nm up to 1 mm while their diameter ranged from 50 nm to hundreds of nanometers. Like Sb-doped nanowires, whisker-like nanorods with smaller diameters and shorter lengths were also observed along with the tendency of the nanowires to fuse. Thus, preformed nanoparticles transformed into nanowires through recrystallization from alkali halide salts exhibited similar characteristics regardless of the nature (n or p type) of the dopant. The morphology of nanowires structure was influenced by a variety of factors, such as precursors, growth temperature, and the moisture content of the grounded mixture before the recrystallization step. Drying the pre-synthesized powder to remove moisture gave a better yield of nanowires and the wires had more uniform dimensional distributions. High-resolution TEM of p-doped SnO_2 nanowires shown in Fig. 5.3d indicated single crystallinity and this was further confirmed by electron diffraction shown in the inset of Fig. 5.3d. The TEM data showed that an inter-planar spacing is about 0.34 nm, which corresponds to the (110) plane of the cassiterite crystal structure of SnO_2.

5.3 Conduction Mechanisms in n- and p-Doped Nanoparticles

Figure 5.4 shows temperature-resistivity behaviors for the p-doped and n-doped SnO_2 thick films. The resistivity for Li-doped SnO_2 (p-type) follows an Efros–Shklovskii variable range hopping conduction mechanism at temperatures below 100°C with a cross over to 2D-Mott variable range hopping conduction at temperatures above 250°C. In a transition region between these two limiting behaviors, the direct current resistivity exhibits an anomalous temperature-independent plateau that depends on the level of Li dopant. We ascribe this anomaly to a carrier inversion phenomena wherein the majority carriers switch from holes to electrons due to Li ion expulsion from the crystalline core and creation of oxygen vacancies generated by loss of oxygen at elevated temperatures [19]. This carrier inversion limits the use of the p-type nanoparticles.

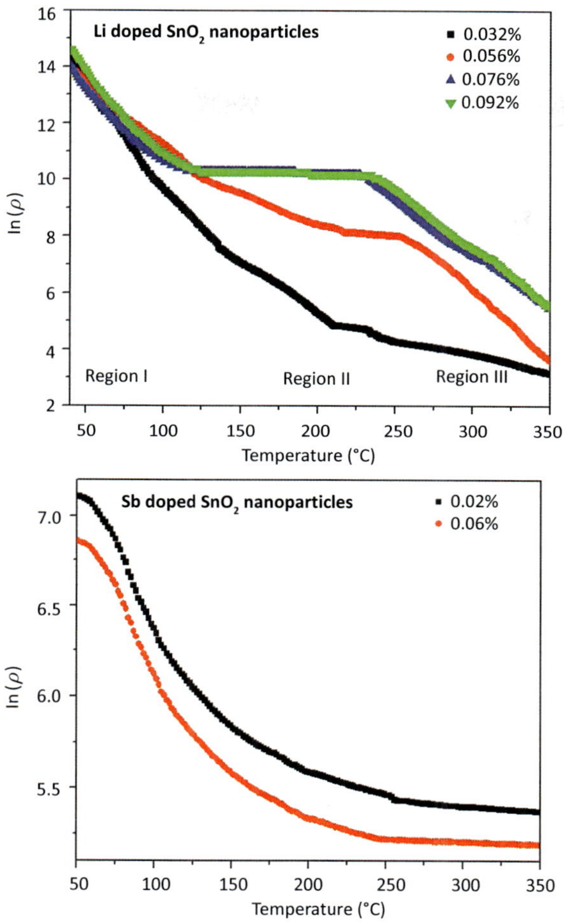

Figure 5.4 Resistivity versus temperature for Li-doped and Sb-doped SnO_2 nanoparticles.

Sb-doped SnO_2 resistivity follows a typical semiconductor conduction behavior, where the resistivity decreases with increasing temperature, which fits the Mott variable range hopping model. It is also important to point out that indefinite doping is not possible at the nanoscale level as chemical complexities arise due to solid–solid nanophase separation. This in turn can affect the carrier transport across nanoparticles/nanowires and hence the conductivity. Carrier concentration and spatial redistribution can also be affected by adsorption of oxidizing/reducing species, which forms the basis of the gas sensors discussed below.

5.4 Sensors for Cl_2 and H_2 Detection

To fabricate the sensor, a simple powder compression technique was employed. For the chlorine sensor, 2 g of Sb-doped SnO_2 were mixed with 0.5 mL of 2% ethyl silicate 40 in acidified ethanol. The magnitude of pressure applied for pellet formation is 5463 kg cm^{-2} to yield a 13 mm (diameter) pellet with a thickness of 800 µm. For the hydrogen sensor, very small amounts (0.1 mg) of palladium nanoparticles were physically compressed onto the surface of p-type tin oxide nanowires pellet. The pellet was heated at 450°C for 30 min. It was then mounted onto a glass substrate and copper contacts attached using conductive silver. The Fermi levels in semiconductive SnO_2 and metallic silver may not necessarily match and hence the possibility of forming Schottky barrier exists. To investigate such a possibility, *I–V* characteristics (shown in Fig. 5.5) were measured at room temperature and the Sb-doped SnO_2 films exhibited a linear dependence between current and voltage. This indicates an ohmic nature of the silver contact. Such linearity might also arise due to an essentially degenerately doped (high carrier concentration) SnO_2, where tunneling barrier becomes small and a quasi-ohmic contact is observed [14].

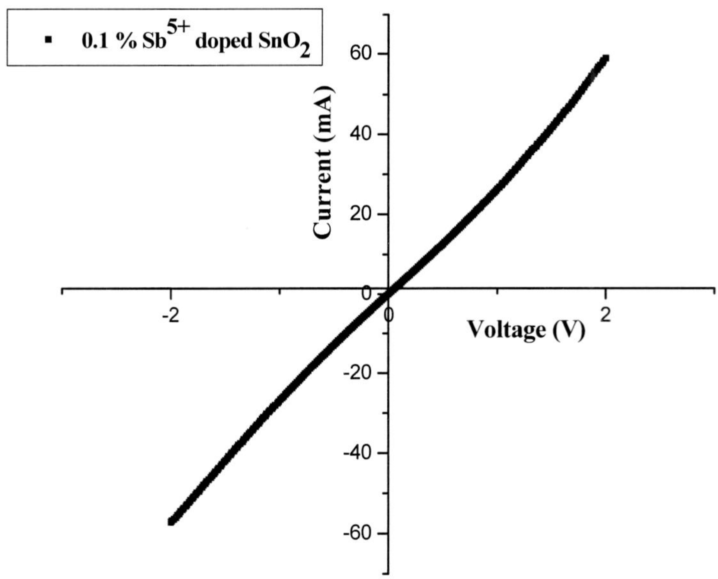

Figure 5.5 *I–V* characteristics of Sb-doped SnO_2 sensor.

5.4.1 Sb-Doped SnO_2 for Chlorine Detection

Typically, gas response (S) of a resistive sensor is defined as the ratio of the change in resistance of the sensor on exposure to the target gas to the original resistance in air; ($S = (R_{gas} - R_{air})/R_{air}$), where R_{gas} and R_{air} are the resistance of the sensor in a target gas medium and in air, respectively. Sensitivity could be defined as the first derivative of response with respect to the concentration of the analyte detected. The time taken for the sensor to attain 90% of the maximum increase in resistance on exposure to the target gas is defined as the response time and the time taken by the sensor to get back to 90% of the original resistance is considered as the recovery time. Selectivity or specificity is defined as the ability of a sensor to respond to certain gas in the presence of other gases. The selectivity coefficient can be calculated as; Selectivity = $S_{target\ gas}/S_o$, where $S_{target\ gas}$ is the response to a targeted gas and S_o is the response to any other non-target gas [14].

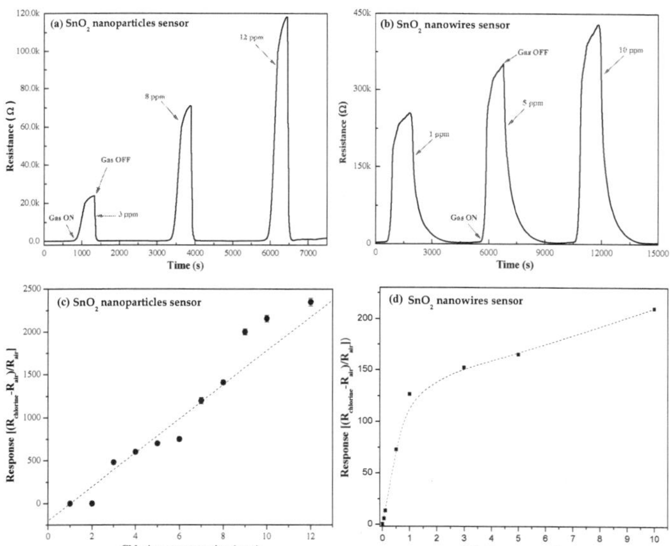

Figure 5.6 (a, b) Change in resistance on exposure to Cl_2 and recovery of nanoparticles and nanowires sensors at three different chlorine concentrations. (c, d) Response versus Cl_2 concentration for nanoparticles and nanowires. The sensors were refreshed by introducing fresh air and heating the sensor after exposure to Cl_2 gas. Reprinted with permission from A. Chaparadza, S. B. Rananavare, *Nanotechnology*, **19**, 245501 (2008).

In order to compare chlorine gas sensor response of different nanostructure materials, we fabricated sensors from SnO_2 nanoparticles and nanowires shown in Figs. 5.2c and 5.3a, respectively. The room temperature changes in resistance on exposure to Cl_2 and recovery results for the Sb-doped SnO_2 nanoparticles and nanowire-based sensors at different chlorine concentrations are shown in Fig. 5.6.

The corresponding temperature dependent response and recovery times for the nanoparticle-based sensor are shown in Fig. 5.7. The response of the sensor is quick (~60 s) but the recovery time is dependent on the temperature. It takes close to 15 min for the sensor to recover at room temperature after exposure to 3 ppm chlorine. The response and recovery times are temperature dependent as expected from analyte adsorption and desorption processes, respectively. Recovery time follows an Arrhenius type behavior (thermally activated behavior, with E_a = 59.5 kJ/mole) while response time does not. Response time exhibits a minimum around 45°C. A crude estimate of adsorption-desorption equilibrium constant is obtained from the ratio of recovery to response times which also shows a thermally activated behavior above room temperature with a free energy of activation of –87 kJ/mole.

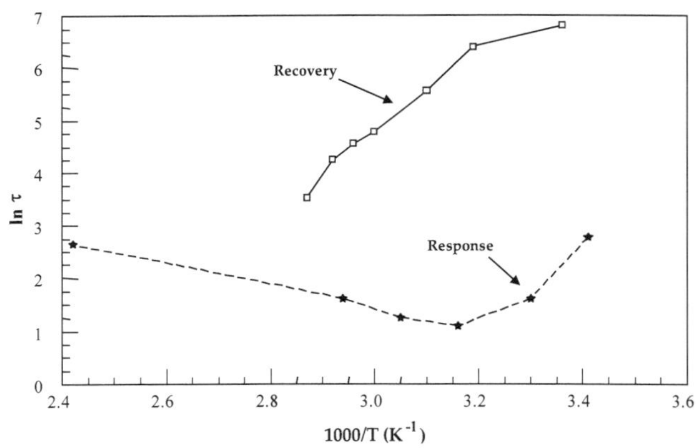

Figure 5.7 Recovery and response time as a function of temperature for nanoparticle-based sensor. A plot of ln τ (time) versus $1/T$ for recovery gives an activation energy of 59.5 kJ/mol. In low-temperature region ($1000/T$), the ratio of recovery/response times (K_{ads}) yielded the free energy of adsorption of –87 kJ/mol. Reprinted with permission from A. Chaparadza, S. B. Rananavare, *Nanotechnology*, **19**, 245501 (2008).

Remarkably, the data presented in Figs. 5.6 and 5.7 show high response (note $S \approx 500$ for 3 ppm of Cl_2 concentration) and rapid response time at much lower temperatures than currently known oxide sensors [14].

In an effort to push the detection limit of these types of sensors, we have also investigated the effect of using Sb-doped SnO_2 nanowires instead of nanoparticles. Figure 5.6b shows the response and recovery of the Sb-doped SnO_2 nanowire-based sensor at three different chlorine concentrations. From the data shown in Fig. 5.6, it can be seen that nanowire-based sensors are capable of pushing the chlorine detection limit lower than the 3 ppm obtained from nanoparticle-based sensors. When the responses for the two different sensors were compared at 1 ppm, nanowire-based sensor had 15 times better response than the nanoparticle-based one. The improvement in response and sensitivity to chlorine detection by nanowires can be attributed to reduced number of hops a carrier has to undergo to reach from one electrode to another. In addition, given the single crystalline nature of nanowires we suspect that charge transport in individual nanowires is dominated by 1-D Luttinger mechanism. As a result SnO_2 nanowires yield much higher sensitivity than nanoparticles at chlorine concentrations less than 3 ppm. Generally, other SnO_2-based sensors exhibit relatively poor sensitivity to Cl_2 gas [20]. Our findings suggest a catalytic effect of Sb for selectivity and morphological changes in enhancing the sensitivity for Cl_2 detection.

5.4.2 Li-doped SnO_2 for Hydrogen Detection

Hydrogen gas sensors have been a subject of intense interest because of the explosive nature of hydrogen when mixed with oxidants. Applications of hydrogen gas sensors range from detection of leaks in hydrogen fueled rocket engines, semiconductor fabrication, petroleum transformation, and chemical substance production [21]. In addition, the rapid development of hydrogen powered fuel cell technologies has also increased the usage of hydrogen dramatically. Figure 5.8 shows the changes in resistance of a Li-doped SnO_2 deposited with palladium at different hydrogen concentrations below its explosive limit (<4%). The corresponding response as a function of concentration is shown in Fig. 5.8b. The resistance of sensor quickly decreases upon exposure to hydrogen followed by a slow increase and stabilization; when exposed to air the resistance returns to the initial resistance of the sample. The sensor does not display any memory effects resulting from the residual hydrogen adsorption on the sensor surface after every air purge cycle, as evidenced by the return to the same baseline resistance after every fresh air purge cycle. The room temperature response time is ~40 s while the recovery time of is about 8 min and the sensor response increases in magnitude with increasing the hydrogen concentration. Even though the sensor shows somewhat limited dynamic range, it is important to point out this detection is at room temperature using a simple resistive device. Further studies of sensitivity of such a sensor are currently underway.

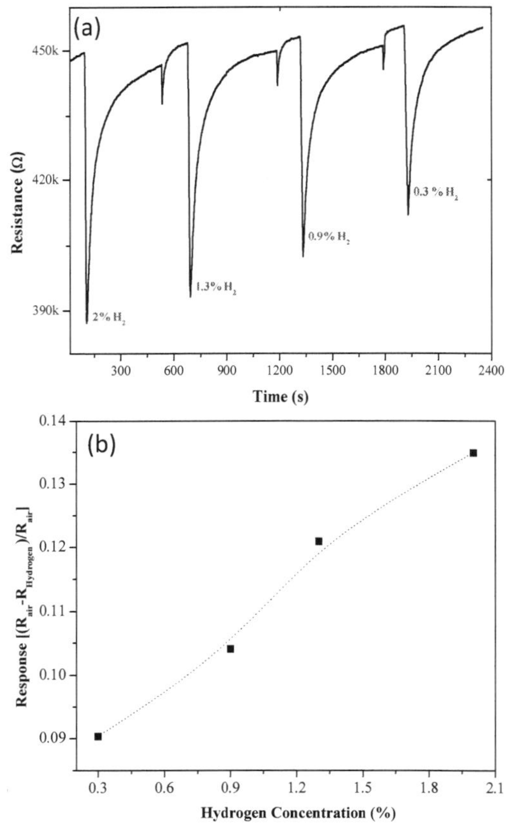

Figure 5.8 (a) Change in resistance on exposure to H_2 and recovery of the sensors at different hydrogen concentrations (b) Response versus H_2 concentration. Note the redefined response.

During the initial stages of hydrogen exposure, the conductivity increases due to the lattice expansion of palladium through nonstoichiometric absorption of hydrogen leading to additional contacts between the Pd nanoparticles [22]. Continuous hydrogen absorption in the second state leads to formation of palladium hydride of higher resistance. This second stage corresponds to a slow rise in the resistance corresponding to a two-phase region between the so-called alpha and beta phases found in palladium/hydrogen phase diagram.

5.5 Conclusions and Future Outlook

Tin oxide as a sensor material has come a long way. We have reported a new class of Cl_2 and H_2 specific oxide-based sensors operable at room temperature. The Cl_2 sensor shows high response and rapid response time at much lower temperatures than currently known oxide sensors. Our doped SnO_2 studies show potential to create selective detector as well as improve the sensitivity of these devices through morphological changes (i.e., nanoparticles → nanowires). Future sensors might include crafting hetero-junctions with different semiconductors, or by using surface derivatization techniques to attach specific chelants or aptamers to enhance selective binding.

References

1. C. Elisabetta, *Analytica Chimica Acta* 2006, **568**, 28–40.
2. A. Ponzoni, E. Comini, G. Sberveglieri, J. Zhou, S. Z. Deng, N. S. Xu, Y. Ding and Z. L. Wang, *Applied Physics Letters* 2006, **88**, 203101–203103.
3. S. Chang Q, *Progress in Solid State Chemistry* 2007, **35**, 1–159.
4. H. Huang, O. K. Tan, Y. C. Lee, T. D. Tran, M. S. Tse and X. Yao, *Applied Physics Letters* 2005, **87**, 163123–163123.
5. E. Comini, G. Faglia, G. Sberveglieri, Z. Pan and Z. L. Wang, *Applied Physics Letters* 2002, **81**, 1869–1871.
6. N. Barsan, M. Schweizer-Berberich and W. Göpel, *Fresenius' Journal of Analytical Chemistry* 1999, **365**, 287–304.
7. J. Q. Hu, Y. Bando, Q. L. Liu and D. Golberg, *Advanced Functional Materials* 2003, **13**, 493–496.

8. H. Kim, J. Lee and C. Lee, *Journal of Materials Science: Materials in Electronics* 2009, **20**, 99–104.
9. S. Mathur, and S. Barth, *Small* 2007, **3**, 2070–2075.
10. J. X. Wang, D. F. Liu, X. Q. Yan, H. J. Yuan, L. J. Ci, Z. P. Zhou, Y. Gao, L. Song, L. F. Liu, W. Y. Zhou, G. Wang and S. S. Xie, *Solid State Communications* 2004, **130**, 89–94.
11. X. L. Ma, Y. Li and Y. L. Zhu, *Chemical Physics Letters* 2003, **376**, 794–798.
12. J. Chan in *Fundamental Study of SnO_2 and Ag Nanoparticles: Growth, Kinetics and Applications* Masters Thesis Portland State University, Portland, 2002.
13. J. C. Chan, N. A. Hannah, S. B. Rananavare, L. Yeager, L. Dinescu, A. Saraswat, P. Iyer and J. P. Coleman, *Japanese Journal of Applied Physics* 2006, **45**, 1300–1303.
14. A. Chaparadza, and S. B. Rananavare, *Nanotechnology* 2008, **19**, 245501.
15. Y. Wang, J. Y. Lee, and T. C. Deivaraj, *The Journal of Physical Chemistry B* 2004, **108**, 13589–13593.
16. A. Chaparadza, S. B. Rananavare and V. Shutthanandan, *Materials Chemistry and Physics* 2007, **102**, 176–180.
17. In *Joint Commission on Powder Diffraction Spectroscopy card No.* 21-1250, Vol. 1980. PA p. 1980.
18. W. Wang, J. Niu and L. Ao, *Journal of Crystal Growth* 2008, **310**, 351–355.
19. A. Chaparadza and S. B. Rananavare, *Nanotechnology* 2010, **21**, 035708.
20. (a) M. D. Mahanubhav, and L. A. Patil, *Sensors and Actuators B: Chemical* 2007, **128**, 186–192; (b) T. Miyata, and T. Minami, *Applied Surface Science* 2005, **244**, 563–567.
21. X. Bevenot, A. Trouillet, C. Veillas, H. Gagnaire and M. Clement, *Sensors and Actuators B: Chemical* 2000, **67**, 57–67.
22. (a) M. Ramanathan, G. Skudlarek, H. H. Wang and S. B. Darling, *Nanotechnology* 2010, **21**, 125501; (b) F. Favier, E. C. Walter, M. P. Zach, T. Benter and R. M. Penner, *Science* 2001, **293**, 2227–2231.

Chapter 6

Titanium Oxide Nano- and Submicron-Structured Coating for Ti and Ti-Related Bio-Implants

Shampa Aich[a] and Banasri Roy[b]

[a]*Department of Metallurgical and Materials Engineering, Indian Institute of Technology, Kharagpur 721302, India*
[b]*Department of Chemical Engineering, BITS-PILANI, Pilani, Rajasthan 333031, India*

saich@metal.iitkgp.ernet.in, banasri.roy@pilani.bits-pilani.ac.in

Titanium oxide (TiO_x) is considered as the best potential candidate to coat the Ti and Ti-based implants and devices because of its superior biocompatibility compared to other known coating materials, such as silicon carbide, titanium nitride, and diamond-like carbon. Use of nanostructured titanium oxide coating can significantly improve the biological activities in vitro and in vivo. Several factors affect the biological activities when a titanium oxide–coated implant comes in contact with the bone tissues and the body-fluid environment. This chapter reviews the important studies involving different synthesis routes of titanium oxide coating on Ti-based biomaterials and how to characterize those coated surfaces.

Oxide Nanostructures: Growth, Microstructures, and Properties
Edited by Avanish Kumar Srivastava

ISBN 978-981-4411-35-6 (Hardcover), 978-981-4411-36-3 (eBook)
www.panstanford.com

This review also includes the important discussion regarding the best biocompatibility of the titanium oxide–coated implant surfaces compared to others and how the different properties of the oxide surface, such as thickness, roughness and porosity, presence of different phases, chemical states, and surface charge influence the biocompatibility (blood compatibility as well as bone compatibility) of the implant surfaces.

6.1 Introduction

Titanium (Ti) and its alloys have widely been used for biomedical devices, such as orthopedic and hard tissue replacement (hip joint, knee joint), dental implants, cardiac and cardiovascular devices (artificial hearts, artificial valves, and stents) (Helsen and Breme, 1998; Branemark et al., 1969; Larsson, 1997; Zhang et al., 1997; Zhang et al., 2001; Walivaara et al., 1994), by utilizing its good mechanical strength (load bearing capacity) (Zhang et al., 2001; Park et al., 2001), good corrosion-, degradation-, wear-, and fatigue resistance (Williams, 1981; Walivaara, 1993; Sunny and Sharma, 1991; Zhang et al., 1998), low specific weight (Kim and Lee, 2009), and low Young modulus close enough to natural bone (10–20 GPa), compared to bioceramics, such as hydroxyapatite (HA, $Ca_{10}(PO_4)_6(OH)_2$) (Velten et al., 2003; Bahadur, 2008), other ceramics and glasses (Martin and Brown, 1999), stainless steel and Co–Cr based alloys (Trentz et al., 1997; Finn et al., 2000). However, success of the implants depends on the surface properties of the implant materials. Pure Ti or Ti based alloys do not make direct bonding with the body cells due to its non-bioactive nature (Park et al., 2010; Park et al., 2007). For bone replacement Ti showed poor ostiointegration (Feng et al., 2003), while for dental implantation both loss of osteointegrity and post implant microbial and bacterial infestation have been reported (Petrini et al., 2006). Moreover, long healing time makes it necessary to improve Ti or Ti alloy quality. Leaching or dissolution of Ti and other metals from alloys and its accumulation in adjacent tissues or its transportation to distant major organs are also possible (Xu et al., 1998; Leng et al., 2002; Sonoda and Kato, 1997).

Therefore, numerous studies have focused toward improving the surface biocompatibility of these otherwise suitable metallic implant candidates. It has been observed that creating a surface oxide layer on the metal can dramatically improve the biocompatibility (Li et al., 2004; Park, 2011; Park et al., 2010; Tao et al., 2009). In bone surgery, where the longevity of the implant could be limited by the weak adhesion between HA and titanium alloy, a titanium oxide coating can be used as an intermediate layer between HA and titanium alloy based implant (Ti–6Al–4V). The presence of an intermediate diffusion layer could improve adhesion, particularly due to the formation of chemical bonding (Jouanny et al., 2010). The protective surface coating of titanium oxide increases the bio-inertness, hardness and wear resistance considerably and acts as a diffusion barrier against transport of metallic ions into the biological tissue. Considerable attentions have been paid to anodic titanium oxide films for implant applications prepared by electrochemical method due to low cost and relative ease of fabrication than those of conventional methods like plasma spray method (Grossm et al., 1998; Oji et al., 1999) or sol–gel technique (Weng et al., 2003; Weng and Baptista, 1998). Oxide films fabricated by anodizing method not only show various surface morphologies, but also contain complicated anion species incorporated from the electrolyte, which can be more effective for bioactive calcium phosphates formation after being implanted in human body (Lee et al., 2006). Also, recent investigation suggests that Ti–O thin film may be a newly developed antithrombotic material. However, its clinical use requires a more complete understanding of the mechanism for its thromboresistance (Leng et al., 2006).

This chapter discusses the synthesis and characterization techniques of titanium oxide coating on Ti and Ti alloy implant surfaces and its related properties and applications. Nanostructure is a structure in nanoscale or nanorange that includes nanofilm, nanograin, or nanoparticles (possess at least one dimension in the nano range 1–100 nm) and which possess high surface area-to-volume ratio that can alter significantly the biological activities in vitro and in vivo (Kalita, 2008). Therefore, discussion will be continued even at nano-and submicron-level for a few cases related to nanostructured coating.

6.2 Synthesis Routes

Several novel processes for surface modification and synthesis of nanostructured titanium oxide coating on Ti and Ti alloys are reported to achieve good biocompatibility and hemocompatibility for the Ti-based implants. These processes include but are not limited to sandblasting (Kononen et al., 1992; Groessner and Tuan, 1992), acid etching (Park et al., 2001; Kononen et al., 1992; Park and Davies, 2000), combination of sandblasting and acid etching (Li et al., 2002), sol–gel technique (Takemoto et al., 2001; Takashima et al., 1999), plasma spraying (Groessner and Tuan, 1992; Gross et al., 1998; Oji et al., 1999), electrochemical methods (Sunny and Sharma, 1991; Walivaara et al., 1994; Lee et al., 2006), magnetron sputtering (Yang et al., 2006; Chen et al., 2006), surface oxidation via heating in different oxidatizing atmosphere (Huang et al., 1994; Takemoto et al., 2004; Feng et al., 2003; Feng et al., 2002), surface oxidation via acid etching and heating (Erriksson et al., 2001; Nygren et al., 1997; Takemoto et al., 2004), plasma immersion ion implantation and deposition (Huang et al., 2003; Leng et al., 2001; Leng et al., 2000; Thorworth et al., 2000), radiofrequency reactive sputtering (Jouanny et al., 2010), direct current (DC) sputtering, ion beam enhanced deposition (Zhang et al., 2001; Zhang et al., 1997; Huang et al., 1998), dual plasma deposition (Leng et al., 2002), DC plasma oxidation (Sonoda and Kato, 1997; Leng et al., 2003), reactive plasma oxidation (Leng et al., 2001), hydrothermal treatment (Park, et al., 2010; Park, et al., 2007; Park et al., 2011; Park et al., 2009), metallo-organic chemical vapor deposition (MOCVD) (Jung et al., 2003), Radio frequency plasma-enhanced chemical vapor deposition (RF-PECVD) (Szymanowski et al., 2005), evaporation (Zhang et al., 1998), and anodic spark deposition method (Angerer et al., 2011).

Plasma immersion ion implantation (PIII) is a method where the ions are implanted into the material using negative high voltage pulses, thus achieving a high implantation rate combined with a simultaneous treatment of arbitrarily shaped objects and a short and cost-effective treatment time (Thorworth et al., 2000). Ion implantation was used at several occasions for Ca, P and other atomic doping purpose (Hideki et al., 1997; Zreiqat et al., 2005). Titanium oxide film fabricated by reactive magnetron sputtering was modified by phosphorus ion implantation and succeeding vacuum annealing. Phosphorus ions were implanted with a dose of about 5×10^{13} ions/cm^2 at room temperature. The implantation energy value used was 35 keV. After ion implantation, annealing was carried out at 200, 400 or 900°C for 20 min in vacuum ($<1 \times 10^{-4}$ Pa). The depth profiles of P atoms implanted in titanium oxide film and the redistribution of P after annealing were measured by means of Rutherford backscattering spectrometry (RBS) (Leng et al., 2006).

The electrochemical methods include anodizing or anodic oxidation (Walivaara et al., 1994; Sunny and Sharma, 1991; Lee et al., 2006; Jaeggi et al., 2006; Oh et al., 2008), micro-arc oxidation (MAO) or plasma electrolysis, also called anodic spark oxidation or plasma electrolytic oxidation (Li et al., 2004; Han et al., 2003; Sul et al., 2005). These electrochemical methods are not only simple, controllable, and cost-effective technique to fabricate oxide layers of various surface morphologies on titanium and its alloys but also contain complicated anion species incorporated from the electrolyte (Kim and Lee, 2009; Hanawa et al., 1997; Sul et al., 2002). A porous layer was formed on the Ti surface after the oxidation (MAO) treatment. The phase and morphology of the oxide layer were dependent on the voltage applied during the oxidation treatment. With increasing voltage, the roughness and thickness of the film increased and the titanium oxide phase changed from anatase to rutile. During the MAO treatment, Ca and P ions were incorporated into the oxide layer. The in vitro cell responses of the specimen were also dependant on the oxidation conditions. With increasing voltage, the ALP (alkaline phosphatase) activity increased, while the cell proliferation rate decreased (Li et al., 2004).

Anodization process at lower voltage is well known for making nanotubular array (Macak et al., 2005; Macak et al., 2005; Taveira et al., 2005; Shankar, 2007). But development of bone apatite for these films are very slow and in order to improve apatite forming ability, the MAO treatment is generally followed by alkali treatment (Wei et al., 2007), hydro-thermal treatment (Ishizawa and Ogino, 1995; Li et al., 2005; Fini, 1999), or sol-gel treatment (Tao et al., 2009; Liu et al., 2005).

On the other hand hydrothermal method reported to be capable of making thin oxide layers of various morphology and incorporating ions in these layers, which can develop osseointegration successfully (Park et al., 2007; Dong et al., 2007). Recently, some researchers have reported about formation of biocompatible robust nanofibre/nanorod titanium oxide based ceramic scaffold (titanate scaffold) on titanium surfaces through a simple, inexpensive, easy scale-up, one-step hydrothermal reaction of alkali (NaOH) with the metal surfaces without using any seeds, templates, titanium oxide powder, or stabilizer (Dong et al., 2006; Dong et al., 2007; Aich et al., 2010). The robust bio-scaffold was able to provide both the requirements of the coating; macroporous to adhere the tissue as well as mechanically enough tough to support the tissue growth and proliferation. These ceramic titanate nanowires can be transformed into their anatase counterparts simply through hydrothermal soft chemical transformation (Mao and Wong, 2006). The morphology of the scaffold (length and diameter of the nanorod and the thickness of the coating) can be well controlled by the strength of the solution and hydrothermal treatment time.

6.3 Characterization Techniques

The titanium oxide-based implants can be characterized using several modern techniques and principles. The major characterization techniques have been sub grouped in three categories; Biological characterization, Physical characterization, and Mechanical characterization describing the characterization techniques related to biological, physical, and mechanical properties/aspects of the implants.

6.3.1 Biological Characterization

The need to perform the biological characterization is to know the degree of bio compatibility of the implants. Two most important aspects of the bio compatibility of the implants are (i) blood compatibility and (ii) bone compatibility. These two can be evaluated by in vitro cell studies as well as in vivo tests. The in vitro hemocompatibility of the titanium oxide films is characterized in terms of clotting time, platelet adhesion and activation, thrombin time and prothrombin time, activated partial thromboplastin time, amount of fibrinogen, and protein adsorption (Erriksson et al., 2001; Huang et al., 1998). The kinetic method for measuring the thrombo-resistant property is used to determine clotting time. One research group did the measurement using blood from a mammal, which was dropped onto the specimens and was allowed to coagulate. After a predetermined time specimens were transferred into distilled water, and incubated. The red blood cells that had not been trapped in a thrombus were hemolyzed, and the free hemoglobin was dispersed in water. The concentration of free hemoglobin in the water was measured colorimetrically by spectrophotometer (Huang et al., 2003). Another group proposed a gravimetric method to measure the amount of thrombus formed instead of the free blood (Imai et al., 1972).

In vitro platelet adhesion testing needs to be performed to investigate the quantity, morphology, aggregation and pseudopodium of the adherent platelets. There are several methods to perform the test. In one of them, the samples were first immersed in platelet-rich plasma collected by centrifuging blood, followed by incubating and finally quantifying by optical microscopy and scanning electron microscopy (SEM) (Park and Davies, 2000; Chen et al., 2006; Takemoto et al., 2004; Huang et al., 2003; Leng et al., 2001). Another method could be using a lactate dehydrogenase based assay (Takemoto et al., 2004; Park and Gemmell, 2001). Platelet activities could be characterized by quantifying the platelet-derived microparticle protein formation and the level of P-selection expression, which involves translation of genetic codes to protein (Park and Gemmell, 2001).

To perform the thrombin time and prethrombin time measurement, platelet poor plasma prepared by centrifuging blood were used and were put into contact with the test samples. The thrombin time measurements were performed by adding thrombin time reagent to platelet poor plasma in a test tube and the thrombin tendencies of both (ppp and ppp mixed with reagent) were measured and compared with a clotting tester. The prethrombin time measurements were carried out by adding platelet poor plasma to prethrombin time reagent in a test tube and then were evaluated similarly with the same clotting tester (Takemoto et al., 2001; Yang et al., 2006; Huang et al., 2003).

Different experimental techniques have been adopted to determine the amount of protein adsorption. In one method, purified blood related proteins (fibrinogen, albumin, globulin, and collagen) of either bovine or other animal or human were labeled with iodine radioisotopes (^{125}I, ^{131}I). The sample was incubated in the protein solution for certain time and finally it was cleaned thoroughly. Amount of protein adsorbed on the surface was estimated by measuring the surface concentration of the radioactive I (Huang et al., 2003; Baszkin et al., 1980). An alternative method is to use chemiluminescence, or fluorescent dye and fluorescence microscope techniques to detect cell adhesion, or antibodies and cell markers (Errikson et al., 2001; Kawahara et al., 1993). Other way could be to dissolve protein in a solvent followed by measuring the concentration (Kawamoto et al., 1997).

For in vivo tests the specimen was implanted in an animal body and left for a few days. Then the implant was removed surgically, cleaned and observed visually with the aid of microscope to check and study the thrombus formation (Zhang et al., 1998; Huang et al., 2003; Huang et al., 1998; Yang et al., 2002).

The bone implantation compatibility is characterized by studying initial bone related cell attachment, cell proliferation, alkaline phosphatase (ALP) activity, and cell gene expression. For initial cell attachment and proliferation study bone cell (osteoblastic cells, mesenchymal stem cells, osteosarcoma cell) from human or other mammal is seeded and cultured on the specimen for a certain time and is then quantified calorimetrically (Park et al., 2007; Li et al., 2004). For morphological evolution of cell spreading the cultured cell on the specimen is fixed and dried critically and then studied microscopically (Park et al., 2010; Feng et al., 2003; Park et al., 2011; Derhami et al., 2000). The ALP activity (an indication of osteoblastic differentiation) is measured spectroscopically (Park et al., 2010; Li et al., 2004; Park et al., 2011; Eriksson et al., 2004). The real-time polymerase chain reaction (PCR), or quantitative real time polymerase chain reaction or kinetic polymerase chain reaction is known to quantify the replication of DNA molecule for measuring cell gene expression. One well known method to study how to detect the samples is performed using non-specific fluorescent dyes that tagged with any double-stranded DNA. In another method the DNA probe is attached with florescent reporter. The PCRs are also capable of measuring messenger RNA level and other gene factors (Park et al., 2010; Park et al., 2011).

For in vivo bone compatibility test the specimen can be implanted in the bone of a mammal in the form of a screw, rod, plate, disc, cylinder, or other convenient shape. After a certain period of time, the implant is surgically removed and the bond strength between the bone and the implant can be inferred from the value of torque necessary to remove it (Park et al., 2007; Li et al., 2004; Park, 2011; Sul et al., 2005; Eriksson et al., 2004; Giavarsei et al., 2003). Surface morphology of the implant can be observed and the histomorphometrical quantification (computer-assisted analysis of images formed by a microscope) can be performed using microscopy techniques (Park et al., 2007; Li et al., 2004; Park, 2011).

6.3.2 Physical Characterization

The important physical properties which should be determined for biocompatible titanium oxide films are structure, morphology, thickness, microstructure, chemical depth profile, corrosion, and chemical composition. Several novel techniques are used to perform the physical characterization of the titanium oxide film, such as, auger electron spectroscopy (AES), spectrophotometry, profilometry, X-ray diffractometry (XRD), X-ray photoelectron spectroscopy (XPS), Raman spectroscopy, environmental scanning electron microscope with field emission (ESEM-FEG), contact angle goniometer, atomic force microscopy (AFM), glow discharge optical emission spectroscopy (GDOES), RBS, ellipsometry, and elastic recoil detection analysis (ERDA).

6.3.2.1 Thickness

In the literature several researchers reported about the various techniques to measure the thickness of the titanium oxide films. Anodic oxide film thickness was measured by Ar sputter etching using Auger electron spectroscopy (AES, PHI 650) (Sul et al., 2002; Sul, 2001). Thickness (*d*) of the film was determined using the following relation.

$$d = v_0 t_\mathrm{d} \tag{6.1}$$

where v_0 is the sputtering rate and t_d is the sputtering time when the amplitude of the oxygen peak is decreased to 50% of its steady-state value at the metal–oxide interface. To obtain precise measurement data the sputtering rate was calibrated using two reference films: SiO_2 and titanium oxide. Thickness can also be identified by observing the interference color of the oxide film by L*a*b system using a spectrophotometer (Sul et al., 2002). Jaeggi et al., reported measurement of Anodic oxide film thickness by ellipsometry as well as GDOPS techniques (Jaeggi et al., 2006). The oxide film obtained from sulfuric acid–based electrolyte was thicker than the film obtained using phosphoric acid based electrolyte. Also, the thickness measured by ellipsometry and GDOPS were in good agreement. Another research group Chen et al., reported the thickness measurement of the titanium oxide film (synthesized by magnetron sputtering, thickness 1.2–1.6 μm) using α-step measurement system (Chen et al., 2006). Cross-sectional SEM measurements can also be used to determine the oxide layer thickness (Tao et al., 2009). Angerer et al., reported the eddy current measurement method using a DualScope MP4 device (Helmut Fischer GmbH & Co. KG, Germany) to measure the oxide film thickness (obtained by anodic spark deposition) (Angerer et al., 2011). In this method, the oxide layer thickness was determined by eddy current measurements and the data were plotted as a function of the applied spark deposition current density.

6.3.2.2 Structural analyses

The exact phase composition and crystallite size of the titanium oxide film can be determined using X-ray diffraction technique. There are three main crystalline phases for titanium oxide; rutile (tetragonal structure), anatase (tetragonal structure) and brookite (orthorhombic structure). These phases can exist in the form of nanomaterials and their stability depends on grain size which can be controlled by process parameters, shapes, and impurities. In the conditions described for a grain size superior to 35 nm, rutile is the stable phase, between 11 and 35 nm, brookite is the stable phase and below 11 nm, anatase is the stable phase. (Jouanny et al., 2010). The X-ray diffraction data obtained from X'Pert Powder Diffractometer (Philips, Netherlands) at 40 kV and 30 mA were analyzed using Reidvelt analysis technique with the help of a structure analysis software for powder diffraction data (Angerer et al., 2011). Crystallite size (t) can be determined using the well known Scherrer formula given as the following (Cullity and Stock, Elements of X-ray Diffraction (2001)).

$$t = \frac{0.9\lambda}{B\cos\theta_B} \tag{6.2}$$

where, λ and θ_B are the wavelength of the radiation used and the Bragg angle, respectively, and B is the angular width (radian) measured at FWHM of the XRD pattern.

Inhomogeneous nature of the titanium oxide phases on the coating surface were determined by Raman Spectroscopic surface analysis method (Angerer et al., 2011). The measurements were performed using a Labram Aramis spectrometer equipped with a BX41 microscope. A diode-pumped solid state laser (532 nm wavelength), with a nominal power of 100 mW was used as the excitation source (exposure time 2 s). The incident beam was focused on the sample surface on a spot of 2 μm diameter by means of a 10× microscope objective lens. The Raman spectra were recorded in the wave number range of 100–800 cm^{-1}. Discrete high rutile content regions were observed in Raman observation at the localized high temperature regions due to discharge (Angerer et al., 2011).

6.3.2.3 Chemical composition and chemical depth profiling

Chemical depth profiling or determination of chemical composition is one of the most important aspects of characterization of the titanium oxide film. The bone responses to the oxidized surface (obtained by anodic oxidation) of the implants are remarkably influenced by type of anodizing electrolytes used (Sul et al., 2003). Bone to metal contact (BMC) was reported to be increased by 186%, 232%, and 272% for S, P, and Ca implants (obtained by anodization of the implants in H_2SO_4, H_3PO_4, and calcium containing mixed electrolyte), respectively compared to the turned implant surfaces (Sul et al., 2003). Chemical depth profiling of the titanium oxide film can be performed in several ways. Several research groups have performed their chemical analyses by GDOES, X-ray Photoelectron Spectroscopy (XPS), and RBS (Jaeggi et al., 2006; Huang et al., 2003). GDOES data revealed small S contamination peak in case of sulfuric acid film and phosphorus peak in case of phosphoric acid oxide film. Hydrogen accumulation at the metal–oxide interface was detected by GDOES and ERDA and is still under investigation for its effect on hydrogen embrittlement sensitive titanium based biomaterials (Jaeggi et al., 2006). Composition (non-stoichiometric titanium oxide (TiO_{2-x})) and the valence state (Ti^{4+}, Ti^{3+}, and Ti^{2+}) of the film were determined with the help of Auger electron spectroscopy (AES, PHI650SAM) and X-ray photoelectron spectroscopy (XPS, XSAM-800) (Huang et al., 2003).

6.3.2.4 Morphology and microstructure

Microstructure, morphology, and various surface topographies of the titanium oxide films can be observed with the help of some well known modern instruments, such as electron microscope and atomic force microscope. A low-magnification (1000×) image of the grain-like morphology of the titanium oxide film with grain diameters in the range of 1–5 μm was observed using ESEM-FEG at 20 kV accelerating voltage (Angerer et al., 2011). Atomic Force Microscopy (AFM, Autoprobe, CP) was used to observe the surface topography of the Ti-O film obtained by PIIID and sputter deposition. AFM results showed how the synthesis temperature affects the surface morphology of the oxide film. The amorphous TiO film synthesized at lower temperature exhibited sharp peaks, whereas surface morphology of rutile Ti-O film synthesized at 800°C presents small flat regions which indicates occurrence of surface diffusion at higher temperature (Huang et al., 2003).

Jouanny et al., reported about the use of electron microscope (SEM & TEM) to observe the significant influence of total chamber pressure (Ar + O_2) on the microstructures (columnar growth with the columns perpendicular to the substrate surface) of the titanium oxide films. As the total pressure increases from 0.17 to 1.47 Pa (O_2 partial pressure decreases from 85 to 35%), the film structure is transferred from rutile single-phase type to anatase type with significant change in columnar film morphology. At higher pressure the columns are straight, wide and extend over the film thickness (~450 nm), whereas at lower pressure the columns are discontinuous and wavy over the whole thickness of the film (~260 nm) (Jouanny et al., 2010). Cross-sectional TEM dark field microscopy of the anatase film helps to reveal the well defined individual columns made of wide well crystallized domains. Corresponding selected area diffraction pattern (SADP) shows rings with some points which confirm the presence of anatase phase. In case of rutile film the TEM micrograph shows less defined and discontinuous columns, constituted of nanograins and with diffused columns boundaries. Corresponding SADP shows continuous and thin rings which result from polycrystalline structure and the indexing confirms presence of rutile phase. High-resolution TEM micrograph of rutile film exhibited an amorphous interlayer (~5 nm) of SiO_2 at the substrate/film interface which was not observed in anatase film (Jouanny et al., 2010).

Aich et al., reported about the observation of the biocompatible robust nanofibre/nanorod titanium oxide based ceramic scaffold (titanate scaffold) on titanium surfaces with the help of SEM (Aich et al., 2012). The height or length of the nanorods was increased from 385 to 555 nm while increasing the reaction time from 2 to 4 h, but the average diameter of the nanorods was decreased from 70 nm to 56 nm, which could be to minimize surface area by nanorods to become more cylindrical and sharp (Aich et al., 2010). According to Aich et al., the average diameter of nanorods and the average height of the nanorods were found to be increased from 37 nm to 6 nm and from 414 nm to 590 nm for the solution strength range of 1.125 M to 2.85 M NaOH (Aich et al., 2012). Also with increase in concentration the tips of the nanowires can get joined, entangled to each other in such a way that they can form pyramid type structure (forming ridges and valleys) to promote tissue growth, adherence and proliferation.

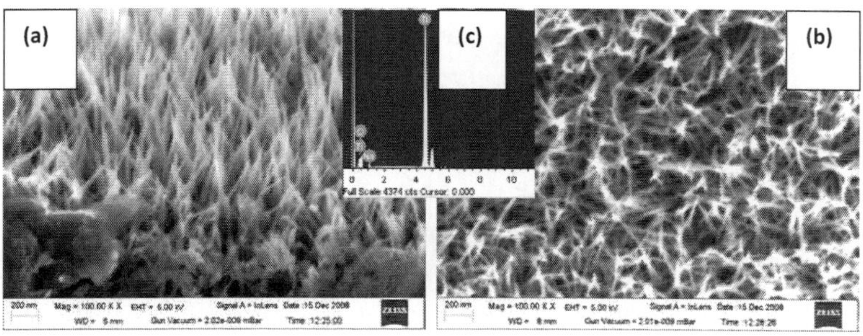

Figure 6.1 FESEM study to check the feasibility of scaffold formation on Ti sheet, magnification 100 K ×, hydrothermal treatment for 2 h in 2 M NaOH (a) plan view (vertical nanowires), (b) top view, (c) EDX profile of the nanowires.

The another research group reported the same time study but at higher temperature (240°C) and the nanorods (50–100 nm diameter) were grown much longer up to 5 micron and eventually the nanowires self-assembled on top into bundles and finally they formed ridges and valleys of 2–10 micron wide and 2–5 micron deep on Ti surface (Dong et al., 2007). The same group also reported the influence of alkali concentration on the morphology of the nanowires (Dong et al., 2007).

6.3.2.5 Surface contact/energy and wettability

The surface–interface tension/energies were determined by contact angle measurement method using JY-82 contact angle goniometer (Huang et al., 2003). Basic hydroxyl group and polar component of surface energy results in better adhesion of osteoblasts on titanium oxide surface compared to the acidic hydroxyl group and the dispersion component of surface energy. Surface energies of different samples were determined by measuring the contact angles (θ) obtained by different sample liquids with the titanium plate surfaces and using the Owens–Wendt–Kaeble's relation:

$$\gamma_L(1+\cos\theta)=2(\gamma_L^d\cdot\gamma_S^d)^{1/2}+2(\gamma_L^p\cdot\gamma_S^p)^{1/2} \tag{6.3}$$

Here, γ^d and γ^p represent the dispersion component and the polar component of the total surface energy (γ).

$$\gamma_L=\gamma_L^p+\gamma_L^d \tag{6.4}$$

The subscripts S and L indicate the solid and liquid surfaces.

6.3.3 Mechanical Characterization

Mechanical properties of titanium oxide thin films, which are used in biomedical applications, are of paramount importance because they need to withstand varieties of loading. In particular, in orthopedic applications the film is subjected to high stresses. The mechanical properties of titanium oxide thin film depends on the microstructure (grain size, phases, porosity) and crystallographic structure (Gaillard et al., 2009), which in turn depends on the substrate and parameters of synthesis. Characterization of mechanical properties of thin film is best carried out using nanoindentation technique (Roa et al., 2007; Gaillard, 2008). In this technique a diamond indenter, with different geometries viz. three and four sided pyramids (Berkovich), wedges, cones, cylinders, filaments, and spheres, indents on a plane surface. The analysis of load versus indentation depth curves implicitly gives the hardness and Young's modulus, even without visualizing the nanoindentation imprint. The analysis is mostly done using the method given by Oliver and Pharr (Oliver and Pharr, 1992). In case of thin film, the load versus indentation depth curves is result of combined mechanical property of the film as well as substrate. In other words the Young's modulus and hardness values obtained are combination of those values of the film and substrate. This is especially true with the value of elastic modulus. Several models have been proposed to extract the Young's modulus of the film from the combined response (Jager, 2002; Bec et al., 2006). However, these models assume the film to be isotropic, which is not a reality.

Depending upon the film thickness and the substrate, estimated Young's modulus varied from 16 to 209 GPa (mostly in the range 116–209 GPa) and the estimated hardness varied from 0.22 to 13 GPa (mostly in the range 4–13 GPa). It appears that wide range is not only due to difference in thickness, substrate and microstructure/crystal structure, but also due to difference in the inferential method for deducing the property of thin film. The value of Young's modulus is lower than that reported for bulk titanium oxide, i.e., 282 GPa (CRC Materials Science and Engineering Handbook, p 510). Gaillard et al., have shown that the measured Young's modulus and Hardness are strong function of the ratio of indentation depth to thickness (Gaillard et al., 2009). In conclusion, deduction of mechanical properties from nanoindentation tests requires further refinement.

6.4 Biocompatibility of Titanium Oxide Coatings

As discussed earlier preventing thrombogenicity (blood clotting at the implant-living tissue interface—blood compatibility) and increasing osseointegration (bone compatibility), which may be defined as the direct structural and functional connection between living bone tissues and the surface of a load-bearing artificial implant, are of same importance for the biocompatibility of an implant. To understand those two important aspects clearly a probable schematic of titanium oxide-body fluid environment has been considered in Fig. 6.2 (Tengvall and Lundstrom, 1992). The bone–titanium interface is covered by a proteoglycan layer (~200–400 Å thick) adjacent to titanium oxide layer. And the outside collagen bundles are observed about 1000 Å apart from the interface. The idea about the thickness and distances are acquired from in vitro experiments and electron microscopy study. Initially, just after the insertion of the implant, ions, proteins and other small molecules are adsorbed into the titanium oxide surfaces. The complex macromolecules of Ti (III) (formed in the partially anaerobic environment due to reduction of oxides) help to facilitate the chemical bonding through inter diffusion of $H_2PO_4^-$, Cl^- ions at the titanium surfaces. Initially, during the first few hours, hemostasis begins to make the blood flow sluggish by arresting the bleeding. In this stage the tissue infection of the inflamed wounds are dominated by polymorphonuclear granulocytes and monocytes (the white blood cells) which are formed in the bone marrow, and eventually enter the blood and migrate into the connective tissue. Later, after a few days, the monocytes and granulocytes differentiate into more divergent macrophages which help to wound healing by facilitating new tissue growth through fibroblast activation (Tengvall and Lundstrom, 1992). The dislocations and impurities are incorporated in the figure to make the oxide realistic.

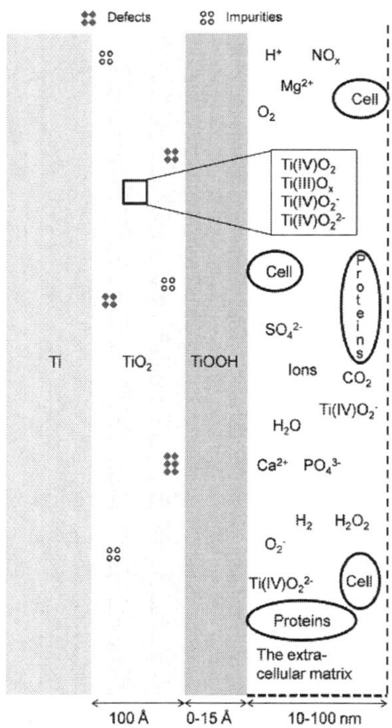

Figure 6.2 A probable schematic of TiO_2 in body fluid environment (adapted from Tengvall and Lundstrom, 1992).

At the initial stage body decides the blood compatibility of the foreign implant and generally bone compatibility consideration takes much longer time period. The OH^- group attached on the titanium oxide surface, in wide range of pH variation, is believed to help bio-adhesion by absorbing proteins (Li, 2003; Dold et al., 1990; Velten et al., 2003). Below we will discuss how titanium oxide coatings on Ti and related alloys affect these two issues widely addressed in the field of biocompatibility of these implants. The important factors influencing those two aspects; blood compatibility and bone compatibility will also be considered for discussion.

6.4.1 Blood Compatibility

Thrombogenicity and the surface induced thrombo-embolism of artificial organs, such as artificial heart valves (AHVs), is a serious problem. Due to this problem, patients with mechanical heart values must undergo lifelong anticoagulation therapy (Huang et al., 2003). Even so, the incidence of thromboembolic complications and bleeding complications has been 1.5–3% per year in the USA (Didisheim, 1996), and 58% of implanted mechanical heart valves have failed within 12 years in China (Stratagem report to the natural science discipline development, 1995). For other organs thrombogenicity could be responsible for implant rejection. Improving the hemocompatibility of these kinds of devices has become a very important task for biomedical material scientists. Interactions between biomaterials and cells mainly depend on characteristics of surface oxide layer including thickness, surface topography, charges, components and chemical states (Sunny and Sharma, 1991; Velten et al., 2003; Park and Davies, 2003; Takemoto, 2001; Yang et al., 2006; Takemoto et al., 2004; Erriksson et al., 2001; Nygren et al., 1997; Park et al., 2001; Nygren, 1997).

6.4.1.1 Blood compatibility of titanium oxide compared to other coating materials

Many different kinds of films have been investigated as candidate materials for the coating of AHVs, such as diamond-like carbon (DLC) (Jones et al., 2000; Tran et al., 1999), silicon carbide (SiC) (Bolz et al., 1990), titanium nitride (TiN) (Dion et al., 1993), and aluminum oxide (Al_2O_3) (Yuhta et al., 1994). However, few of the coatings have been reported to possess better hemocompatibility than low-temperature isotropic pyrolytic carbon (LTIC), which has shown the highest level of hemocompatibility as an AHV material for long time of clinical applications, even though its hemocompatibility is still far from ideal (Huang et al., 2003). From the beginning of the ninth decade of the last millennium, people started to notice that rutile type titanium oxide ceramics and TiO_2 and TiO_x prepared by thermal oxidation and ion beam assisted deposition generally have blood compatibility better than that of LTIC (Huang et al., 1998; Ebert and Schaldach proceedings, 1982; Huang et al., 1994; Liu et al., 1996). Over the years extensive research has shown that titanium oxide coating has the best blood compatibility (Zhang et al., 1998; Huang et al., 1998; Ebert and Schaldach proceedings, 1982; Xianghuai et al., 1996; Wang et al., 2000). A few attempts have also been made to coat the LTIC and other coating materials with TiO_2 layer to improve the blood compatibility further (Zhang et al., 1998; Xianghuai et al., 1996; Wang et al., 2000).

6.4.1.2 Effect of thickness

Surface modification or making thin solid oxide film is a new approach for improving the hemocompatibility of mechanical heart valves (Sunny and Sharma, 1991; Zhang et al., 1998; Huang et al., 1998; Ebert and Schaldach proceedings, 1982; Huang et al., 1994). Ti is a highly oxidative metal which naturally forms oxide layer on its surface. However, this naturally formed oxide layer is very thin (about 10 nm), porous (because of O_2 diffusion through the layer), and has a high density of defects (because of the low-temperature process). Better biocompatibility has been reported by improving the quality and thickness of titanium oxide films (Sunny and Sharma, 1991; Zhang et al., 1998; Huang et al., 2003; Huang et al., 1998; Ebert and Schaldach proceedings, 1982; Huang et al., 1994).

Sunny and Sharma (Sunny and Sharma, 1991) reported that adsorption and desorption of fibrinogen and albumin on the surface decreased with increase in thickness of the titanium oxide layer grown on Ti by anodic oxidation process. But the ratio of absorbed albumin/fibrinogen proteins increased ~5 times as the thickness of titanium oxide film was increased from zero nm to ~41 nm, which is a clear indication of lower thrombogenicity. Since no further significant improvement of the ratio of absorbed albumin/fibrinogen could notice as the thickness of the oxide layer increased, they considered the ~41 nm as the optimum thickness for the titanium oxide layer. Huang et al., reported that the clotting time increased 1.5 times when the thickness of the titanium oxide film was increased from 10 to 250 nm by a thermal oxidation process (Huang et al., 1994).

6.4.1.3 Effect of chemical nature

The proteins present in blood can be divided in to two groups; good protein such as albumin, which discourage thrombogenicity and bad protein such as fibrinogen or globulin, which encourage thrombogenicity. Blood clotting or thrombus formation proceeds through two pathways: one is the route in which several proteins (factors) are activated in a cascade manner, and the other is associated with platelet activation (Huang et al., 1998). Detail mechanism of how exactly blood interacts in presence of an artificial organ is not fully clear yet. One easy way to explain the blood behavior is that as soon as blood comes in contact with the foreign body, electrolytes, water, blood cells, and other kinds of protein get adsorbed on the surface. Nygren et al., demonstrated that a complex biofilm can form in the blood-implant materials interface within a few seconds of blood exposure (Nygren et al., 1997). If the surface property triggers the denaturization process of fibrinogen or globulin, these proteins can decompose and eventually leads to coagulation and/or platelet activation (Baier and Dutton, 1969). During decomposition the fibrinogen decomposes into fibrin monomer and fibrinpeptides (Baurschmidt and Schaldach, 1977). How this triggering happens? Gutmann and Keyzer proved that fibrinogen has electronic band structure, similar to that of an intrinsic semiconductor with a band gap of 1.8 eV; the Fermi level is situated at ~0.9 eV below and above the conduction band and valence band, respectively. So, when the electronic charge transfer happen from the occupied valence bond level of fibrinogen to the free states of the material surface, denaturization of the protein starts (Gutmann F and Keyzer H, 1986).

TiO_2 is a high band gap material with low carrier density. For the non-stoichiometric (deprived of oxygen, Ti^{2+}, Ti^{3+} are present in the lattice) or n-type TiO_2 (doped with donor ions such as Ta^{5+}, P^{5+}), work function is much lower than that of the fibrinogen as a consequence of Fermi level closer to the conduction band. Figure 6.3a represents the schematic of the electronic band structure of TiO_2 and fibrinogen side by side. When fibrinogen comes in contact with TiO_2, an n-n′ isotype heterojunction is formed, as shown in Fig. 6.3b (Zhang et al., 1998). With increasing carrier density the Fermi level pushes closer to the conduction band. Thus the charge transfer from the protein to the TiO_2 gets inhibited and anticoagulation effect increases.

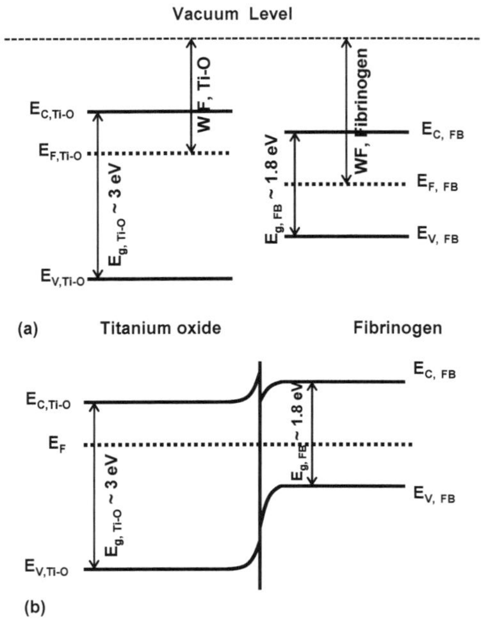

Figure 6.3 The energy band diagram of titanium oxide and fibrinogen (a) before adsorption and (b) after adsorption (Adapted from Zhang et al. (1998)).

One method, which may overcome thrombogenicity and thrombo-embolism problem, is seeding with endothelial cells on vascular prostheses, because of non-thrombogenic properties of the cells in natural state in contact with blood (Stansby, 1994; Zilla, (1999). Sanborn et al., found that endothelial cells can form acting stress fibers and focal adhesion complexes required for survival on hydrophilic plasma polymers. This suggests that these materials could be exploited for seeding endothelial cells (Sanborn, 2002).

6.4.1.4 Effect of phase

Amorphous, anatase, and rutile titanium oxide phases demonstrated better biocompatibility compared to pristine titanium oxide (Sunny and Sharma, 1991; Leng et al., 2002; Nan et al., 1994; Bolz et al., 1990). However, the role of titanium oxide phases as well as crystallinity differences on biocompatibility is not clear yet. Generally rutile phase is chosen over anatase or amorphous titanium oxide for biomedical applications due to the fact that the dissolution of Ti metal ions from rutile is one order of magnitude lower than that from anatase (Wisby et al., 1991). Huang et al., demonstrated that number and aggregation of platelets adhered on rutile surface was less compared to amorphous titanium oxide surface, which is a sign of enhanced anticoagulation property of rutile (Huang et al., 2003). On the other side, Huang et al., argued that absence of grain boundary related local states could be helpful for increasing biocompatibility of amorphous titanium oxide (Huang et al., 1998).

6.4.1.5 Effect of surface

The thrombogenic tendency of the materials was observed to be directly proportional to the amount of the fibrinogen adsorbed on the surface. As adsorption of protein is an important stage, the wettability and surface tension (energy) of the implant should have an influential effect on hemocompatibility of titanium oxide films and this has been shown experimentally (Nygren et al., 1997; Baszkin et al., 1980; Brash and Uniyal, 1979). Brash and Uniyal reported less adsorption of albumin and fibrinogen on hydrophilic surface compared to the hydrophobic surface due to hydration sheath on the surface. On that basis hydrophilic surface should be more blood compatible, while some research works, based on both in vitro and in vivo studies showed higher thombogenic tendency for the hydrophilic surfaces (smaller contact angel) than the hydrophobic ones (larger contact angel) (Brash and Uniyal, 1979).

Takemoto et al., observed that the hydrophilicity of the titanium oxide coating can influence platelet adhesion only up to certain level, but composition and thickness of the oxide layer affect platelet adhesion more. Thicker titanium oxide layers formed on Ti substrates by heating-oxidation displayed less platelet adhesion than thin oxide layers on untreated Ti substrates (Takemoto et al., 2004).

Huang et al., demonstrated that the adsorption of protein on the surface depends on the interfacial energy between blood and titanium oxide surface rather than the surface tension of the implant (Huang et al., 2003). Reportedly smoother surface enhance the adherence, growth, shape and proliferation of endothelial cells on Ti-O films (Chen et al., 2006). Erriksson et al., reported that titanium oxide coating grown by acid etching and surface oxidation via heating can be considered to be similarly hydrophilic irrespective of thickness and roughness. For these similarly hydrophilic samples, the film topography rather than thickness showed more distinctive effect on the protein cell responses. Films with rougher surfaces showed more leukocytic cell adhesion and surface coverage than the smoother films due to better anchorage, while the thicker films work in opposite way (Erriksson et al., 2001). Park et al., observed that rougher titanium oxide surface adhere and activate higher density of platelets obtained from human blood than the smoother films (Park et al., 2001; Park and Davies 2000).

6.4.2 Bone compatibility

Successful adhesion and attachment of the bone cells with the implant are two initial phases of osseointegration. Cell adhesion can be explained as formation of physico-chemical linkages between cells and materials by virtue of ionic forces, and van der Waals forces. The attachment phase occurs in the longer term involving various biological molecules: extracellular matrix proteins, cell membrane proteins, and cytoskeleton proteins which interact together to induce signal transduction, promoting the action of transcription factors and consequently regulating gene expression (Anselme, 2000). Morphological features such as surface roughness, porosity, surface energy, and wettability are observed to be the important factors governing the cell-implant interaction.

6.4.2.1 Effect of roughness and porosity

It is generally accepted that rough, textured and submicron scaled porous surfaces could stimulate cell attachment, differentiation and the formation of extracellular matrix (Groessner and Tuan, 1992; Anselme, 2000; Anselme et al., 2000). Li et al., observed that the proliferation rate of MG-63 human osteosarcoma cell line on the somewhat porous titanium oxide layer (oxidized at the relatively low voltage of 190 V) to be 10 times higher than that of the untreated titanium oxide. As the porosity and roughness of the films increased with increasing voltage beyond 190 V cell proliferation rate decreased, but the ALP activity of the cells increased with the increasing applied voltage (Li et al., 2004).

How the nano-porous and nanocrystalline nature of titanium oxide affect cell-implant interaction is not very clear. Titanium oxide gel, either made directly from Ti metal using NaOH or by sol-gel route, with nanoporous surface structure is reported to have better bone apatite formation capability than the larger pore titanium oxide coating. One possible cause for it could be that the higher energy defective oxide structures in the nanopores are better nucleation sites (Li et al., 1994; Kim et al. 1999). Due to increased long-term osteoblast adhesion coupled with decreased cell motility as well as enhanced proliferation, synthesis of alkaline phosphatase and deposition of calcium-containing mineral has been more observed on nanocrystalline (grain sizes less than 100 nm) titanium oxide surfaces compared to the conventional titanium oxide layer with larger grain size (Webster et al., 2000).

6.4.2.2 Effect of surface energy and wettability

Effect of surface energy and wettability on blood compatibility of titanium oxide is already discussed. Consistent to this, it has been observed that while the initial interactions of the bone cells with hydrophilic and hydrophobic implants are different, long-term responses are very similar (Eriksson et al., 2004). Surface energy would affect cell behaviors and cells are more easily attached and spread on surfaces with high surface energy (Zhao et al., 2005; Rupp et al., 2006; Zhao et al., 2007; Anselme, 2000). There was also an opposite opinion that the cell attachment increased with the hydrophobicity of surfaces (De Santis D et al., 1996). Park et al., thought that increasing wettability enhances biocompatibility of titanium oxide layer (Park, 2011; Park et al., 2010; Park et al., 2009). Feng et al., observed that the specific surface area as well as the surface energy increased with the surface roughness, which could be increased by heat treatment. The rougher surface means higher density of thermodynamically high energy sharp points and edges, which in turn can provide more favorable places for cell attachment and proliferation. The surface roughness, surface energy, and amount of the hydroxyl group attached on the surface depend on the environment of oxidation heat treatment. For the sample oxidized in water vapor all three were found out to be higher than the sample treated in air or oxygen. The results indicated that polar component

(γ_p) of the surface energy and the basic hydroxyl $(OH)_b$ groups of the total $(OH)_s$ groups attached with the surface have stronger influence on the amount and the interaction energy of the osteoblasts attachment on titanium surfaces, compared to the acidic hydroxyl $(OH)_a$ groups in $(OH)_s$ groups and the dispersion component (γ_d) of the total surface energy (Feng et al., (2003)).

6.4.2.3 Using seeds

Not only the inorganics (mostly Ca and P as calcium hydroapatite), but also there are other trace elements which are necessary for healthy bone functions and formation (Zreiqat et al., 2005). Deficiency of Mg, Zn, Mn, and Cu can lead to the development of osteoporosis (Smith and Nisbet, 1972; Komori, 1997; Herzberg, 1990). Human skeleton contains trace amounts of strontium. Recent studies show that strontium positively affects bone metabolism to promote bone formation and to decrease bone resorption, leading to normalized bone density (Imanishi, and Nishizawa, 2005; Ammann, 2005; Meunier et al., 2002; Compston, 2005; Ego et al., 2006).

P ions are incorporated into the titanium oxide during processing, which act as preferential nucleation sites for calcium phosphate in the biological fluid. Using hydroxyapatite or related calcium phosphate seeds on the titanium oxide layer is an interesting method to encourage osteoconductivity of the implants. The incorporated Ca and P ions were even crystallized into hydroxyapatite or other calcium phosphates by hydrothermal treatment (Han et al., 2003; Ishizawa et al., 1995; Ishizawa et al., 1995).

Tao et al., modified the surface of Ti–24Nb–4Zr–7.9Sn alloy by MAO and incorporated Ca^{2+} ion in the layer which led to improved bioactivity in simulated body fluid. They argued that Ca^{2+} can attract the negatively charged protein, which in turn increased cell differentiation and propagation (Tao et al., 2009). Park et al., incorporated Ca hydrothermally on the titanium oxide layer on the Ti–6Al–4V alloy surfaces. These Ca-enriched implants showed a considerable apatite deposition, which results in significant increase in cell viability and removal of torque forces, and better bone-to-implant contact compared with those for untreated Ti–6Al–4V implants (Park et al., 2007). Oh et al., used sodium silicate solution as an additive during anodization of titanium oxide (titanium oxide) films by an electrochemical method for preferential nucleation of calcium phosphate (Oh et al., 2008). Preliminary results on Sr/P ion incorporation in Ti oxide surface showed improved implant osseointegration in both cortical and cancellous bone by increasing bone apposition, due to its surface properties combining micro-topography, P/Sr chemistry and superior wettability (Park (2011)). But Mn incorporation notably decreased cellular attachment, spreading, proliferation, alkaline phosphatase activity, and osteoblast phenotype gene expression compared with the bare Ti surface ($p < 0.05$). The results indicate that the Mn-incorporation into the surface Ti oxide layer has no evidence of any beneficial effect on osteoblastic cell function, but instead, actually impaired cell behavior (Park et al., 2011).

6.4.2.4 Effect of phase

In photo catalyst and photovoltaic field, anatase is clearly considered as a better candidate than rutile due to favorable band structure, but in biomedical field there is controversy. Jouanny et al., studied structural and mechanical properties of titanium oxide thin films and reported that although the rutile phase is ~2.5 times harder than the anatase phase, the anatase phase has better potential as a titanium oxide intermediate layer between hydroxyapatite and titanium alloy, because of its reduced elastic modulus which is comparable with those of hydroxyapatite (around 60–100 GPa) and Ti-6Al-4V (140 GPa) (Jouanny et al., 2010). Feng et al., thought that anatase to rutile transformation is a thermal treatment phenomenon. Both phases demonstrated bioactivity under appropriate conditions and bioactivity of titanium oxide would not depend on the phase of the oxide (Feng et al., 2002).

6.5 Conclusions

In this chapter, we discussed how the titanium oxide surface coating on Ti and Ti alloy based biomedical implants influences the biomedical, physical, and mechanical properties of the implant surface. By adopting a special synthesis route (as required) and by controlling the process parameters of a specific synthesis technique we can control the morphology, chemistry (specific ion incorporation which can help to improve biocompatibility), phase formations (anatase or rutile), and crystallinity of the titanium oxide layer on the implant surface. The effects of some critical aspects such as chemical nature of doped and undoped oxides, thickness, surface morphology, and surface properties of the oxide layer on biocompatibility are well researched so far and are well discussed here also. But, some aspects like effect of nanoscaled structure and phase are not discussed thoroughly here and need more attention. Titanium oxide is a multifunctional ceramic oxide, which has several applications in other biomedical fields also, such as treatment of cancerous growth, dental implant field. Lack of time and space did not allow us to cover these aspects in this chapter, but will be considered elaborately in future.

References

Aich, S., Sekhar, C., and Mishra, M. (2010). Nanostructured bio-scaffold for bone implants, stents: A biomedical evolution, *Supplemental Proceedings: Volume 2: Materials Characterization, Computation, Modeling and Energy*, TMS (The Minerals, Metals and Materials Society), pp. 617–624.

Ammann, P. (2005). Strontium ranelate: a novel mode of action leading to renewed bone quality, *Osteoporos Int.*, **16**, pp. 11–15.

Angerer, P., Kiss, N., Löcker, C.H., Gavrilović, A., Artner, W., Godja, N., and Nauer, G.E. (2011). Titanium oxide layers on aluminium substrates produced by the anodic spark deposition process, *Mater. Chem. Phys.*, **128**, pp. 28–31.

Anselme, K., Bigerelle, M., Noel, B., Dufresne, E., Judas, D., Iost, A., and Hardouin, P. (2000). Qualitative and quantitative study of human osteoblast adhesion on materials with various surface roughness, *J. Biomed. Mat. Res.*, **49**, pp. 155–166.

Anselme, K. (2000). Osteoblast adhesion on biomaterials, *Biomaterials,* **21**, pp. 667–681.

Bahadur, R., Chul, K., Kim, K., Khil, M., Kim, H., and Kim, I. (2008). Synthesis of hydroxyapatite crystals using titanium oxide electro spun nanofibers, *Mater. Sci. Eng. C,* **28**, pp. 70–74.

Baier, R.E., and Dutton, R.C. (1969). Initial events in interaction of blood with foreign surfaces, *J. Biomed. Mat. Res.*, **3**, pp. 191–206.

Baszkin, A., and Lyman, D.J. (1980). The interaction of plasma-proteins with polymers. I. Relationship between polymer surface-energy and protein adsorption-desorption, *J. Biomed. Mat. Res.*, **14**, pp. 393–403.

Baurschmidt, P., and Schaldach, M. (1977). The electrochemical aspects of the thrombogenicity of a material, *J. Bioengng.*, **1**(11), pp. 261–278.

Bec, S., Tonck, A., and Loubet, J.L. (2006). A simple guide to determine elastic properties of films on substrate from nanoindentation experiments, *Philos. Mag.*, **86**(33–35), pp. 5347–5358.

Bolz, A., and Schaldach, M. (1990). Artificial heart valves: improved blood compatibility by PECVD a-SIC coating, *Artif. Organ,* **144**(4), pp. 260–269.

Brånemark, P.I., Bryonia, U., Adell, R., Hansson, B.O., Lindstrom, J., and Olsson, A. (1969). Intra-osseous anchorage of dental prostheses. I. Experimental studies. *Scand. J. Plast. Reconstr. Surg.*, **3**, pp. 81–100.

Brash, J.L., and Uniyal, S. (1979). Dependence of albumin-fibrinogen simple and competitive adsorption on surface properties of biomaterials, *J. Polym. Sci., Polym. Symp.*, **66**, pp. 377–389.

Chen, J. Y., Leng, Y. X., Zhang, X., Yang, P., Sun, H., Wang, J., Wan, G. J., Zhao, A. S., Huang, N., and Chu, P. K. (2006). Effect of tantalum content of titanium oxide film fabricated by magnetron sputtering on the behavior of cultured human umbilical vein endothelial cells (HUVEC), *Nucl. Instrum. Method Phys. Res. Section B.*, **242**, pp. 26–29.

Christensen, F. B., Dalstra, M., Sejling, F., Overgaard, S., and Cody Bünger, C. E. (2000). Titanium-alloy enhances bone-pedicle screw fixation: mechanical and histomorphometrical results of titanium-alloy versus stainless steel, *Eur. Spine J.*, **9**, pp. 97–103.

Compston, J. (2005). Prevention of vertebral fractures by strontium ranelate in postmenopausal women with osteoporosis. *Osteoporos Int.*, **16**, pp. 4–6.

Demuth, P., Irvine, C., Darrell, J., Raymond, S., Hammond, E., and Paula, T. Multilayer coating compositions, coated substrates and methods thereof, United States Patent Application 20120027837.

Derhami, K., Wolfaardt, J. F., Wennerberg, A., and Scott, P. G. (2000). Quantifying the adherence of fibroblasts to titanium and its enhancement by substrate-attached material, *J. Biomed. Mat. Res.*, **52**, pp. 315–322.

De Santis, D., Guerriero, C., Armato, U., Nocini, P. F., Richards, G., and Gotte, P. (1996). Adult human bone cells from jaw bones cultured on plasma-sprayed or polished surfaces of titanium or hydroxyapa-tite discs, *J. Mater. Sci. Mater. Med.*, **7**, pp. 21–28.

Didisheim, P. (1996). Substitute heart valves: Do we need better ones?, *Biomater. Forum.*, **18**(5), pp. 14–16.

Dion, I., Rouais, F., Trut, L., Baquey, C., Monties, J. R., and Havlik, P. (1993). TiN coating: surface characterization and hemocompatibility, *Biomaterials,* **14**(3), pp. 169–176.

Dold, J. M., Schimdt, M., and Steinemann, S. G. (1990). XPS study of retrieved titanium and Ti alloy implants, in clinical implant materials. In *Advances in Biomaterials* (Heimke, G., Soltesz, U., and Lee, A. J. C., ed.), Elesevier, Amsterdam, Chapter 9, pp. 69–74.

Dong, W., Zhang, T., Epstein, J., Cooney, L., Wang, H. Li, Y., Jiang, Y., Cogbill, A., Vardan, V., and Tian, Z. R. (2007). Multifunctional nanowire bioscaffolds on titanium, *Chem. Mater.*, **19**, pp. 4454–4459.

Dong, W., Zhang, T., McDonald, M., Padilla, C., Epstein, J., and Ryan, Z. (2006). Biocompatible nanofiber scaffolds on metal for controlled release and cell colonization, *Nanomed. Nanotech. Biol. Med.,* **4**(2), pp. 248–252.

Ego, S., Bruno, V., Claude, B., Jean, P. A., Jutta, S., Jean, M., Kaufman, K. H., Alfredo, R., Varela, C., Fiore, Kim, Brixen, J., Yves, R., and Steven, B. (2006). Strontium ranelate reduces the risk of vertebral and nonvertebral fractures in women eighty years of age and older. *J. Bone Min. Res.*, **21**, pp. 1113–1120.

Eriksson, C., Lausmaa, J., and Nygren, H. (2001). Interactions between human whole blood and modified TiO_2-surfaces: Influence of surface topography and oxide thickness on leukocyte adhesion and activation. *Biomaterials,* **22**(14), pp. 1987–1996.

Eriksson, C., Nygren, H., and Ohlson, K. (2004). Implantation of hydrophilic and hydrophobic titanium discs in rat tibia: cellular reactions on the surfaces during the first 3 weeks in bone. *Biomaterials,* **25**, pp. 4759–4766.

Feng, B., Chen, J. Y., Qi, S. K., He, L., Zhao, J. Z., and Zhang, X. D. (2002). Characterization of surface oxide films on titanium and bioactivity, *J. Mater. Sci. Mater. Med.*, **3**(5), pp. 457–464.

Feng, B., Weng, J., Yang, B. C., Qu, S. X., and Zhang, X. D. (2003). Characterization of surface oxide films on titanium and adhesion of osteoblast, *Biomaterials,* **49**, pp. 129–137.

Feng, B., Weng, J., Yang, B. C., Qu, S. X., and Zhang, X. D. (2003). Characterization of surface oxide films on titanium and adhesion of osteoblast, *Biomaterials,* **24**, pp. 4663–4670.

Feng, Z., Xianghuai, L., Yingjun, M., Nan, H., Yu, C., Zhihong, Z., Zuyao, Z., Anqing, C., and Zhenbin, J. (1998). Artificial heart valves: improved hemocompatibility by titanium oxide coatings prepared by ion beam assisted deposition, *Surf. Coatings Technol.*, **103**, pp. 146–150.

Fini, M., Cigada, A., Rondelli, G., Chiesa, R., Giardino, R., Giavaresi, G., Nicoli, A.N., Torricelli, P., and Vicentini, B. (1999). In vitro and in vivo behaviour of Ca- and P-enriched anodized titanium, *Biomaterials,* **20**, pp. 1587–94.

Liu, F.T., Wang, F., Shimizu, T., Igarashi, K., and Zhao, L. (2005). Formation of hydroxyapatite on Ti–6Al–4V alloy by microarc oxidation and hydrothermal treatment, *Surf. Coatings Technol.*, **199**, pp. 200–224.

Gaillard, Y., Jimenez-pique, E., Soldera, F., Mücklich, F., and Anglada, M. (2008). Quantification of hydrothermal degradation in zirconia by nanoindentation, *Acta Mater.*, **56**, pp. 4206–4216.

Gaillard, Y., Rico, V., Pique, E., and Elipe, A. (2009). Nanoindentation of TiO_2 thin films with different microstructures. *J. Phys. D: Appl. Phy.*, **42**, pp. 145305.

Giavaresi, G., Fini, M., Cigada, A., Chiesa, R., Rondelli, G., Rimondini, L., Torricelli, P., Aldini, N.N., and Giardino, R. (2003). Mechanical and histomorphometric evaluations of titanium implants with different surface treatments inserted in sheep cortical bone, *Biomaterials,* **24**, pp. 1583–194.

Groessner-Schreiber, B., and Tuan, R.S. (1992). Enhanced extracellular matrix production and mineralization by osteoblasts cultured on titanium surfaces in vitro, *J. Cell. Sci.,* **101**(1), pp. 209–217.

Gross, K.A., Chai, C.S., Kannangara, G.S.K., Ben-Nissan, B., and Hanley, L. (1998). Thin hydroxyapatite coatings via sol–gel synthesis, *J. Mater. Sci. Mater. Med.*, **9**(12), pp. 839–843.

Grossm, K.A., Chai, C.S., Kannangara, G.S.K., Nissam, B.B. (2006). Effects of microstructure of anodic titania on the formation of bioactive compounds, *Mater. Chem. Phys.,* **98**, pp. 39–43.

Gutmann, F., and Keyzer, H. (1986). *Modern bioelectrochemistry,* (Plenum Press, New York).

Han, Y., Hong, S.H., and Xu, K.W. (2003). Structure and in vitro bioactivity of titania-based films by micro-arc oxidation, *Surf. CoatingsTechnol.*, **168**, pp. 249–258.

Helsen, J. A., and Breme, H. J. (1998). *Metals as Biomaterials*, (John Wiley & Sons, New York).

Herzberg, M., Foldes, J., Steinberg, R., and Menczel, J. (1990). Zinc excretion in osteoporotic women, *J. Bone Miner Res.,* **5**(3), pp. 251–257.

Hanawa, H., Persons, D. A., and Nienhuis, A. W. (1997). Mobilization and mechanism of transcription of integrated self-inactivating lentiviral vectors, *J. Virol.,* **79**(13), pp. 8410–8421.

Huang, N., Yang, P., Cheng, X., Leng, Y., Zheng, X., Cai, G., Zhen, Z., Zhang, F., Chen, Y., Liu, X., and Xi, T. (1998). Blood compatibility of amorphous titanium oxide films synthesized by ion beam enhanced deposition, *Biomaterials,* **19**, pp. 771–776.

Huang, N., Chen, Y. R., Luo, J. M., Yi, J., Lu, R., Xiao, J., Xue, Z. N., and Liu, X. H. (1994). In vitro investigation of blood compatibility of Ti with oxide layers of rutile structure, *J. Biomater. Appl.*, **8**(4), pp. 404–412.

Huang, N., Yang, P., and Leng, Y. X. (2003). Hemocompatibility of titanium oxide films, *Biomaterials,* **24**(13), pp. 2177–2187.

Huimin, H., and Xianghuai, L. (2000). Influence of IBED parameters on the composition and structure of titanium oxide films deposited on LTI-carbon, *Mater. Lett.*, **45**, pp. 1–5.

Imai, Y., Nose, Y. (1972). A new method for evaluation of antithrombogenicity of materials. *J. Biomed. Mat. Res.,* **6**, pp. 165–172.

Imanishi, Y., and Nishizawa, Y. (2005). Strontium ranelate as a new therapeutic agent for osteoporosis, *Clin. Calcium,* **15**, pp. 25–28.

Ishizawa, H., and Ogino, M. (2004). Characterization of thin hydroxyapatite layers formed on anodic titanium oxide films containing Ca and P by hydrothermal treatment, *J. Biomed. Mat. Res.,* **29**, pp. 1071–1079.

Jager, I.L. (2002). Comment on: Effects of the substrate on the determination of thin films mechanical properties by nanoindentation by Saha and Nix [*Acta Mater.* 2002; **50**: 23], *Scr. Mater.,* **47**, pp. 429–432.

Jaeggi, C., Kern, P., Michler, J., Patscheider, J., Tharian, J., and Munnik, F. (2006). Film formation and characterization of anodic oxides on titanium for biomedical applications, *Surf. Interf. Anal.*, **38**, pp. 182–185.

Jones, M. I., McColl, I. R., Grant, D. M., Parker, K. G., and Parker, T. L. (2000). Protein adsorption and platelet attachment and activation, on TiN, TiC, and DLC coatings on titanium for cardiovascular application, *J. Biomed. Mat. Res.,* **52**(2), pp. 413–421.

Jouanny, I., Labdi, S., Aubert, P., Buscema, C., Maciejak, O., Berger, M. H., Guipont, V., and Jeandin, M. (2010). Structural and mechanical properties of titanium oxide thin films for biomedical application, *Thin Solid Films*, **518**, pp. 3212–3217.

Jung, O. J., Kim, S. H., Cheong, K. H., Li, W., and Saha, S. I. (2003). Metallorganic chemical vapor deposition and characterization of TiO_2 nanoparticles, *Bull. Korean Chem. Soc*, **24**, pp. 49–54.

Kalita, S.J. (2008). Nanostructured biomaterials. In *Functional Nanostructures: Processing, Characterization and Applications* (Sudipto Seal, ed.) Springer Science + Business Media, New York, pp. 168–219.

Kawahara, D., Kimura, Y., Nakamura, M., and Kawahara, H. (1993). Studies on the tissue adhesive capabilities to titanium by dynamic wettability Tst and cell attachment, invitro, *Clin. Mater.,* **14**, pp. 229–233.

Kawamoto, N., Mori, H., Terano, M., and Yui, N. (1997). Blood compatibility of polypropylene surfaces in relation to the crystalline-amorphous microstructure, *J. Biomater. Sci. Polym. Ed.,* **8**(11), pp. 859--877.

Kim, H. J., and Lee, K. H. (2009). Dependence of morphology of nanostructured titanium oxide on fluoride ion content, *Electrochem. Solid State Lett.,* **12**(3), pp. 10–12.

Kim, H. M., Miyaji, F., Kokubo, T., and Nishiguchi, S. (1999). Graded surface structure of bioactive titanium prepared by chemical treatment, *J. Biomed. Mat. Res.,* **45**, pp. 100–107.

Kim, J. H., Lee, H. S., Goak, J. C., Seo, Y. H., Kim, K. B., Park, K. S., and Lee, N. S. (2010). Effect of micro and nanoparticle inorganic fillers on the field emission characteristics of photosensitive carbon nanotube pastes, *Appl. Surf. Sci.,* **258**, pp. 2636–2642.

Komori, T., Yagi, H., Nomura, S., Yamaguchi, A., Sasaki, K., Deguchi, K., Shimizu, Y., Bronson, R. T., Gao, Y. H., Inada, M., Sato, M., Okamoto, R., Kitamura, Y., Yoshiki, S., and Kishimoto, T. (1997). Targeted disruption of Cbfa1 results in a complete lack of bone formation owing to maturational arrest of osteoblasts, *Cell,* **89**(5), pp. 755–764.

Könönen, M., Hormia, M., Kivilahti, J., Hautaniemi, J., and Thesleff, I. (1992). Effect of surface processing on the attachment, orientation, and proliferation of human gingival fibroblasts on titanium, *J. Biomater. Res.,* **26**, pp. 1325–4131.

Larsson, C., Emanuelsson, L., Thomsen, P., Ericsonm, E. L., Aronsson, B. O., Kasemo, B., Lausmaa, J., and Mater, J. (1997). Bone response to surface modified titanium implants—studies on the tissue response after 1 year to machined and electropolished implants with different oxide thicknesses, *Sci. Mater. Med.,* **8**, pp. 721–729.

Lee, J.-H., Kim, S.-E., Kim, Y.-J., Chi, C.-S., and Oh, H.-J. (2006). Effects of microstructure of anodic titania on the formation of bioactive compounds, *Mater. Chem. Phys.,* **98**, pp. 39–43.

Leng, X. L., Chen, J. Y., Yang, P., Sun, H., and Huang, N. (2003). Structure and properties of passivating titanium oxide films fabricated by DC plasma oxidation, *Surf. Coatings Technol.,* **166**, pp. 176–182.

Leng, Y. X., Chen, J. Y., Zeng, Z. M., Tian, X. B., Yang, P., Huang, N., Zhou, Z. R., and Chu, P.K. (2000). Properties of titanium oxide biomaterials synthesized by titanium plasma immersion ion implantation and reactive ion oxidation, *Thin Solid Films,* **377**, pp. 573–577.

Leng, Y. X., Huang, N., Yang, P., Chen, J. Y., Sun, H., Wang, J., Wan, G. J., Tian, X. B., Fu, R. K. Y., Wang, L. P., and Chu, P. K. (2002). Structure and properties of biomedical TiO_2 films synthesized by dual plasma deposition, *Surf. Coating Technol.*, **156**, pp. 295–300.

Li, D., Ferguson, S. J., Beutler, T., Cochran, D. L., Sittig, C., Hirt, H. P., and Buser, D. (2002). Biomechanical comparison of the sandblasted and acid-etched and the machined and acid-etched titanium surface for dental implants, *J. Biomed. Mat. Res.*, **60**(2), pp. 325–332.

Li, J. (1993). Behaviour of titanium and titania-based ceramics in vitro and in vivo, *Biomaterials*, **14**, pp. 229–232.

Li, L. H., Kim, H. W., Lee, S. H., Kong, Y. M., and Kim, H. E. (2005). Biocompatibility of titanium implants modified by microarc oxidation and hydroxyapatite coating, *J. Biomed. Mat. Res.*, **73A**, pp. 48–54.

Liu, X., Zheng, Z., Zhou, Z., Huang, N.,Yang, P., Cai, U., and Chen, Y. (1996). Surface modification of titanium based biomaterials by ion beam. *J. Biomater. Appl.*, **10**, pp. 330–337.

Li, P. J., Ohtsuki, C., Kokubo, T., Nakanishi, K., Soga, N., and Degroot, K. (1994). The role of hydrated silica, titania, and alumina in inducing apatite on implants, *J. Biomed. Mat. Res.*, **28**, pp. 7–15.

Li, L.-H., Kong, Y.-M., Kim, H.-W., Kim, Y.-W., Kim, H.-E., Heo, S.-J., and Koak, J.-Y. (2004). Improved biological performance of Ti implants due to surface modification by micro-arc oxidation, *Biomaterials*, **25**(14), pp. 2867–2875.

Macak, M., Tsuchiya, H., and Schmuki, P. (2005). High-aspect-ratio TiO_2 nanotubes by anodization of titanium, *Angew. Chem. Int.*, **44**, pp. 2100–2102.

Macak, J. M., Tsuchiya, H., Taveira, L., Aldabergerova, S., and Schmuki, P. (2005). Smooth Anodic TiO_2 Nanotubes, *Angew. Chem. Int.*, **44**, pp. 7463–7465.

Mao, Y., and Wong, S. S. (2006). *J. Am. Chem. Soc.*, **128**, pp. 8217–8226.

Martin, R. I., and Brown, P. W. (1995). Mechanical properties of hydroxyapatite formed at physiological temperature, *J. Mater. Sci. Mater. Med.*, **6**, pp. 138–143.

Meunier, P. J., Slosman, D. O., Delmas, P. D., Sebert, J. L., Brandi, M. L., Albanese, C., Lorenc, R., Pors-nielsen, S., Devernejoul, M. C., Roces, A., and Reginster, A.J. (2002). Strontium ranelate: dose-dep endent effects in established postmenopausal vertebral osteoporosis—A 2-year randomized placebo controlled trial, *J. Clin. Endocrinol. Metab.*, **87**, pp. 2060–2066.

Nygren, H., Eriksson, C., and Lausmaa, J. (1997). Adhesion and activation of platelets and polymorphonuclear granulocyte cells at TiO_2 surfaces, *J. Lab. Clin. Med.,* **129**, pp. 35–46.

Nygren, H., Tengvall, P., and Lundstrom, I. (1997). The initial reactions of TiO_2 with blood, *J. Biomed. Mat. Res.,* **34**(4), pp. 487–492.

Oh, H.-J., Lee, J.-H., Kim, Y.-J., Suhc, S.-J., Lee, J.-H., and Chi, C.-S. (2008). Surface characteristics of porous anodic TiO_2 layer for biomedical applications, *Mater. Chem. Phys.,* **109**, pp. 10–14.

Oji, M. O., Wood, J. V., and Downes, S. (1999). Effects of surface-treated cp Ti and Ti–6Al–4V alloy on the initial attachment of human osteoblast cells, *J. Mater. Sci. Mater. Med.,* **10**, pp. 869–872.

Oliver, W. C., and Pharr, G. M. (1992). An improvement technique for determining hardness and elastic modulus using load and displacement sensing indentation experiment, *J. Mater. Res.,* **7**, pp. 1564–1583.

Park, J.-W., Kim, Y.-J., Jang, J.-H., and An, C.-H. (2010). In vitro biocompatibility of magnesium-incorporated submicro-porous titanium oxide surface produced by hydrothermal treatment, *Appl. Surf. Sci.,* **6**, pp. 1661–1670.

Park, J.-W., Kim, Y.-J., and Jang, J.-H. (2011). Surface characteristics and in vitro biocompatibility of a manganese-containing titanium oxide surface, *Appl. Surf. Sci.,* **22**, pp. 230–234.

Park, J.-W., Kim, Y.-J., and Jang, J.-H. (2011). Surface characteristics and in vitro biocompatibility of a manganese-containing titanium oxide surface, *Appl. Surf. Sci.,* **258**, pp. 977–985.

Park, J.-W., Park, K.-B., and Suh, J.-Y. (2007). Effects of calcium ion incorporation on bone healing of Ti–6Al–4V alloy implants in rabbit tibiae. *Biomaterials*, **28**, pp. 306–3313.

Park, J., Cynthia, H.G., and Davies, J.E. (2001). Platelet interactions with titanium: modulation of platelet activity by surface topography, *Biomaterials*, **22**, pp. 2671–2682.

Park, J. Y., and Davies, J. E. (2000). Red blood cell and platelet interactions with titanium implant surfaces, *Clin. Oral. Impl. Res.*, **11**, pp. 530–539.

Petrini, P., Arciola, C. R., Pezzali, I., Bozzini, S., Montanaro, L., Tanzi, M. C., Speziale, P., and Visai, L. (2006). Antibacterial activity of zinc modified titanium oxide surface, *Int. J. Artif. Organs,* **29**(4), pp. 434–442.

Roa, J. J., Capdevila, X. G., Martinez, M., Espiell, F., and Sagarra, M. (2007). Nanohardness and Young's modulus of YBCO samples textured by the Bridgman technique, *Nanotechnology,* **18**, pp. 385701.

Rupp, F., Scheideler, L., Olshanska, N., de Wild, M., Wieland, M., and Geis-Gerstorfer, J. (2006). Enhancing surface free energy and hydrophilicity thorough chemical modification of microstructured titanium implant surfaces, *J. Biomed. Mat. Res. A.*, **76**, pp. 323–334.

Shackelford, J. F., and Alexander, W. (2001). Selecting structural properties. In *CRC Material Science and Engineering Handbook* (Shackelford, J. F., and Alexander, W., ed.), CRC Press, Boca Raton, FL, 3rd ed., Chapter 10, p. 510.

Shankar, K., Mor, G. K., Fitzgerald, A., and Grimes, C. A. (2007). Cation effect on the electrochemical formation of very high aspect ratio TiO_2 nanotubes in formamide-water mixtures, *J. Phys. Chem. C.*, **111**, pp. 21–26.

Sanborn, S. L., Gurunathan, M., Roger, E. M., and Kandice, K.-M. (2002). Endothelial cell formation of focal adhesions on hydrophilic plasma polymers. *Biomaterials*, **23**, pp. 1–8.

Smith, B. S., and Nisbet, D. I. (1972). Biochemical and pathological studies on magnesium deficiency in the rat. II. adult animals, *J. Comp. Pathol.*, **82**(1), pp. 37–46.

Sonoda, T., and Kato, M. (1997). Effects of discharge voltage on Ti-O film formation on Ti–6Al–4V alloy by reactive DC sputtering, *Thin Solid Films*, **303**, pp. 196–199.

Stansby, G., Berwanger, C., Shukla, N., and Hamilton, G. (1994). Endothelial cell seeding of vascular grafts: status and prospects, *Cardiovasc. Surg.*, **2**, pp. 543–548.

Sul, Y. T., Johansson, C. B., Petronis, S., Krozer, A., Jeong, Y., Wennerberg, A., and Albrektsson, T. (2002). Characteristics of surface oxides on turned and electrochemically oxidized pure titanium implants up to dielectric breakdown: oxide thickness, micropore configurations, surface roughness, crystal structure, and chemical composition, *Biomaterials*, **23**, pp. 491–501.

Sul, Y. T. (2003). The significance of the surface properties of oxidized titanium to the bone response: special emphasis on potential biochemical bonding of oxidized titanium implant, *Biomaterials*, **24**, pp. 3893–3907.

Sul, Y.-T., Johansson, C., Byon, E., and Albrektsson, T. (2005). The bone response of oxidized bioactive and non-bioactive titanium implants. *Biomaterials*, **26**, pp. 6720–6730.

Sul, Y. T., Johansson, B. C., Jeong, Y., and Albrektsson, T. (2001). The electrochemical oxide growth behaviour on titanium in acid and alkaline electrolytes, *Med. Eng. Phys.*, **23**, pp. 329–346.

Sunny, M. C., and Sharma, C. P. (1991). Titanium-protein interaction: change with oxide layer thickness. *J. Biomater. Appl.*, **5**(6), pp. 89–98.

Szymanowski, A., Gazicki-Lipman, M., Jakubowski, W., and Klimek, L. (2005). Plasma enhanced CVD deposition of titanium oxide for biomedical applications, *Surf. Coatings Technol.*, **200**, pp.1036–1040.

Takemoto, S., Tsuru, K., Hayakawa, S., Osaka, A., and Takashima, S. (2001). Highly blood compatible titania gels. *J. Sol–Gel Sci. Technol.*, **21**, pp. 97–104.

Takemoto, S., Yamamoto, T., Tsuru, K., Hayakawa, S., Osaka, A., and Takashima, S. (2004). Platelet adhesion on titanium oxide gels: effect of surface oxidation. *Biomaterials*, **17**, pp. 3485–3492.

Takashima, S., Takemoto, S., Hayakawa, S., and Osaka, A. (1999). In *Bioceramics*, (Ohgushi, H., Hastings, G. W., and Yoshikawa, T., ed.) World Scientific, Nara City, Japan, **12**, p. 551.

Tao, X. J., Li, S. J., Zheng, C. Y., Fu, J., Guo, Z., Hao, Y. Z., Yang, R., and Guo, Z. X. (2009). Synthesis of a porous oxide layer on multifunctional biomedical titanium by micro-arc oxidation, *Mater. Sci. Eng. C*, **29**, pp. 1923–1934.

Taveira, L. V., Macak, J. M., Tsuchiya, H., Dick, L. F. P., and Schmuki, P. (2005). Initiation and growth of self-organized TiO_2 nanotubes anodically formed in $NH_4F/(NH_4)_2SO_4$ electrolytes. *J. Electrochem. Soc.*, **152**, pp. B405–B410.

Tengvall, P., and Lundström, I. (1992). Physico-chemical considerations of titanium as a biomaterial, *Clin. mater.*, **9**(2), pp. 115–134.

Thorwarth, G., Mändl, S., and Rauschenbach, B. (2000). Plasma immersion ion implantation using titanium and oxygen ions, *Surf. Coatings Technol.*, **128**, pp. 116–120.

Tran, H. S., Puc, M. M., Hewitt, C. W., Soll, D. B., Marra, S. W., Simonetti, V. A., Cilley, J. H., and DelRossi, A. J. (1999). Diamond-like carbon coating and plasma or glow discharge treatment of mechanical heart valves, *J. Invest. Surg.*, **12**(3), pp. 133–140.

Trentz, O. A., Zellweger, R., Amgwerd, M. G., and Uhlschmid, G. K. (1997). Testing bone implants in cell lines and human osteoblasts, *Unfallchirurg*, **100**(1), pp. 39–43.

Tsyganov, I., Maitz, M. F., Wieser, E., Prokert, F., Richter, E., and Rogozin, A. (2003). Structure and properties of titanium oxide layers prepared by metal plasma immersion ion implantation and deposition, *Surf. Coating. Technol.*, **174–175C**, pp. 591–596.

Velten, D., Schenk-Meuser, K., Biehl, V., Duschner, H., and Breme, J. (2003). Characterization of thermal and anodic oxide layers on beta- and on near-beta- titanium alloys for biomedical application, *Zeitschrift fu"r Metallkunde*, **94**, pp. 667–675.

Walivaara, B., Lundstrom, I., and Tengvall, P. (1993). An in vitro study of H_2O_2-treated titanium surfaces in contact with blood plasma and a simulated body fluid, *Clin. Mater.*, **12**, pp. 141–148.

Walivaara, B., Aronsson, B., Rodahl, M., Lausmaa, J., and Tengvall, P. (1994). Titanium with different oxides: in-vitro studies of protein adsorption and contact activation, *Biomaterials*, **15**(10), pp. 827–834.

Wang, X., Zhang, F., Changrong, Li., Zheng, Z., Chen, L., Wang, H., and Liu, X. (2000). Influence of IBED parameters on the composition and structure of titanium oxide films deposited on LTI-carbon, *Mater. Lett.*, **45**, pp. 1–5.

Webster, T., Ergun, C., Doremus, R., Siegel, R., and Bizios, R. (2000). Enhanced functions of osteoblasts on nanophase ceramics. *Biomaterials*, **21**, pp. 1803–1810.

Wei, D. Q., Zhou, Y., Jia, D. C., and Wang, Y. M. (2007). Characteristic and in vitro bioactivity of a microarc-oxidized TiO_2-based coating after chemical treatment, *Acta Biomater.*, **3**, pp. 817–827.

Weng, W., Baptista, J. L. (1998). Alkoxide route for preparing hydroxyapatite and its coatings, *Biomaterials*, **19**, pp. 125–131.

Weng, W., Zhang, S., Cheng, K., Qu, H., Du, P., Shen, G., Yuan, J., and Han, G. (2003). Sol–gel preparation of bioactive apatite films, *Surf. Coatings Technol.*, **167**, pp. 292–296.

Williams, D. F. (1981). Electrochemical aspects of corrosion in the physiological environment. In *Fundamental Aspects of Biocompatibility* (Williams, D. F., ed.), CRS Press, Boca Raton, Florida, Chapter 1, pp. 12–41.

Wisby, A., Gregson, P. J., and Pater, L. M. (1991). Effect of surface treatment on the dissolution of titanium-based implant materials, *Biomaterials*, **12**, pp. 470–473.

Liu, X., Zheng, Z., Zhou, Z., Huang, N., Yang, P., Cai, G., and Chen, Y. (1996). Surface modification of titanium based biomaterials by ion beam, *J. Biomater. Appl*, **10**(4), pp. 330–337.

Xu, J., Xu, M., Rossi, C. G., Gavril, W. P., and Pan, Y.-X. (2011). Identification and characterization of seven new exon 11-associated splice variants of the rat mu opioid receptor gene, OPRM 1, *Mol. Pain*, **7**, pp. 9–23.

Yang, P., Huang, N., Leng, X. Y., Chen, J. Y., Sun, H., Wang, J., Chen, F., and Chu, P.K. (2002). In vivo study of Ti-O thin film fabricated by PIII, *Surf. Coatings Technol.,* **156**, pp. 284–288.

Yang, P., Leng, Y. X., Zhao, A. S., Zhou, H. F., Xu, L. X., Hong, S., and Huang, N. (2006). Blood compatibility improvement of titanium oxide film modified by phosphorus ion implantation, *Nucl. Instrum Methods Phys. Res. B,* **242**, pp. 15–17.

Yuhta, T., Kikuta, Y., Mitamura, Y., Nakagane, K., Murabayashi, S., and Nishimura, I. (1994). Blood compatibility of sputter-deposited alumina films, *J. Biomed. Mater. Res.*, **28**(2), pp. 217–224.

Zhang, F., Jin, S., Mao, Y., Zheng, Z., Chen, Y., and Liu, X. (1997). Surface characterization of titanium oxide films synthesized by ion beam enhanced deposition, *Thin Solid Film*, **310**, pp. 29–33.

Zhang, F., Zheng, Z., Chen, Y., Mao, Y., and Liu, X. (1998). Effect of Xe+ bombardment on the formation of highly oriented rutile-type titanium oxide films, *Thin Solid Film*, **312**, pp. 1–3.

Zhang, X., Wu, W. T., Wang, M. C., and Man, H. C. (2001). In-situ synthesis and wear performance of TiC particle reinforced composite coating on alloy Ti–6Al–4V, *Surf. Coatings Technol.*, **138**, pp. 95–100.

Zhao, G., Schwartz, Z., Wieland, M., Rupp, F., Geis-Gerstorfer, J., Cochran, D. L., and Boyan, B.D. (2005). High surface energy enhances cell response to titanium substrate microstructure, *J. Biomed. Mat. Res.,* **74**, pp. 49–58.

Zhao, G., Raines, A. L., Wieland, M., Schwartz, Z., and Boyan, B. D. (2007). Requirement for both micron- and submicron scale structure for synergistic responses of osteoblasts to substrate surface energy and topography, *Biomaterials*, **28**, pp. 2821–2829.

Zilla, P., Deutsch, M., and Meinhart, J. (1999). Endothelial cell transplantation, *Seminars in Vascular Surgery*, **12**, pp. 52–63.

Zreiqat, H., Valenzuela, S. M., Nissan, B. B., Roest, R., Knabe, C., Radlanski, R.J., Renz, H., and Evans, P.J. (2005). The effect of surface chemistry modification of titanium alloy on signaling pathways in human osteoblasts, *Biomaterials*, **26**, pp. 7579–7586.

Chapter 7

Metal Oxide Nanostructured Films for Photovoltaic Applications

S. K. Tripathi

Department of Physics, Panjab University, Chandigarh 160014, India

surya@pu.ac.in

7.1 Introduction to Nanotechnology

Nanotechnology is current a booming trend in science and technology. Nanotechnology refers to a field of applied science which deals with matters on an atomic and molecule scale and involves developing devices within that size. Nanostructures which correspond to the structures having at least one physical dimension between 1 and 100 nm have been constantly gaining interest for their peculiar and fascinating properties [1]. Nanostructured material includes atomic clusters, layered (lamellar) films, filamentary structures and the bulk nanostructured materials. Recently, various nanostructures have been under research, such as carbon nanotubes, nanowires, nanobelts, nanorings and nanorods [2].

Oxide Nanostructures: Growth, Microstructures, and Properties
Edited by Avanish Kumar Srivastava

ISBN 978-981-4411-35-6 (Hardcover), 978-981-4411-36-3 (eBook)
www.panstanford.com

As we know that nanomaterials have the structural features in between of those of atoms and the bulk materials. Because of the very fine grain sizes and consequent high density of interfaces, nanocrystalline materials exhibit a variety of properties that are different and often considerably improved in comparison with those of conventional coarse-grained materials. These include increased strength/hardness, improved ductility/toughness, reduced density, reduced elastic modulus, higher electrical resistivity, increased specific heat, higher coefficient of thermal expansion, lower thermal conductivity and superior soft magnetic properties. Microstructural features on the nanometre scale have been known to be useful in enhancing the properties and the performance of materials. At present, the broad field of nanostructured materials includes (i) nanoparticles, (ii) nanocrystalline materials and (iii) nanodevices. The potential applications for the various kinds of nanoscale materials include dispersions and coatings, high-surface-area materials, functional nanostructures (optoelectronic devices, biosensors, nanomachines, etc.) and bulk nanostructured materials for structural or magnetic applications.

Among the problems of modern physics two challenges to the semiconductor physics can be emphasized. The first one is related to the continuous miniaturization of the operational scale of semiconductor devices and the second one to the development of low-cost energy sources. The synthesized nanomaterials are extremely useful in both present and future semiconductor industry. Recently, electronic devices are being made in smaller and smaller sizes and of better performance. According to a report by the National Renewable Energy Lab (NREL), the worldwide electrical energy consumption will increase from currently 10 terawatts to 30 terawatts by 2050. Semiconductor solar cells are efficient solar energy converters. Solar electricity is a steadily growing energy technology today and solar cells have found markets in variety of applications ranging from consumer electronics and small scale distributed power systems to centralized megawatt scale power plants. The wide range of novel solar energy converters with unique properties are based on the physical ideas on the nanometre scale as well as on the use of novel materials.

7.1.1 Metal Oxide Nanomaterials

Metal oxide semiconductor has been widely studied due to it varied properties and applications. Metal oxide semiconductor films have received considerable attention in recent years due to their optical and electrical properties. Some of them are good candidates for transparent conductive oxide films. Oxide semiconductors are preferential in photo-electrochemistry because of their exceptional stability against photo-corrosion on optical excitation in the bandgap [3]. Furthermore, because of large bandgap (>3 eV) of the oxide semiconductors, it cannot absorb visible light and thus provides the transparency of the semiconductor electrode for the large part of the solar spectrum. In addition to TiO_2, semiconductors used in porous nanocrystalline electrodes in dye-sensitized solar cells include, for example, ZnO, CdSe, CdS, WO_3, Fe_2O_3, SnO_2, Nb_2O_5 and Ta_2O_5 [4]. Among them, TiO_2 is one of the metal oxide semiconductors suitable for use in optoelectric devices. Undoped and doped oxide thin films have a wide range of applications as an important semiconductor material such as solar cells and electrical, piezoelectric or luminescent devices and also as gas sensors and chemical sensors. It is well known that chemical doping greatly influences the electronic and optical properties of oxide thin films. Al, In and Ga-doped oxide thin films are one of the promising transparent conductive oxides for advanced applications such as displays and electro chromic devices [5].

Metal oxides are wide-bandgap materials with a large degree of disorder and numerous defects in the forbidden gap. The scattering of free charge carriers by phonons and defects in such materials with small average scattering time may lead to the spatial localization of carriers within the mean free path. The very low value of the charge carrier mobility in metal oxides [6] implies localization and a mean free path of the order of the inter-atomic distance. Such scattering decreases the uncertainty in a particle's position and hence increases its momentum (energy) uncertainty. The electronic states in the forbidden gap of metal oxides films are crucial for the charge carrier transport. The conductivity and mobility of nanoporous TiO_2 films change over several orders of magnitude depending on the ambient conditions (vacuum, oxygen, water) by means of the change the electronic states distribution or dielectric screening of the barriers.

7.1.2 Titanium Dioxide as a Material

Titanium dioxide became the semiconductor of choice. It is a low-cost, widely available, non-toxic and biocompatible material, and as such is even used in health care products as well as domestic applications such as paint pigmentation [7]. Over recent years, it was shown that titanium dioxide, because of its suitable electronic and optical properties, is one of the materials that could be incorporated in to construction of DSSCs to improve their performance. Various TiO_2 nanostructures can be prepared on optically transparent and electrically conducting oxide substrates using various methods.

Anatase is widely reported as the most active crystalline phase for photocatalytic applications. However, to utilize anatase for high temperature applications, it is necessary to delay the anatase-to-rutile transformation temperature. This can be achieved through the addition of dopants [8]. It is necessary to develop visible light active TiO_2 in order to utilize the full solar spectrum and also was to develop a high temperature stable visible light active anatase photocatalysts. Addition of metal has been well reported to influence early phase transformation of anatase to rutile. It has been reported that silver causes an increase in nucleation sites and oxygen vacancies which can all result in early phase transformation. Metal doping can improve visible light absorption of metal oxide thin films due to surface plasmon absorption [9].

TiO_2 served as a model system to study the properties and factors that control the photoconversion efficiency of DSSCs. The electron transport occurs through nanocrystalline TiO_2 through trapping and detrapping events; however, exact nature of these trap states are not thoroughly quantified. Research efforts are going on not only to quantify the trap states in mesoporous metal oxides but new mesoporous architectures also to increase the conversion efficiency of metal oxide-based photovoltaics.

7.2 Crystal Structure of TiO_2

TiO_2 has been used in a wide range of applications such as paint, pigment, water splitting, hydrogen and polluting gas sensors, oxygen gas sensor, and biosensors [10,11]. Oxygen vacancies and the titanium interstitial as well as the decreased recombination probability with increased surface area make this attractive material for the application in DSSCs [12]. Important crystalline forms of TiO_2 anatase, rutile and brookite are shown in Fig. 7.1. Anatase appears as pyramid-like crystals and is stable at low temperatures, whereas needle-like rutile crystals are dominantly formed at high temperature processes. Single crystals of TiO_2 also have rutile structure. The third crystalline form of TiO_2, brookite, is difficult to produce and is therefore not of practical interest for the DSSCs. The band-gaps of the crystalline forms are 3.2 eV (absorption edge at 388 nm) for anatase and 3.0 eV (absorption edge at 413 nm) for rutile.

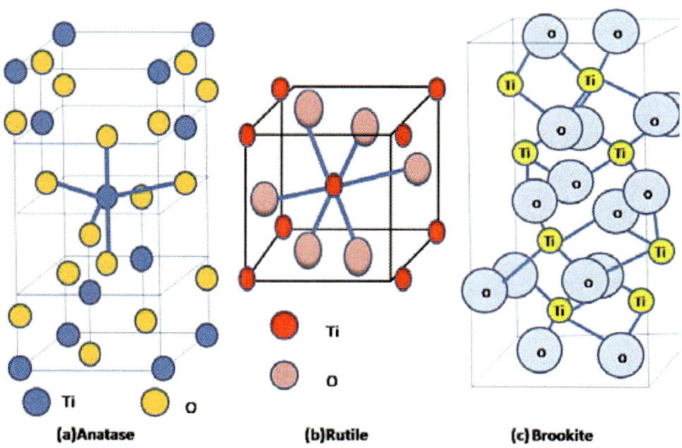

Figure 7.1 Crystalline polymorphs of TiO_2 (a) anatase, (b) rutile, (c) brookite, where small red sphere: O^{2-}, big grey sphere: Ti^{4+}.

TiO_2 also exists in many other metastable forms such as monoclinic, tetragonal, orthorhombic and cubic. The electronic state of Ti in TiO_2 is Ti^{4+} ($3d^0$), in which the valence band of TiO_2 is composed of the hybridized 2p orbital of oxygen and 3d orbital of Ti. However, the conduction band is completely composed of pure 3d orbital of Ti. Therefore, the different parity of valence band and the conduction band of TiO_2 decreases the transition of electrons to the valence band, which decrease the electron hole (e^-–h^+) recombination probability. Among the polymorphs of TiO_2, it is known that rutile is the most stable and common. However, the Fermi level of anatase is 100 mV higher than that of rutile, which leads to higher open circuit potential (V_{OC}). The larger surface area of anatase is responsible for efficient dye loading, due to which the higher photocurrent, and higher photovoltaic performance effect is achieved.

Anatase has been the phase of interest since it is the primary crystal formed in the usual colloidal preparation method of the nanostructured TiO_2 electrodes. However, low light harvesting efficiency with low absorption in red region and the lack of efficient electron transfer through the ITO/TiO_2 interface increases the back electron transfer to the redox ion in electrolyte limits the conversion efficiency of DSSCs. The enhanced grain boundary density of spherical TiO_2 nanoparticles and random electron diffusion increase the loss of conversion efficiency with the increase of carrier recombination [13]. Therefore, broad work has been performed to minimize the recombination. For example, the carrier leakage of direct electron acceptance from the TiO_2 film can be avoided through prevention of hole mediation to excited dye by addition of 4-tert-butylpyridine into redox solution [14]. Other ways are the use of TiO_2 blocking layer by mixing the rutile TiO_2 with anatase TiO_2 and the fabrication of one-dimensional (1-D) nanostructures such as nanowires, nanotubes and nanorods. The TCO substrate and TiO_2 are important routes of the back electron transfer to electrolyte. Since the Fermi level of FTO is close to the redox Fermi level at open circuit potential, so the back electron transfer from FTO to the I^-/I^{3-} redox couple is negligible. So, nanocrystalline TiO_2 is direct route of recombination of I^-/I^{3-} redox couple. Focus of research is to prevent this back electron transfer by using rutile/rutile mixed TiO_2 with an upper layer of mesoporous anatase TiO_2 film. The compact TiO_2 layer acts as a shielding layer against the ionic penetration. It also stimulate provoke different microstructure.

7.3 Electron Transport in TiO_2

Electron transport through any material is controlled by its band structure. So, it is usually linked to the bonding characteristics. The ionic compound TiO_2 exhibits strong electron–phonon coupling (polarons). Small polarons correspond to localized and self-trapped electrons are formed for a sufficiently strong electron–phonon interaction [15]. The electron transport usually occurs through thermally activated hopping from one electronic state to the other. Weaker coupling strength corresponds to large polarons having spatially extended wave functions that exhibit band-type behaviour. In the bulk semiconductor, electron transport takes place by this way. While in the sintered nanocrystalline film, the electron transport is dependent on the extent of particle connectivity, properties of the individual nanoparticle or electronic coupling between the particles [16]. The TiO_2 particles considered for DSC is typically in the 10–25 nm range, i.e. much higher than TiO_2 exciton Bohr radius (1.5 nm). Therefore, the nanocrystalline film composed of 10 nm particles can be considered as a network of bulk crystals having a band of extended electronic states that arise from the high surface area of nanocrystals. The surface states in nanomaterials lie within the bandgap usually arise from deviation in bond angles, unsaturated bonds and presence of impurities and bond angle from that of the bulk materials. However, in the case of TiO_2, for the particles in the 10–25 nm range, the nature and origin of the surface states have not been understood so far.

There is a junction capacitor in the metal oxide–electrolyte interface. So, the electron transport in DSSC through the mesoporous electrode is coupled with the electrolyte. The capacitive charging at this junction further retards the electron mobility. Figure 7.2 shows directional electron diffusion in the 1-D nanostructures. By frequency-resolved modulated photocurrent/photovoltage spectroscopy, a comparative study of electron transport and recombination in DSSCs between spherical nanoparticles (NPs) and TiO_2 nanotubes (TNTs) has been made [17].

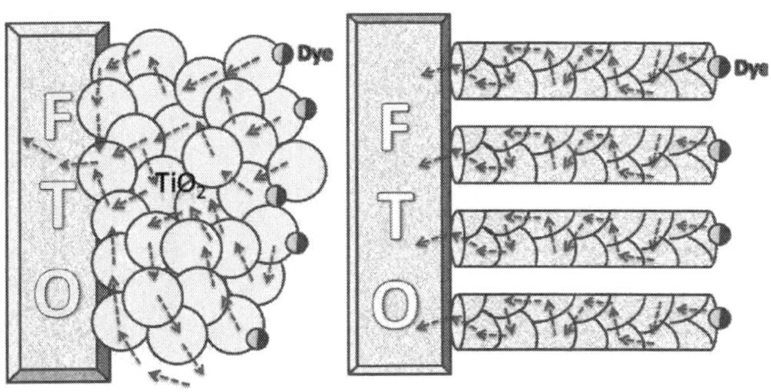

Figure 7.2 Random electron diffusion through spherical TiO_2 nanoparticles (C), and 1-D nanostructures of TiO_2 with directional electron diffusion (D).

In contrast to NPs photoelectrode, the rate of recombination in TNTs is much slower. Therefore, charge collection efficiency in TNTs photoelectrode is increased compare to NPs photoelectrode. The internal and external dye loading with stronger internal light-scattering effects of TNTs exhibits higher photoconversion efficiency. However, charge collection efficiency and the dye loading of TNTs is strongly dependent on the nanotube arrays. Bundling of nanotube creates additional pathway via intertube connection. It decreases the internal surface area of the films accessible to dye molecules and increases the charge recombination, hence decreases the photoconversion efficiency. The high photocurrent density and photocurrent efficiency can be achieved by better-aligned nanotube arrays. Supercritical CO_2 drying process is developed for preparing crack-free and bundle-free nanotubes because capillary stress creates the bundling and microcracks of TNTs during the solvent evaporation [18].

The other 1-D nanostrucutres are nanorods, nanowires, etc. Recently, TiO_2 nanorods have been prepared and showed improved photoconversion performance compared with the naocrystalline counterparts due to the increased rate of electron transport arisen from the high crystallinity of the nanorod. The region for the poor extent of dye loading and the poor light harvesting of nanorods based DSSC is due to the low specific surface area of the nanorod. TiO_2 nanorod (4 nm × 20 nm) of high specific surface area has been synthesized from the necking of truncated NPs by oriented attachment approach to facilitate the improved dye loading [19]. So, the improved conversion efficiency is achieved due to higher dye adsorption ability of nanorod by improving its roughness factor (169 μm^{-1}) compare to nanoparticles (115.6 μm^{-1}).

7.4 Introduction to Photovoltaics

The wide range of novel solar energy converters with unique properties is based on the use of the novel materials and on the physical ideas on the nanometre scale. Among novel photonic devices, one can select, for example, organic light-emitting diodes or dye-sensitized solar cells. These devices are based on materials, e.g. porous Si, the organic polymer poly(p-phenylene vinylene) (PPV) or porous TiO_2. The extremely high radiative recombination efficiency of organic materials is combined with the poor transport properties of disordered polymer chains in the organic light emitting diodes [20]. As the particle size decreases, the surface-to-volume ratio increases in the porous materials.

Although, the propagation of charge carriers through the porous matrix is the limiting factor for the performance of DSSCs, the wide surface area of a porous matrix enlarges the illumination area and hence increases the light harvesting efficiency. Scientists are trying to understand the transport properties of the materials through nanostructured metal-oxide networks in order to improve the efficiency. A high localization of the mean free path of charge carriers via scattering on the intrinsic defects or grain boundaries takes place in nanoporous materials [21]. When the photon strikes on the photoanode, photovoltage arises due to separation of excess charge carriers. The photovoltage techniques were numerously used for the investigation of crystalline semiconductors. The value of diffusion length of minority carriers and surface potential lifetime, parameters of surface states such as energy position, density, occupancy and capture cross section for electrons and photons can be determined from the transient and spectral photovoltage measurements. For the investigation of transport properties of porous materials, the contactless approach is the main advantage of the photovoltage technique. The biocompatibility of porous TiO_2 and porous Si and the comparable size of characteristic dimensions allow combining porous semiconductors with biomolecular and biological objects and applying the PV technique to the investigation of the charge carrier generation and their spatial separation in these systems.

Direct utilization of solar radiation to produce electricity is close to an ideal way to utilize the nature's renewable energy flow. With PV cells, power can be produced near the end user of the electricity, thus avoiding transmission losses and costs, and the solar panels themselves operate without noise, toxic and greenhouse gas emissions, and require very little maintenance. Furthermore, the huge theoretical potential and the very high practical potential of the solar electricity make it attractive for large-scale utilization. Despite the significant development over the past decades, the high cost of solar cells has remained a limiting factor for the implementation of the solar electricity in a large scale. The standard silicon solar cell technology has matured to a stage where costs reductions are mostly foreseen only by the economies of scale. Costs calculations of the thin film photovoltaic technologies on the other hand place them more or less in the same line with the standard silicon technology [22]. There is a prevailing need for the development of new materials and concepts for the photovoltaic conversion, to lower the price of the PV power production.

7.4.1 Solar Irradiation

The intensity of solar radiation in the earth's distance from the sun is approximately 1353 kW/m^2. The spectrum of solar radiation is similar to black body at 6000 K temperature. The available solar irradiation in a certain place depends on the latitude, the altitude and the climatic type in a yearly basis, and on the season, the time of day and the weather conditions in a specific time. The total yearly solar irradiation on horizontal surface is 700–1000 kWh/m^2 in North Europe, 900–1300 kWh/m^2 in Middle Europe, 1300–1800 kWh/m^2 in South Europe, 1800–2300 kWh/m^2 in the equator and 2000–2500 kWh/m^2 in the so-called "solar belt," i.e. between 20° and 30° latitude [23].

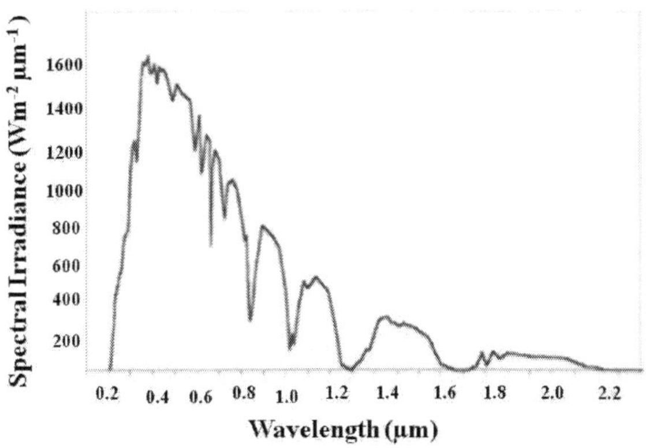

Figure 7.3 Standard AM1.5 global solar spectrum. The data are from http://www.pv.unsw.edu.au/am1.5.html.

The yearly solar irradiation falling to the same horizontal area in North Europe conditions (about 1000 kWh/m^2/year), converted to solar electricity with total PV system efficiency of 10% amounts to about 12000 kWh/m^2, i.e. about 65% of the yearly electricity consumption of the examined typical Finnish family.

7.4.2 Photovoltaic Characterization

A photovoltaic cell is a device, which converts incident photon energy to electrical energy. Generation of electricity depends upon the capability of the PV device under illumination to produce voltage over an external load and current through the load at the same time. This is characterized by the current-voltage (I–V) curve of the cell at certain illumination and temperature. When the cell is short circuited under illumination, the maximum current called short circuit current (I_{SC}), is generated, while under open circuit conditions, the voltage is at its maximum, called the open circuit voltage (V_{OC}). The maximum power point (MPP) is the point in the I–V curve corresponds to maximum power which gives maximum product of current and voltage. One more important characteristic of the solar cell performance is the fill factor (FF), defined as

$$\mathrm{FF} = P_{\mathrm{MAX}}/I_{\mathrm{SC}}\cdot V_{\mathrm{OC}} = I_{\mathrm{m}}\cdot V_{\mathrm{m}}/I_{\mathrm{SC}}\cdot V_{\mathrm{OC}}.$$

So maximum power can be given by

$$P_{\mathrm{MAX}} = V_{\mathrm{OC}}\cdot I_{\mathrm{SC}}\cdot \mathrm{FF}.$$

While the physical principles behind the operation of different types of photovoltaic cells are generally different, the I–V curve of well performing cells are similar and can be characterized and compared with each other in terms of FF, V_{OC} and I_{SC}.

The energy conversion efficiency of the solar cell is defined as the power produced by the cell (P_{MAX}) divided by the power incident on the representative area of the cell (P_{light}), i.e.

$$\eta = P_{\mathrm{MAX}}/P_{\mathrm{light}}$$

The efficiency of the solar cell depends on the temperature of the cell, on the total light intensity and the spectral distribution of the intensity at AM 1.5 global standard solar spectrum and temperature 25°C. Generally, special solar simulator light sources are used for the standard measurements.

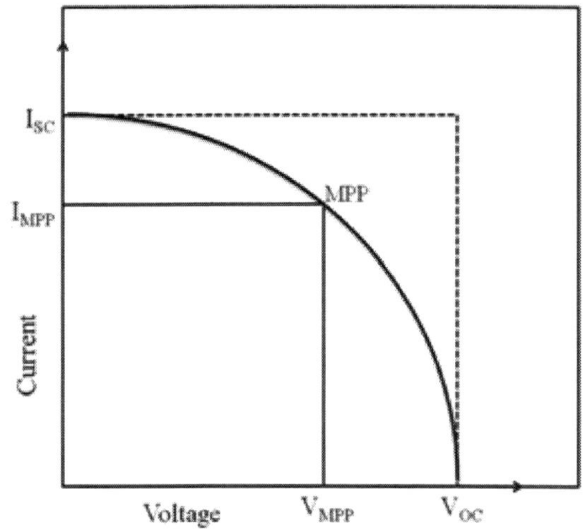

Figure 7.4 I–V curve of a PV cell showing the open-circuit voltage (V_{OC}), short-circuit current (I_{SC}), maximum power point (MPP), and the current and voltage at the MPP: I_{MPP}, V_{MPP}.

7.5 Dye-Sensitized Solar Cell

The most familiar PV system is the dye-sensitized nanostructured solar cell developed by Prof. Grätzel in 1991. This photoelectrochemical solar cell is based on a TiO_2 nanoparticle photoelectrode sensitized with a light-harvesting metallo-organic dye and is on the verge of commercialization offering an interesting alternative for the existing thin film solar cell and silicon based solar cells. At the same time the research activity as well as the industrial interest around the technology is growing fast.

Their low fabrication cost and high solar-to-electrical power conversion efficiency make them very attractive for producing electricity in large scales [24]. In some aspects, they are superior conventional semiconductor solar cell based on the p–n junction diodes. In contrast to the conventional systems, where the semiconductor assumes both the task of light absorption and charge carrier transport, the two functions are separated by in DSSCs. Semiconductor in DSSCs provide the path for electron transport and the photoelectrons are generated by a photosensitive dye via photo-induced electron injection from dye into conduction band of solid. The interfaces between the electrolyte, semiconductor and dye play an important role in charge separation. A schematic presentation of the operating principles of the DSSC is shown in Fig. 7.5.

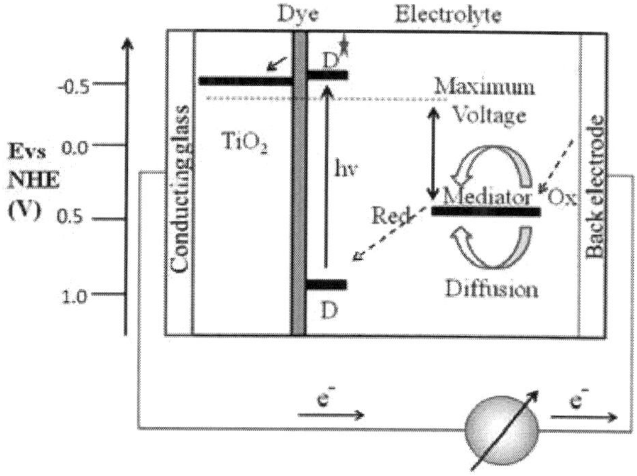

Figure 7.5 Operation principle and energy-level diagram of the dye-sensitized nanocrystalline solar cell.

Photo-excitation of the sensitizer (S) is followed by electron injection into the conduction band of the mesoporous oxide semiconductors. The dye molecule is regenerated by the redox system, which itself is regenerated at the counter electrode by electrons passed through the load. Potentials are referred to the normal hydrogen electrode (NHE). The open-circuit voltage of the solar cell corresponds to the difference between the redox potential of the mediator and the Fermi level of the nanocrystalline film. Dye-sensitized solar cells (DSSCs) comprise transparent conducting glass electrode coated with porous nanocrystalline TiO_2, ZnO or SnO_2 thick film, dye molecule attached to surface of nanocrystalline metal oxide. When the dye-sensitized nanocrystalline solar cell was first presented, perhaps the most puzzling phenomenon was the highly efficient charge transport through the nanocrystalline TiO_2 layer. The mesoporous electrodes are very much different compared to their compact analogs because (i) the electron conductivity of the film is very small; (ii) individual particle size of nanocrystalline particles does not support a built-in electrical field; and (iii) the electrolyte surrounds all the particles effectively, decouples the particles and screen any existing electric field with in a nm. Charge transport in mesoporous systems is under keen debate today. However, the "effective" electron diffusion coefficient is expected to depend on a number of factors such as space charge compensation by ionic motion in electrolyte and trap filling. Therefore, research attempt will continue fractal-like assembly of nanoparticles. A desirable morphology of the films would have the mesoporous nanotubes or nanorods aligned in parallel to each other and vertically with respect to the conducting glass slide. This would facilitate charge collection efficiency avid grain boundaries and allow the junction to be formed under better control. It was reported that the solar energy conversion efficiency of DSSCs can achieve 11% by using a titanium dioxide (TiO_2) nanoparticle film as the photoanode. Porous TiO_2 or ZnO structures and Ruthenium complexes have been mostly used as semiconductor materials and sensitizers in DSSCs, respectively [25]. Another group of new type of solar cells grabbing increasing industrial and research attention is the purely organic solar cells based on organic semiconducting polymers, macromolecules, dyes, pigments or liquid crystalline materials.

The depletion of fossil fuel resources, increased demand for energy and global warming of planet caused by green house effect forces us to seek it from sustainable renewable energy sources and concerns on environment demands that resources should be clean as well. Metal oxide semiconductors, which are stable and less costly and environment friendly materials, are used in photovoltaics either as photoelectrode in dye solar cells (DSSCs) or to build metal oxide p–n junctions. The d-block binary metal oxides viz. TiO_2, ZnO, SnO_2 and Nb_2O_5 are the best candidates as photoelectrode due to the dissimilarity in orbitals constituting their conduction band and valence band which decreases the probability of charge recombination and enhances the carrier lifetime in these materials. Various I-D nanostructures such as nanowires, nanotubes, and nanofibres have been explored to enhance the efficiency of photoelectrochemical cell. Metal oxide thin films for dye-sensitized solar cell.

7.5.1 Metal Oxide Thin Films for Dye-Sensitized Solar Cell

Metal oxide nanostructured films have sufficiently large surface area for dye adsorption. However, these nanostructures limit the photo conversion efficiency by enhancing electron recombination probability and slowing electron transport rate which essentially depends on the process of electrons diffusion to the anode through particles [26]. The electron diffusion rate can be increased by replacing nanoparticle film by vertical aligned single crystalline nanowire or nanorod film which can provide a direct path to anode for further increasing the electron diffusion rate. Benefits of aligned I-D nanostructures are their mesoporosity and large surface area, which allow large adsorption of dye molecules and larger density of states in metal oxides than the molecular orbital of dye allows faster injection of electrons from the dye to the metal oxide. The trapping and detrapping of electrons at the surface atomic states in the electronic band causes inefficient electrical transport in nanostructured metal oxide semiconductors. The large fraction of surface atoms increase trap density in nanostructured materials which reduces the kinetic energy of the flowing electrons, i.e. decrease in charge collection efficiency which ultimately results in inferior cell performance. Binary metal oxides such as TiO_2, ZnO, Nb_2O_5, Fe_2O_3, ZrO_2, Al_2O_3 and CeO_2 and ternary compounds such as $SrTiO_3$ and Zn_2SnO_4 have been tested for their use as photoelectrodes in DSSC. Among them the d-block binary metal oxides, viz., TiO_2, ZnO and Nb_2O_5 are the best candidates as photoelectrode due to the dissimilarity in orbitals constituting their conduction band and valence band. Various morphologies and approaches, such as doping, composites, core/shell structures, have been explored to enhance the electrical transport properties. One-dimensional morphologies play a great role in enhancing the electrical transport properties. One-dimensional nanowires synthesized through template-assisted method are able to increase the recombination time and allowed enhancement in charge collection efficiency, but their reduced surface area adversely affects the dye loading. A lot of effort is being undertaken to synthesize mesoporous metal/non-metal doped TiO_2 and composites with high crystallinity to realize high-performance DSSC [27].

7.5.2 TiO_2 Photoelectrode with Scattering Layer

Use of scattering (SC) TiO_2 particles (films or mixed with nanocrystalline particles) help to develop very thin nanocrystalline. TiO_2 film of high surface area reduces the use of dye without affecting the cell efficiency. Figure 7.6 shows the light confined effect on nanocrystalline TiO_2 particles. It is clear that SC TiO_2 particles scattered the incident longer wavelength light and prolonged the optical path length. Hence, the transmitted visible light can be used for photon conversion. The light scattering property is strongly dependent on the refractive index between the active TiO_2 film, on the relative size of the scattering particles and on scattering layer [28].

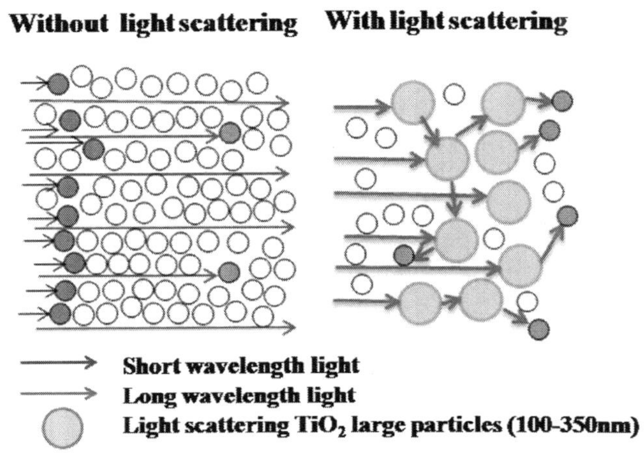

Figure 7.6 Conceptual scheme of light-confined effect.

A study of the performance of different thickness (4 μm, 8μm) of nanocrystalline TiO_2 film of high surface area with and without scattering layer shown that the 4 μm nanocrystalline TiO_2 thin film with SC layer has a comparable performance with 8 μm thick TiO_2 film without SC layer [29]. It concludes that it is better to use thin film of TiO_2 with SC layer for the industrial production of large DSSC modules, which will minimize the cost of dye by lesser consumption in the production procession. Other metal oxides such as ZnO, Al_2O_3, ZrO_2 and TiO_2-rutile form has been also explored as effective SC layer on top of the active TiO_2 film.

7.5.3 Metal-Doped Titania (TiO_2) Photoelectrode

Doping of metal ions has a considerable effect on the electrical surface state modification of TiO_2, which leads to significant changes in electrical conductivity, particle aggregation, shifting of Fermi level potential, dye absorption characteristics, and charge transfer kinetics of TiO_2.

Researchers have suggested that the optimized level of metal ion doping into TiO_2 differs mainly with respect to the tuning of the flat band of TiO_2, hence the performance of DSSCs. Nb doping level of 2.5–7.5 mol% on doped anatase TiO_2 (Nb–TiO_2) has positive effect on the I_{SC} with positive shift of flat band potential [30]. Enhanced electron injection, increased charge transfer kinetics as well as the improvement of dye loading up to $5.7 \pm 0.2 \times 10^{-8}$ (mol cm^{-2}) at a doping level of 7.5 mol% compared to the undoped state is the region of increase in I_{SC}. The TiO_2 doped by various metals such as tungsten (W), neodymium (Nd) also exhibited the similar phenomenon, while Al doped TiO_2 showed an opposite effect [31]. The performance of the DSSC of compact Nb-doped TiO_2 layer with an upper mesoporous TiO_2 layer is greater as compared to the compact TiO_2 layer. Figure 7.7 describes the band diagram of two compact layers and the charge transfer kinetics.

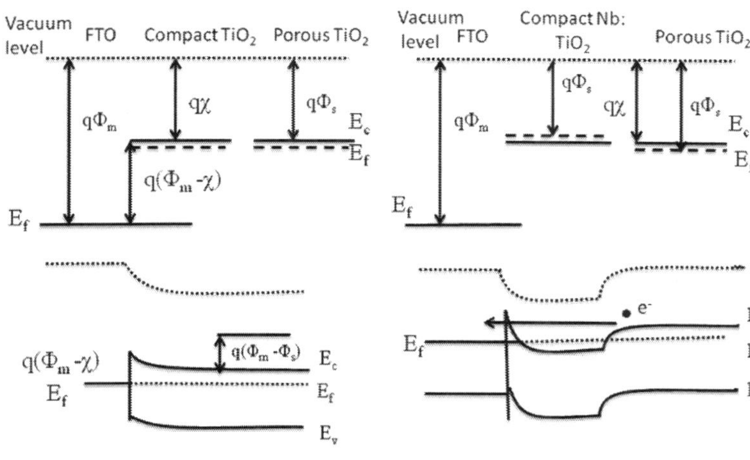

Figure 7.7 Schematic illustration of band diagram of compact TiO_2 and compact Nb–TiO_2 layer before (upper) and after (under) contact with FTO.

It is clear that the Fermi level of compact TiO_2 and Nb-doped TiO_2 layers is below and above of the conduction band (ECB) respectively. This is attributed to the non-degeneracy of compact TiO_2 and the degeneracy of Nb-doped TiO_2. When compact TiO_2 and FTO are in contact, i.e. semiconductor–metal contact, Schottky barrier or depletion region, $q(\Phi_m - \Phi_s)$ is form and TiO_2 (Φ_s) and by the difference of work function of FTO (Φ_m). Electrons have to overcome this barrier in order to be transferred from compact TiO_2 to FTO. As a result, the electron injection efficiency is decreased and the back electron transfer is increased. On the other hand, the contact between FTO and the compact degenerate Nb-doped TiO_2 layer form a very narrow depletion region. Electrons injected rapidly from compact Nb-doped TiO_2 to FTO by tunnelling lead to ohmic contact characteristics [32]. Therefore, the incorporation of compact Nb-doped TiO_2 layer reduces the interfacial charge transfer resistance and results in the enhancement in efficiency of DSSC. Doped metal ions have a strong effect for the tuning of flat band potential, as well as the electron injection efficiency.

7.5.4 Core–Shell Composite of Titania (TiO_2) and Other Metal Oxides for Photoelectrode

Coating of TiO_2 with different metal hydroxides or oxides of wide bandgap such as $Mg(OH)_2$, $Zn(OH)_2$, MgO, ZnO and $SrTiO_3$ has received much attention for the enhancement in the efficiency of DSSCs. Metal hydroxide or oxide coating on TiO_2 employed a core–shell structure is very useful for the efficient photoconversion. First, the wide bandgap of shell metal oxide increases the surface resistance of TiO_2. So, it minimizes the electron–hole recombination by reducing the back electron transfer to the electrolyte. Also, the shell layer increases the dye absorption due to increased surface area leading to high photoconversion and finally tuning the conduction band of core structure. Figure 7.8 shows the core–shell structure of metal oxide.

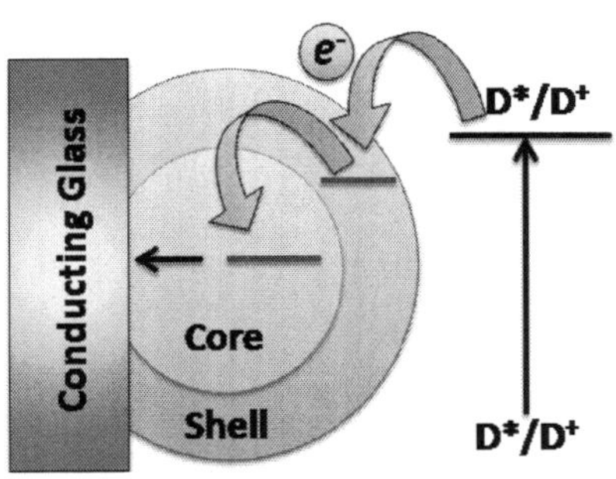

Figure 7.8 Core–shell matrix.

The use of a thin metal hydroxide or oxide layer (shell) allows the formation of an energy barrier at the metal oxide/electrolyte interface. It reduces the recombination rate and, hence, improves the cell performance. Injected electrons do not face any energy barriers during the electron diffusion to the FTO due to favourable core–shell structure [33]. Figure 7.9 shows the energy band tuning of TiO_2 photoelectrode. A bilayer of Nb_2O_5 coated nanoporous TiO_2 photoelectrode has been prepared [34]. The conduction band of shell materials Nb_2O_5 is 100 mV negative than the core materials TiO_2. Thus, the Nb_2O_5 layer acts as an energy barrier, therefore, confined the electrons to the core TiO_2 particles. It reduces the recombination and improves the efficiency by increasing the J_{sc}. The increase of the photovoltage is due to the shift of the Fermi level of TiO_2 towards negative direction by Nb_2O_5.

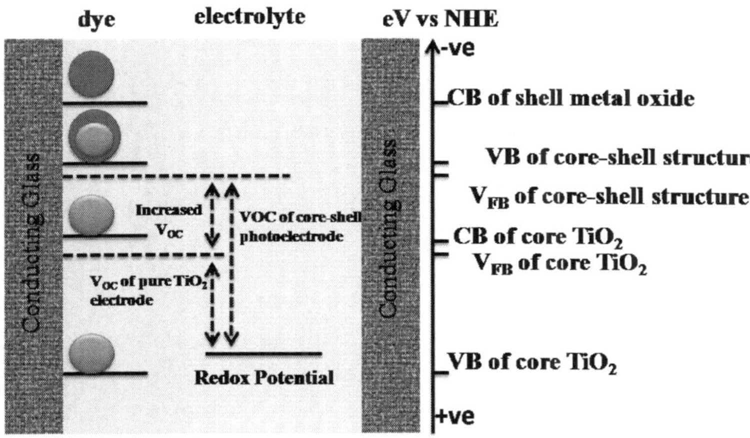

Figure 7.9 Schematic illustration of energy band tuning of TiO_2 photoelectrode.

Diamant et al. reported the development of $SrTiO_3$ coated TiO_2 photoelectrode. The conduction band of $SrTiO_3$ is 200 mV negative than that of TiO_2. The increment of J_{SC} by the reduced recombination as well as the efficient electron injection to FTO is attributed to suitable band position of core–shell composite. Increment of V_{OC} and I_{SC} results in the overall DSSC performance. Non- metal oxides such as $CaCO_3$ with high band energy (6.0 eV) for the formation of favourable core–shell with TiO_2 and high isoelectric point caused the enhancement of dye loading. The increased photovoltage is also attributed to the phenomenon described above with the more negative conduction band edge of $CaCO_3$ compare to that of TiO_2. Overall, photo conversion increased by 24.4% with this approach [35].

7.5.5 TiO_2 Coupled with Other Semiconductors

For achieving better charge separation in a light energy conversion system the fabrication, design, and tailoring of the semiconductor heterojunction has received significant attention. The vectorial charge transfer from one semiconductor to another with suitable band edge positions that is thermodynamically favourable can increase the lifetime of the charge carriers and promoting the interfacial charge transfer and catalytic efficiency [36]. To enhance the charge separation and photo catalytic activity the coupling of TiO_2 wide-bandgap semiconductor like SnO_2 has been reported [37]. Both TiO_2 and SnO_2 are n-type semiconductors with bandgap energies greater than 3.0 eV and exhibit a strong absorption threshold in the UV region. The CB edges of anatase TiO_2 and SnO_2 are situated at 0.34 and +0.07 V, respectively, while the VB edge of SnO_2 (+3.67 V) is much more positive than that of anatase TiO_2 (+2.87 V).

Composite films of SnO_2–TiO_2 fabricated on transparent conductive glass substrate showed superior activity for the degradation of Rhodamine B (RhB) due to the combined effects of appropriate phase composition of anatase (73.4%) and rutile (26.3%), better crystallization, low sodium content and slower recombination rate of charge carriers. Electrons flow into the SnO_2 underlayer after the UV exposure, while the hole diffuses oppositely to the TiO_2 overlayer. Consequently, more number of holes reach the TiO_2 surface to cause oxidation reactions, while electrons are consumed for dioxygen reduction at the band edge of SnO_2 film. VB edge of TiO_2 inhibits the interfacial hole transfer from TiO_2 to SnO_2, thereby increasing the hole density in the TiO_2 VB, and oxidizes the organic contaminant. Double-layered SnO_2–TiO_2 film with both the components exposed to the surface prepared by plasma-enhanced chemical vapour deposition (PECVD) showed superior activity for phenol degradation. Accumulation of photogenerated electrons on SnO_2 has been confirmed by doing the systematic studies [38]. TiO_2 composite film under applied bias and formation of SnO_2–TiO_2 heterojunction interface result in higher quantum efficiency and better photocatalytic activity for Acid Orange 7 degradation. The activity increased at the mass ratio of SnO_2 to TiO_2 (2:1) and decreased at lower dosage of SnO_2 than TiO_2 [39]. Using ball-milling technique and H_2O as a disperser, SnO_2–TiO_2 has been prepared which showed higher activity for the degradation of monocrotophos, compared to the SnO_2.

7.6 Synthesis Techniques

The synthesis technique is a key factor in determining the effectiveness of the device as studies have shown that TiO_2 production and physical form are among the most important factors in determining the overall efficiency of the device. There are many methods of synthesising titanium dioxide, such as hydrothermal combustion synthesis, gas-phase method, microwave synthesis and sol-gel processing. Here is the brief overview of these synthesis techniques.

7.6.1 Hydrothermal Synthesis

Hydrothermal synthesis is carried out in a vessel under pressure known as autoclave with the reaction in aqueous solution. The temperature of the autoclave can be raised above the boiling point of water, reaching the vapour saturation pressure. This technique is widely used for the preparation of TiO_2 nanoparticles which can be easily obtained through hydrothermal treatment of peptized precipitates of a titanium precursor with water. The hydrothermal method can be useful to control particle morphology, crystalline phase, grain size and surface chemistry through regulation of the solution composition, reaction temperature, pressure, solvent properties, aging time and additives.

7.6.2 Combustion

Combustion process involves a rapid heating of a solution containing redox groups. In this synthesis technique, highly crystalline particles with large surface areas are obtained. During combustion, the temperature reaches approximately 650°C for one or 2 min, which makes the material crystalline. Since the time is so short, the transition from anatase phase to rutile phase is forbidden.

7.6.3 Gas Phase Methods

Gas phase method is an ideal method for the synthesis of thin films. It can be carried out chemically or physically. Chemical vapour deposition (CVD) is a widely used industrial technique that can coat large areas in a short space of time [40]. In this technique, titanium dioxide is formed from a chemical reaction or decomposition of a precursor in the gas phase.

Physical vapour deposition (PVD) is another thin film deposition technique. Films are formed without a chemical transition from precursor to product in the gas phase. For the preparation of TiO_2 thin films a focused beam electrons, are produced from a tungsten wire heated by a current, which heats the titanium dioxide material. This is known as electron beam (E-beam) evaporation. Titanium dioxide films deposited with E-beam evaporation have superior characteristics over CVD grown films such as conductivity, smoothness, presence of contaminations and crystallinity. Reduced TiO_2 powder (heated at 900°C in a hydrogen atmosphere) is necessary for the required conductance needed to focus an electron beam on the TiO_2 [41].

7.6.4 Microwave Synthesis

TiO_2 materials are also synthesized using microwave radiation method. The advantage of this technique is that it eliminates the use of high temperature calcination for extended periods of time and allow for fast, reproducible synthesis of crystalline TiO_2 nanomaterials. TiO_2 colloidal nanoparticles prepared within 5 min using microwave radiation have been studied [42]. By combining microwave synthesis and hydrothermal method, high-quality rutile rods have been generated; also, TiO_2 hollow and open-ended nanotubes were synthesized by reaction of anatase and rutile crystals in NaOH solution.

7.6.5 Sol-Gel Processing

Sol-gel processing is a very common chemical approach to produce high-purity materials shaped as fibres, powders, monoliths, thin film coatings and self-supported bulk structures [43]. This method has several advantages over other synthesis techniques such as homogeneity, purity, stoichiometric control, composition, ease of preparation and ease of introducing dopants and the ability to produce porous powders or thin film coatings. There are two possible routes for carrying out sol-gel synthesis, the alkoxide route and the non-alkoxide route.

The alkoxide route involves hydrolysis of a metal alkoxide, followed by condensation. The hydrolysis/condensation reactions typically form a three-dimensional polymeric structure that, upon calcination, will result in anatase or rutile titanium dioxide crystals depending on the calcination temperature.

In the non-alkoxide route, inorganic salts such as $TiCl_4$ are used as the starting material and require the removal of the inorganic anion to produce the required oxide such as titanium dioxide. However, halides often remain in the final oxide material and are difficult to remove.

Acknowledgements

The author is thankful to Ms. Mamta Rani and Ms. Gurvir Kaur for helping in the preparation of this chapter.

References

1. Tang, C. C., Fan, S. S., Li, P., Liu, Y. M. and Dang, H. Y. (2001). Synthesis of boron nitride in tubular form, *Mater. Lett.*, **51**, pp. 315–319.
2. Lu, W. and Lieber, C. M. (2006). Semiconductor nanowires, *J. Phys. D: Appl. Phys.*, **39**, pp. R387-R406.
3. Luque, A. and Hegedus, S. (2010). *Handbook of Photovoltaic Science and Engineering*, 2nd Ed. (John Wiley & Sons).
4. Papadopoulou, E. L., Pagkozidis, A., Barberoglou, M., Fotakis, C. and Stratakis, E. (2010). Electrowetting Properties of ZnO and TiO_2 Nanostructured Thin Films, *J. Phys. Chem. C*, **114**, pp. 10249–10253.

5. Eun, L. K., Mingsong, W., Tung, K. E. and Hong, H. S. (2009). Structural, electrical and optical properties of sol–gel AZO thin films, *Current Applied Physics*, **9**, pp. 683–687.
6. Breckenridge, R. and Hosler, W. (1953). Electrical Properties of Titanium Dioxide Semiconductors. *Phys. Rev.*, **91**, pp. 793–802.
7. Liu, J., Wang, Y. and Sun D. (2012). Enhancing the performance of dye-sensitized solar cells by benzoic acid modified TiO_2 nanorod electrode. *Renew. Energ.*, **38**, pp. 214–218.
8. Li, Y., Jia, L., Wu, C., Han, S., Gong, Y., Chia, B., Pu, J. and Jian, L. (2012). Mesoporous (N, S)-codoped TiO_2 nanoparticles as effective photoanode for dye-sensitized solar cells, *J. Alloys Comp.*, **512**, pp. 23–26.
9. Wei, N. N., Han, T., Deng, G. Z., Li, J. L. and Du, J. Y. (2011). Synthesis and characterizations of three-dimensional ordered gold- nanoparticle-doped titanium dioxide photonic crystals, *Thin Solid Films*, **519**, pp. 2409–2414.
10. Fujishima, A. and Honda, K. (1972). Electrochemical photolysis of water at a semiconductor electrode, *Nature*, **238**, pp. 37–38.
11. Rahman, M.M., Ahammad, A.J.S., Jin, J.H., Ahn, S.J. and Lee, J.J. (2010). A Comprehensive Review of Glucose Biosensors Based on Nanostructured Metal-Oxides, *Sensors*, **10**, pp. 4855–4886.
12. Li, J. G., Ishigaki, T. and Sun, X. (2007). Anatase, Brookite, and Rutile Nanocrystals via Redox Reactions under Mild Hydrothermal Conditions: Phase-Selective Synthesis and Physicochemical Properties, *J. Phys. Chem. C*, **111**, pp. 4969–4976.
13. Benkstein, K. D., Kopidakis, N., Van de Lagemaat, J. and Frank, A. J. (2003). Influence of the Percolation Network Geometry on Electron Transport in Dye-Sensitized Titanium Dioxide Solar Cells, *J. Phys. Chem. B*, **107**, pp. 7759–7767.
14. Hattori, R. and Goto, H. (2007). Carrier leakage blocking effect of high temperature sputtered TiO_2 film on dye-sensitized mesoporous photoelectrode, *Thin Solid Films*, **515**, pp. 8045–8049.
15. Jase, R., Thavasi, V. and Ramakrishna, S. (2009). Metal Oxides for Dye-Sensitized Solar Cells, *J. Am. Ceram Soc.*, **92**, pp. 289–301.
16. Bisquert, J. (2008). Physical Electrochemistry of Nanostructured Devices, *Phys. Chem. Chem. Phys.*, **10**, pp. 49–72.
17. Zhu, K., Neale, N. R., Miedaner, A. and Frank, A. J. (2006). Enhanced Charge-Collection Efficiencies and Light Scattering in Dye-Sensitized Solar Cells Using Oriented TiO_2 Nanotubes Arrays, *Nano Lett.*, **7**, pp. 69–74.

18. Zhu, K., Vinzant, T. B., Neale, N. R. and Frank, A. J. (2007). Removing Structural Disorder from Oriented TiO_2 Nanotube Arrays: Reducing the Dimensionality of Transport and Recombination in Dye-Sensitized Solar Cells, *Nano Lett.*, **7**, pp. 3739–3746.

19. Kang, S. H., Choi, S. H., Kang, M. S., Kim, J. Y., Kim, H. S., Hyeon, T. and Sung, Y. E. (2008). Nanorod-Based Dye-Sensitized Solar Cells with Improved Charge Collection Efficiency, *Adv. Mater.*, **20**, pp. 54–58.

20. Lebedev, E., Dittrich, T., Petrova-Koch V., Karg, S. and Brütting, W. (1997). Charge Carrier Mobility in poly(p-phenylenevinylene) studied by time of flight technique, *Appl. Phys. Lett.*, **71**, pp. 2686–2688.

21. Tiong, K.K., Amirtharaj, P.M., Pollak, F.H. and Aspnes, D.E. (1984). Effects of As^+ ion implantation on the Raman spectra of GaAs: "Spatial correlation" interpretation, *Appl. Phys. Lett.*, **44**, pp. 122–124.

22. Frantzis, L. (2000). Opportunities for Cost Reductions in Photovoltaic Modules, *16th European Photovoltaic Solar Energy Conference*, (Glasgow, UK).

23. Peippo, K. (1992). Aurinkosähköjärjestelmien käytön ja mitoituksen perusteita, NEMO-raportti **23**, ISSN 0785-644X, ISNB 951-22-1342-7, Teknillinen korkeakoulu, in Finnish.

24. Gratzel, M. (2001). Photoelectrochemical Cells, *Nature*, **414**, pp. 338–344.

25. Yanagida, M., Komatsuzaki, N. O., Kurashige, M., Sayama, K. and Sugihara, H. (2010). Optimization of Tandem structured dye-sensitized solar cells, *Sol. Energ. Mater. Sol. Cells*, **94**, pp. 297–302.

26. Liu, R., Yang, W. and Qiang, L. (2012). Enhanced efficiency for dye-sensitized solar cells using a surface-treated photo-anode, *J. Power Sour.*, **199**, pp. 418–425.

27. Kim, D. Y., Kim, J., Kim, J., Kim, A., Lee, G. and Kang, M. (2012). The photovoltaic efficiencies on dye sensitized solar cells assembled with nanoporous carbon/TiO_2 composites, *J. Ind. Eng. Chem.*, **18**, pp. 1–5.

28. Arakawa, H., Yamaguchi, T., Takeuchim, A. and Agatsuma, S. (2006). Efficiency improvement of dye-sensitized solar cell by light confined effect, *Conference Record of the 2006 IEEE 4th World Conference on Photovoltaic Energy Conversion*, pp. 36–39, (New York).

29. Hore, S., Vetter, C., Kern, R., Smit, H. and Hinsch, A. (2006). Influence of scattering layers on efficiency of dye-sensitized solar cells, *Sol. Energy Mater. Sol. Cells*, **90**, pp. 1176–1188.

30. Lü, X., Mou, X., Wu, J., Zhang, D., Zhang, L. Huang, F., Xu, F. and Huang, S. (2010). Improved-Performance Dye-Sensitized Solar Cells Using Nb-Doped TiO_2 Electrodes: Efficient Electron Injection and Transfer, *Adv. Funct. Mater.*, **20**, pp. 509–515.

31. Ko, K. H., Lee, Y. C. and Jung, Y. J. (2005). Enhanced efficiency of dye-sensitized TiO_2 solar cells (DSSC) by doping of metal ions, *J. Colloid Interface Sci.*, **283**, pp. 482–487.

32. Lee, S., Noh, J. H., Han, H. S., Yim, D. K., Kim, D. H., Lee, J. K., Kim, J. Y., Jung, H. S. and Hong, K. S. (2009). Nb-Doped TiO_2: A New Compact Layer Material for TiO_2 Dye-Sensitized Solar Cells, *J. Phys. Chem. C*, **113**, pp. 6878–6882.

33. Chappel, S., Chen, S. G. and Zaban, A. (2002). TiO_2-Coated Nanoporous SnO_2 Electrodes for Dye-Sensitized Solar Cells, *Langmuir*, **18**, pp. 3336–3342.

34. Chen, S. G., Chappel, S., Diamant, Y. and Zaban, A. (2001). Preparation of Nb_2O_5 Coated TiO_2 Nanoporous Electrodes and Their Application in Dye-Sensitized Solar Cells, *Chem. Mater.*, **13**, pp. 629–4634.

35. Lee, S., Kim, J. Y., Youn, S. H., Park, M., Hong, K. S., Jung, H. S., Lee, J. K. and Shin, H. (2007). Preparation of a Nanoporous $CaCO_3$-Coated TiO_2 Electrode and Its Application to a Dye-Sensitized Solar Cell, *Langmuir*, **23**, pp. 11907–11910.

36. Serpone, N., Maruthamuthu, P., Pichat, P., Pelizzetti, E. and Hidaka, H. (1995). Exploiting the interparticle electron transfer process in the photocatalysed oxidation of phenol, 2-chlorophenol and pentachlorophenol: chemical evidence for electron and hole transfer between coupled semiconductors, *J. Photochem. Photobiol. A: Chem.*, **85**, pp. 247–255.

37. Cao, Y., Zhang, X., Yang, W., Du, H., Bai, Y., Li, T. and Yao, J. (2000). A Bicomponent TiO_2/SnO_2 Particulate Film for Photocatalysis, *Chem. Mater.*, **12**, pp. 3445–3448.

38. Vinodgopal, K. and Kamat, P. V. (1995). Enhanced Rates of Photocatalytic Degradation of an Azo Dye Using SnO_2/TiO_2 Coupled Semiconductor Thin Films, *Environ. Sci. Technol.*, **29**, pp. 841–845.

39. Shifu, C., Lie, C., Shen, G. and Gengyu, C. (2006). The Preparation of coupled SnO_2/TiO_2 photocatalyst by ball milling, *Mater. Chem. Phys.*, **98**, pp. 116–120.

40. Chen, X. and Mao, S. S. (2007). Titanium dioxide Nanomaterials: Synthesis, Properties, Modifications, and Applications, *Chem. Rev.*, **107**, pp. 2891–2959.

41. Van de Krol, R., Goossens, A. and Schoonman, J. (1997). Mott-Schottky Analysis of Nanometer-Scale Thin-Film Anatase TiO_2, *J. Electrochem. Soc.*, **144**, pp. 1723–1727.

42. Corradi, A. B., Bondioli, F., Focher, B., Ferrari, A. M., Grippo, C., Mariani, E. and Villa, C. (2005). Conventional and Microwave-Hydrothermal Synthesis of TiO_2 Nanopowders, *J. Am. Ceram. Soc.*, **88**, pp. 2639–2641.

43. Keshmiri, M., Troczynski, T. and Mohseni, M. (2006). *J. Hazard. Mater. B*, **128**, pp. 130–137.

Chapter 8

Nanostructured Materials as Nanoprobes for Bioimaging Applications

S. D. Geethanjali and A. Vadivel Murugan

Centre for Nanoscience and Technology,
School of Green Energy Technologies,
Pondicherry University (A Central University),
Dr. R. V. Nagar, Kalapet, Pondicherry 605014, India

avmrajeshwar@gmail.com

8.1 Overview

Nanostructured materials as bionanoprobes have gained much interest in recent times owing to their ultra small size and functional properties, which make them effective tools for bioimaging applications. In the last decade, semiconducting nanoparticles such as cadmium sulphide (CdS), cadmium selenide (CdSe), cadmium telluride (CdTe) and lead selenide (PbSe) have been studied as potential candidates for making nanoprobes. However, either due to their intrinsic cytotoxicity or due to their lack of target specificity and uncontrolled drug release, most of them pose several challenges that need to be solved. Hence, it is of prime importance to further

Oxide Nanostructures: Growth, Microstructures, and Properties
Edited by Avanish Kumar Srivastava

ISBN 978-981-4411-35-6 (Hardcover), 978-981-4411-36-3 (eBook)
www.panstanford.com

deepen the search for the nanoprobes that are non-toxic as well as highly photoluminescent. In this chapter, we discuss some of the robust nanostructured materials that can function as effective bionanoprobes with regard to their synthesis, characterization, functional properties and bioimaging applications.

8.2 Introduction

Nanobiotechnology is an imminent interdisciplinary field that is a perfect fusion of material science and biotechnology. The nanostructured materials that are used in biotechnological applications have exceptional prospects like high luminescence and other optoelectronic and magnetic novel properties which make them excellent probes for biological systems since they are of the same size as that of these biomolecules [1]. Moreover, a basic understanding of these nanostructured materials is necessary to manipulate and fabricate these nanoparticles into robust nanoprobes. By this way, we can solve many daunting challenges that our humankind faces, such as cancer and AIDS. This will further help to improve the healthcare industry [2]. Nanoprobes are nanostructured materials of size 10–100 nm which are very essential tools of bioimaging. Many researchers demonstrate that the advantages of using nanomaterials for bioimaging are aplenty since they are target specific, provide better biofunctionalization and absorption, have lower retention and circulation time with better luminescence [3]. Moreover, they can be amicably tuned for controlled drug release and minimal side effects. This area of research is said to increase many fold in the coming years due to the tremendous application of these nanoprobes. It is indeed a very challenging and overwhelming task to comprehensively state all the applications of nanostructured materials as nanoprobes. In this present chapter, we address the synthesis, functional characteristics and the emergence of nanomaterial-based therapeutics which is likely to accelerate future biomedical imaging and further improve health care.

In addition, this chapter further embarks on a broad discussion on the conventional nanoprobes that have been used in the past decades focussing mainly on their design, biofunctionalization and imaging strategy. We further proceed by extensively discussing the newer generation nanoprobes especially the III–V semiconductor quantum dots, lanthanide-based quantum dots and carbon-based quantum dots. This section is also designed to specifically concentrate on the design, synthesis, biofunctionalization, cytotoxicity studies, etc., with the main focus on the upconversion that forms an essential part of the applicability of these lanthanide-based quantum dots. Further, the steps to yield the quantum efficiency are also discussed. Specific applications may not be fully discussed but some examples are given nevertheless for better understanding. The major limitations of currently used nanoprobes are highlighted with a view of suggesting the direction of research in this area to create effective bioimaging tools. In conclusion, this chapter will discuss the indispensable necessities for designing nanostructured materials as nanoprobes for superior bioimaging applications.

8.3 Nanoprobes for Bioimaging Applications

8.3.1 Nanostructured Materials as Nanoprobes

Xiaogang Liu et al. have extensively studied therapeutic crystalline nanomaterials which include gold nanoparticles with different shapes, magnetic nanoparticles, carbon-based nanomaterials, quantum dots and lanthanide-based up conversion nanoparticles for bionanoprobe applications [4].

The intrinsic properties of nanomaterials play an important role in influencing their therapeutic applications. Accordingly, their size, shape, composition, nature of crystallinity, biological interaction, drug delivery routes and in vivo targeting completely influence the bioimaging strategy.

8.3.1.1 Size of the nanoprobe

The intake of any nanomaterial into the biological system depends on the size of the nanomaterial. Small sized nanoprobes are transported much rapidly across cell membranes with better surface modification and functionalization to target specific diseased tissues. The size basically depends on the construct of the nanoprobe. They can be either small drug molecules that self assemble (micellar structures) to function as both bioimaging and drug carriers or those that are hybrid nanostructured materials where the drug moieties are encapsulated or surface bound to nanomaterial carriers. The carrier molecules can be organic, inorganic-crystalline or polymeric that is sensitive to stimulus which specializes in harnessing the dual advantage of both magnetic and luminescence properties with specific drug delivery [5–7]. The entire size of the nanoprobe must be below 150 nm for effective clearance from the body. It is common knowledge that a 20 nm nanoparticle has a high surface area to mass ratio than that of bulk which makes it efficient to attach, absorb and bear drug molecules, biomolecules or probes [8]. Moreover, by controlling the size, we can actually control the circulation of the nanoprobes in the blood stream and further we can induce accumulation of the probes at cancer tissues better that bulk or large sized particles. Along with size, surface properties and charge on the nanoparticle also influences cancer destruction. Neutral and negatively charged particles with size less than 150 nm have been proven to destroy cancer tissues. Furthermore, 10–100 nm sized nanoparticles can easily diffuse through the extracellular matrix and thus prove to be effectual cancer destruction agents [9]. Scientific studies conducted in mice have proved that sub 100 nm nanoprobes were successfully up taken while larger particles accumulated in the stomach lymphoid tissue when injected directly into the blood [10]. In addition, the removal also depends on the size of the nanomaterial. If the nanomaterial is allowed to accumulate in a part of the biological system, then it may lead to dangerous consequences like mutation or toxicity that may eventually lead to cancer. This has been noted in scientific literature pertaining to iron poisoning in hepatic injury where iron poisoning is mainly observed in liver and gastrointestinal tract [11]. Recently, Choi et al. reported that the renal clearance is about 5.5 nm [12]. In case of nanoprobes used as direct imaging tools, different sized particles have diverse optical properties with different emission and absorption wavelengths. Interestingly, the same materials when size tuned, exhibits different luminescent properties. This aspect is currently exploited in size-dependent emission of quantum dots and lanthanide-based nanoprobes.

8.3.1.2 Nanoparticle shape

Shape is also an important aspect of nanoprobe design as the shape directly implicates a particular application with a specific required wavelength. Also surface plasmonic nanomaterial optical properties are greatly influenced by shape [13]. Depending on shape, the optical properties also vary. Spherical gold nanoparticles have low surface area when considering irregularly shaped non-spherical nanoparticles. Owing to this, the non-spherical particles have more binding capacity due to their coarse edges and uneven surface. Carbon nanotubes have high carrier density for different drugs with good retention time and controlled drug release.

8.3.1.3 Nanoparticle composition

Materials composition has direct influence on cellular toxicity. It also influences the various optoelectronic, magnetic properties which single them out towards their particular application as therapeutic agents. The surface reactivity if the nanomaterials are a consequence of their composition which completely determines their retention time, binding efficiency with the target receptor, interfaces with the biological systems and eventually the cellular toxicity. For example, polymeric nanoparticles are found to be more biocompatible than cadmium-based quantum dots. Also silica-based nanomaterials are more suitable for drug carriers as they are more stable and comparatively less toxic. The current research has suggested lanthanide-based nanomaterials to be non-toxic and effective replacements for cadmium-based quantum dots. Additionally, along with composition crystallinity also plays an important role in photoluminescence. In photodynamic therapies, effective light emissions are necessary to bring about destruction of diseased tissues. This is possible only with highly crystalline nanomaterials. Hexagonal phase sodium yttrium fluoride ($NaYF_4$) has high upconversion rate that the cubic phase. Also by doping, the luminescent can be substantially increased [14].

8.3.1.4 Nanomaterial functionalization

Most of the materials that have great potential as nanoprobes fail to impress as biologically safe agents for non-invasive applications. Hence, to make these materials biologically safe, functionalization is necessary. Surface engineering or interior modification which consists of encapsulating in a biopolymer, attaching receptor recognizing moieties, doping with lanthanide ions, etc., are some of the tactics to improve biocompatibility of nanoprobes as shown in Fig. 8.1. Moreover, the main aim of functionalization is to render the nanoparticle chemically inert and biologically safe. For example, nanomaterials can be made water soluble by functionalizing them with thiolated DNA as in the case of gold nanomaterials. Another interesting example is the coating of single-layer silica on the quantum dots that renders it non-invasive and biologically safe [15]. This also protects the core from unwanted interaction with biomolecules like proteins, etc., which results in the quantum dot core detachment from the shell. However the choice of biofunctionalization agent greatly depends on the type of biomolecule used with the nanoparticle core which can be Au, CdS, CdSe, CdTe, etc., Generally cysteine residues are used to prevent nanomaterial swelling when they are exposed to the aqueous environments and polyethylene glycol (PEG) and oligoethylene (OEG) glycol are used as linker molecules to enhance the stability and biocompatibility [16–19].

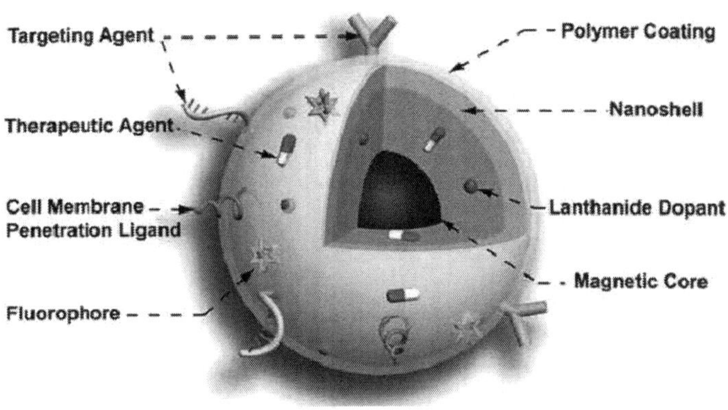

Figure 8.1 Nanomaterial functionalization through surface engineering and interior modification for targeted delivery of therapeutic agents (Reprinted permission from [4] and copyright 2011 from the Royal Society of Chemistry).

The composition of these nanoprobes greatly influences toxicity. Cadmium compounds still pose serious issues of toxicity even when they are effectively biofunctionalized. This is due to the fact that when cadmium compounds come in contact with aqueous environments they split up into Cd^{2+} ions that damage the cellular components. They destabilize DNA which will lead to cell death or mutation that may lead to cancer. Hence, choosing the right composition is necessary for the non-invasive drug delivery. For example, silica which is inherently non-toxic can be coated to quantum dots with controlled layer thickness to achieve biocompatibility. The main aim of surface modification is to develop an exceedingly specific targeting ligand that can be effectively used for drug therapy. By bioengineering these nanoprobes we can make them compatible to the internal environment since most of the nanomaterials do not posses positive interaction with the host. These nanoprobes effectively bind only with their specific receptors on the diseased cells and are imbibed into the cells through endocytosis. By this receptor mediated transport we can actually increase the cellular uptake rate of nanomaterials with better retention time and controlled drug release.

8.3.1.5 Nanoprobe–biomolecule interaction

The direct interaction of a nanoprobe with a biomolecule is crucial in determining the control over the cellular processes and environments when a nanoprobe enters a biological system. All physical aspects like size, shape, composition, etc., equally play important roles to determine the biological interaction. First, the nanomaterials must be sub 10 nm because only then it can effectively interact with the biomolecules like proteins and penetrate membranes [20].

Nanoprobes that enter the biological systems are functionalized with proteins or DNA either by covalent or other non-covalent interactions [21–26]. Of this covalent attachment seems to be favoured for bioconjugation [27]. Chemisorptions and heterobifunctional linkers are two direct, facile methods for attaching nanomaterials with biomolecules. Chemisorption can be brought about by attaching cysteine residues onto the core of nanomaterials like Au, ZnS, CdS, and CdSe/ZnS hybrid [28]. Here cysteine greatly prevents quantum dot swelling that is a common degenerative effect that is faced by nanomaterials in biological systems. This effect usually leads to core detachment that will cause generation of toxicity. In case of heterobifunctional linkers, OEG or PEG is used in the linker to improve the stability of the attached biomolecules and minimize non-specific adsorption of other materials. Non-covalent interactions include electrostatic interactions, groove binding, intercalation, and complementary single-strand DNA binding [29]. Recently Zheng et al. worked on biotin and glutathione-functionalized gold nanoparticles functionalized with tri (ethylene glycol)-terminated thiols [30]. The ligand they bind with is streptavidin and glutathione -S- transferase and can be detected with time-resolved fluoroimmunoassay. Another biomolecule that can be used is histidine which when functionalized on gold nanoparticles showed great promise to study protein carbohydrate mechanisms.

8.3.1.6 Drug delivery route and in vivo targeting

Irrespective of the physical properties and composition, the most challenging factor for the usage of nanoprobes is the exact method of drug delivery. Since the sizes of these probes are very small, they may easily get deposited or entrapped inside normal tissues and may lead to the development of cancerous lesions. Hence, the drug delivery route must be pre determined and properly executed. Of these the pre-oral route offers the most easiest solution but the limitation is that there is a chance of nanomaterial deposition in the digestive tract mucosa that might lead to cancer and also nanoprobe swelling can happen due to interaction with digestive enzymes. Here there is also a chance if non-specific binding of nanomaterials with body tissues [31]. The second most attractive route is the pulmonary route that provides rapid intake of nanomaterial [32]. But safety concerns still challenge this mode of delivery. Other routes include topical application on the skin and parenteral injection, etc., once the drug loaded nanoparticle is delivered into the body the drug is absorbed by the cancerous tissue and undergoes degeneration [33]. In addition to this, recently developed nanoprobes are being tagged with specific antibodies that help them to specifically attach to the antigen in the cancerous tissues.

8.3.2 Conventional Nanoprobes

Conventional nanoprobes have been demonstrated to have basically two different structures. The first group generally has small nanomaterials that self assemble to form micellar structures which make them biocompatible in the essence that they have both hydrophobic and hydrophilic group that help them to balance the conditions of the biological environment. This type of nanoprobes needs no further biofunctionalization. On the other hand, the second group of nanoprobes has an inorganic or organic crystalline core that is encapsulated in a polymer matrix with effective biofunctionalization with subsequent drug load. This type of nanohybrids offers the synergistic effect of both luminescent properties and drug targeting. The traditionally used nanohybrid probes include the noble metal nanomaterials, magnetic nanomaterials, etc.

8.3.2.1 Gold-based nanomaterials

Spherical gold Nanomaterials, Noble metal nanomaterials demonstrate a very distinguishing optical property called as the surface plasmon resonance. Surface plasmons are coherent electron oscillations that are stimulated by the interaction between the conduction band electrons of metal nanoparticles and an electromagnetic field in form of light. Size, shape, dielectric constant, refractive indices are some of the main factors that influence the surface plasmon effect of nanomaterials. The light absorbed by the nanomaterials is converted to thermal energy or heat that can be used to increase the temperature at the site if binding of these nanomaterials so as to destroy the cancerous tissues. Furthermore, noble metals have been found to be highly biocompatible and can be this used for photodynamic therapy.

Although many noble metals have been scientifically studied, gold with its high stability, oxidation resistance, biocompatibility and easy biofunctionalization with DNA interconnects via ligand exchange step is the most investigated nanoprobe for biological application. Also gold has the ability of size-dependent tunable optical absorption with NIR absorption and visible emission. This ability coupled with the surface plasmon effect offers very high level of heating effect of about five times more than other photo-absorbing probes [34].

The gold nanoprobes can be of varying sizes and shapes. Spherical gold nanoparticles can be easily prepared with high monodispersity and small size that can be stored for months without agglomeration. The gold nanoparticles have short retention time with minimal side effects. Thermal ablation is also an important aspect of gold nanomaterials. This property involves the generation of heat at the site of cancer tissue and gradually shrinks the cancer by increasing the temperature of the gold nanoparticles. In addition, Kim et al. have recently published a work on pH-induced Aggregation of sub 10 nm gold nanoparticles that have effectively destroyed cancer cells through thermal ablation [35]. This method of treatment has also been reported to destroy pathogenic organisms like *T. gondii* by Pissuwan et al. [36]. Spherical gold nanoparticles are effective drug carriers that can carry multiple drugs at the same time with controlled drug release thus bringing about parallel therapeutic applications. Rotello et al. have demonstrated that 5-fluorouracil; an anticancer drug functionalized on a gold nanoparticle carrier through a nitrobenzyl group can be fine tuned to release the drug by photolytic cleavage of the nitrobenzyl group upon photo absorption [37]. This study proves that by prolonged and controlled drug release, the viability of cancer cells can be reduced. Also spherical gold nanomaterials are resistant to nuclease action and can couple with DNA effectively. Mirkin et al. demonstrated that Gene regulation in some instances can be controlled by RNA-gold nanoparticles bioconjugates [38]. In spite of their tremendous effectiveness as drug carriers, gold nanoparticles have been shown to disrupt regulated cell cycle mainly affecting DNA and thus destabilizing them in tumourous tissues. The gold nanoparticles specifically target the cell nucleus with membrane disruption. But efficient drug release needs drug delivery without bilayer membrane disruption. By effectively balancing the amount of membrane penetration to the drug release greatly determines the pharmacokinetics of the drug. In order to overcome this negative aspect Stellacci et al. developed ordered amphiphilic structure of some cell penetrating [39]. Peptides with ribbon like repeating domains of hydrophobic and hydrophilic groups. This is effective to prevent cell membrane disruption but one drawback was particles with identical hydrophobic content only exhibited moderate cell-penetrating abilities.

Gold nanorods exhibit both longitudinal surface plasmon and transverse surface Plasmon. The longitudinal surface plasmon provides a much larger extinction coefficient than the transverse mode due to oscillation of electrons of the nanorod. Gold nanorods have one great advantage to be used as nanoprobes because when fine tuned these nanorods can have optical excitation peaks between 700–900 nm which is actually the so called "medical spectral window" associated with minimum absorption of light in most human tissues [40]. EI-Sayed et al. have reported conjugated anti-epidermal growth factor receptor (anti-EGFR) to gold nanorods and have demonstrated cancer tissue destruction malignant epithelial cell lines (HOC313 clone 8 and HSC3) when exposed to NIR irradiation [41]. Gold–silver hybrid nanorods have also been studied for anticancer therapy. In addition, the application of gold nanorods in gene regulation therapy is also being currently investigated.

Gold nanoshells are efficient plasmonic absorbers with good plasmon resonance and peak absorption efficiency. They can be tuned as effective nanoprobes by controlling the shell thickness as illustrated in Fig. 8.2. The most commonly investigated gold nanoshell construction is thin layer of gold-Shell formed on a silica or polymeric nanocore. Thermal ablation, tumour imaging and thermal-triggered drug delivery are some of the major application areas of the plasmon resonance of these gold nanoshells. In 2003, Halas et al. first demonstrated the usage of gold nanoshells on silica templates in photothermal cancer therapy of breast carcinoma cells [42]. On subjecting the nanoshells to NIR radiation, the temperature in the vicinity of the cancerous tissues increased to about 50°C, thus destroying the cancer cells.

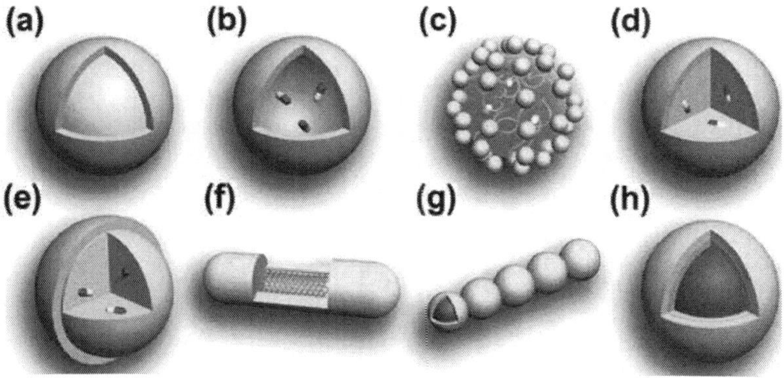

Figure 8.2 Various forms of therapeutic gold nanoshell structures. (a) Silica/polymer–gold core–shell particle. (b) Gold nanoshell particle without the core. (c) Gold cluster-coated liposome particle. (d) Polymer–gold core–shell particle. (e) Janus-type core–shell particle. (f) Gold-coated carbon nanotube. (g) Gold coated-metal oxide nanoparticle chain. (h) Multi-layered gold nanoshell particle (Reprinted permission from [4] and copyright 2011 from the Royal Society of Chemistry).

Gold hollow nanoparticles are a different category of nanoshell organization that can offer high capability of drug load and induce strong photothermal effects because of their strong surface plasmon absorption band centred in the NIR region. Romanowski and co-workers have demonstrated a liposome encapsulated shell shaped array of discrete gold clusters [43]. When NIR radiation is applied, the gold clusters absorb the light and convert it to heat by surface plasmon resonance. This will lead to phase-transition by the liposome that will gradually release the drug to the cancerous tissues.

The combination of chemotherapy and photothermal treatment may lead to substantial improvement of therapeutic effectiveness. Many different systems have been reported in the recent years based on the nanoshell structures. Doxorubicin coupled gold nanoshells have been extensively studied as drug delivery agents. Gold plated carbon nanotubes are being studied as drug nanoprobes for targeting lymphatic vessels in mice. To increase the effectiveness of drug delivery of these nanomaterials, Yu et al. have engineered hydrophilic magnetic shell nanoprobes [44]. They have the basic construct of a cobalt core and porous gold nanoshell. These structures specialize as non-viral vectors for gene delivery agents and transfection vectors. Other core structures have also been extensively studied namely an iron oxide core, a dielectric polymer middle layer and a gold shell. Yeh et al. have proved that gold nanoshells are having greater photothermal conversion efficiency than gold nanorods [45]. But gold nanoshells have low absorption efficiency due to electron scattering that resulted in broadening of plasmon resonance. Furthermore, the tuning of shell thickness in nanoshells can pose a challenge as the acceptable biological clearance size is about 5.5 nm [12].

Nanocages: Gold nanocages as shown in Fig. 8.3 have hollow interiors and porous walls [46–48]. Xia et al. have demonstrated that nanocages have cubic shape with pores formed in the clipped corners of the cubes [49]. Gold nanocages absorb in the NIR region and can be synthesized by galvanic replacement reaction among silver nanocube precursors and $HAuCl_4$ (or $HAuCl_2$) in an aqueous solution. The nanocages are basically encapsulated with a monolayer of smart polymers composed of poly(*N*-isopropylacrylamide) and its derivatives. These nanocages also exhibit photo thermal effect. When the heat is applied owing to the NIR radiation, the heat is dissipated into the polymer which undergoes structural change thus releasing the drug. When the heat absorption stops, the polymer retracts back to their original, extended conformation and thus turning off the drug release. Thus by using these gold nanocages, we can efficiently control drug release.

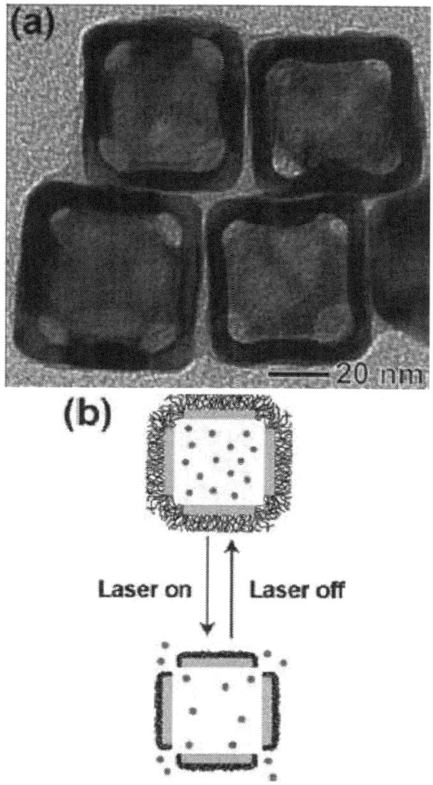

Figure 8.3 (a) TEM image of polymer-coated Au nanocages prepared by Xia and co-workers. The photo shows a high-magnification TEM image of a gold nanocage (b) Schematic representation of laser light-responsive confirmation change of the polymer that leads to the release of the drug molecules encapsulated in the nanocage. (Reprinted with permission from [46], *Nat. Mater.*, 2009, **8**, 935–939, Copyright 2009, Nature Publishing Group).

8.3.2.2 Semiconductor quantum dots

Semiconductor quantum dots are nanocrystalline imaging probes of group II–VI or III–V atoms from the periodic table of elements. These nanoprobes are smaller that the exciton Bohr radius and have size that is typically between 1–10 nm [50]. They exhibit quantum confinement effects that induce them to have different optical properties than their bulk counterparts. Thus by varying the size and intrinsic composition, we can manipulate these nanoprobes to absorb in the NIR and emit in the visible region. The most researched nanoparticles in this field are cadmium-based nanomaterials, especially cadmium sulphide (CdS) and zinc selenide (ZnSe), which can emit in the blue to near UV, while cadmium selenide (CdSe) can emit light across the entire visible spectrum. Alivisatos et al. proved that quantum dots can be made water soluble and can be biofunctionalized and biotagged [51]. They also have high luminescent properties that make them effective bioimaging tools. However due to their intrinsic toxicity their application in biological systems has been considerably reduced. Although quantum dots exhibit slightly low quantum yields of 40–50%, they have high molar extinction coefficients (105–106 M^{-1}) that compensates for this defect. These quantum dots are also resistant to photo blinking and photobleaching.

8.3.2.3 Photodynamic therapy

Although the applications of quantum dots has been demonstrated from the 1990s, the proper usage of quantum dots in cancer tissues has been demonstrated only in 2003 by Samia et al. they demonstrated that quantum dots when irradiated by UV can give rise to triplet energy transfer to molecular oxygen which stimulated the generation of reactive oxygen species [52]. This reactive oxygen species denature DNA thus destroying cancer cells. Alternatively, quantum dots can act as cofactors to promote the effect of conventional photosensitizing agents used in photodynamic therapy (PDT) via Förster resonance energy transfer (FRET) Quantum dots present several advantages over conventional molecular photo sensitizers. They have persistent production of singlet oxygen ($^1O_2$) under extended and repetitive irradiation, large absorption cross-section, and the possibility to generate $^1O_2$ under excitation in the NIR spectral region where biological tissues are optically transparent. Surface passivation is one aspect that can be exploited to improve $^1O_2$ generation by reducing defect sites and improve quantum yield. Concerning toxicity, a single layer of silica can be coated on the core to make it biocompatible. This can also be achieved by using polymer layers. If this major disadvantage is overcome then these quantum dots can become major nanoprobes for photodynamic therapy. The two possible pathways of quantum dot-induced generation of singlet oxygen are shown in Fig. 8.4.

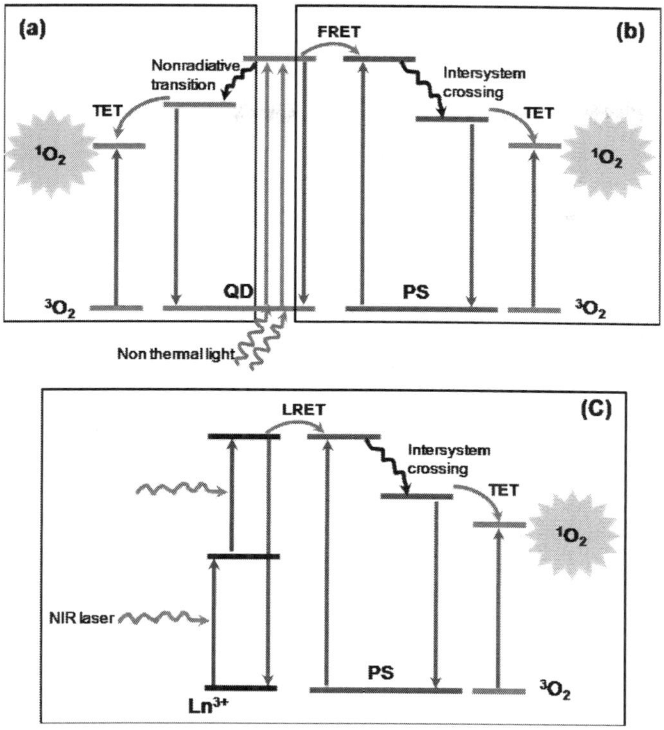

Figure 8.4 (a) Two possible pathways of quantum dot-induced generation of singlet oxygen. (a) The quantum dot (QD) directly interacts with molecular oxygen through triplet energy transfer (TET) to generate singlet oxygen. QD with energy of triplet state larger than 0.97 eV (the energy of singlet oxygen) is required to activate triplet oxygen. (b) QD acts as a cofactor to promote the effect of a conventional photosensitizer (PS) through FRET (c) Proposed mechanism of UC nanoparticle-induced generation of singlet oxygen. The incident NIR photons are absorbed by the Ln^{3+}-doped UC nanoparticles, which subsequently activate a conventional photosensitizer (PS) via a LRET process.

8.3.3 Oxide-Based Bioimaging Probes

8.3.3.1 Iron oxide–based magnetic bioimaging probes

Drug coated magnetic nanoparticles can be easily focused onto a specific target site by manipulating the external magnetic field. Furthermore, they can effectively be used as contrast agents in MRI (magnetic resonance imaging) and hyperthermic cancer treatment by subjecting the nanoparticles to magnetic AC field that heats the nanomaterial. Hence, iron oxide-based magnetic nanomaterials promise exciting new opportunities as bioimaging and drug delivery agents.

Magnetic iron oxide-based nanoparticles, such as magnetite (Fe_3O_4), maghemite (Fe_2O_3) and cobalt ferrite ($CoFe_2O_4$), are the members of the ferrite family with spinel structure and superparamagnetic properties with very small sizes of about 10 nm. Superparamagnetism arises as a result of magnetic anisotropy, i.e. the spins are aligned along a preferred crystallographic direction. If the sample is made up of smaller particles, the total magnetization decreases with decreasing particle size [109]. Hence, proper fabrication of these magnetic particles is necessary for effective biofunctionalization.

Based on their size, they can be Superparamagnetic iron oxide (SPIO) (>50 nm) or nanocrystalline super paramagnetic iron oxide (NCSPIO) (<50 nm). SPIO are effective in imaging liver and spleen tumours, while USPIO can be used for lymphography and angiography [110]. In order to make them highly biocompatible and prevent aggregation, biofunctionalization with polymers such as dextran, dextran derivatives, albumin, poly(vinyl alcohol), or PEG is highly essential. Saeed et al. have demonstrated that PEG coated iron oxide has long circulation time and the magnitude and duration of vascular signal enhancement was 0.05 mM/kg which was twice the magnitude than commercially available gadolinium-based contrast agents [111]. Harris et al. have concluded that hydrophilic triblock copolymers coated iron nanoparticles with carboxylic acid groups in their central segments and with poly(ethylene oxide) (PEO) tails are highly stable with good magnetization ability [112].

8.3.3.2 Rare earth oxide–based nanoprobes

Rare earth (RE)-doped nanophosphors have attracted considerable attention because of their wide range of species and excellent luminescent properties [113]. The emission wavelengths of RE elements are fixed and independent of the particle size, which allows their synthesis and use within a certain size range that is determined mainly by the particular application.Y_2O_3, Gd_2O_3, and $NaYF_4$ nanocrystals have been known to have strong luminescence for biological systems. Oxide-based rare earth nanomaterials are more environment-friendly, cost-effective than fluoride-based rare earth nanoprobes. Body-centred cubic crystallographic Yttrium oxide can be doped with other rare earth–based ions to ensure photostability and biocompatibility [114]. Moreover, Eu and Tb ions may also be used as imaging agents because of their characteristic red and green emission [115–119]. But the main problems that restrict the usage of these rare earth oxide–based probes are agglomeration, irregular shape and size. Furthermore, the precursors can be non–environmental friendly, costly and may lead to toxic by-products. Eu/Tb co-doped Y_2O_3 can be colour tuned from green to red by varying the dopant ion concentration. L-929 fibroblastic cells incubated with Eu/Tb-codoped Y_2O_3 at concentrations less than 62 ppm did not show any cytotoxicity [120].

8.3.3.3 Silica-based nanoprobes

Silica nanoparticles (<50 nm diameter) are potential nanoprobes for drug delivery, gene therapy, and molecular imaging. Biofunctionalized silica nanoprobes showed remarkable biocompatibility with long circulation time and targeted drug delivery [121]. Silica nanoparticles in combination with magnetic and fluorescent particles have also been extensively researched.

Silica as a coating

Silica can be used as a coating for various imaging probes to make them biocompatible and stable. Some of the characteristics that make silica an effective coating are as follows:

- An effective barrier to quench any fluorophores by the magnetic cores. Here the quenching ability can be controlled by manipulating the thickness of the silica core.
- Silica shell is inert, optically transparent, easily functionalized, prevents toxicity of the bare nanomaterial core, prevents particle agglomeration and prevents desorption.

8.3.3.4 Zinc oxide (ZnO)–based nanoprobes

ZnO is a direct band gap (3.37 eV) semiconductor with a relatively high exciton binding energy (60 meV) which has found various applications in optical, electronic and sensing devices [122]. Conventional ZnO QDs are not stable in water because water molecules are able to attack the luminescent centres on the surface and destroy them rapidly [123]. Jana et al. reported ZnO QDs capped with 3-aminopropyltrimethoxysilane was highly stable [123]. Xiong et al. demonstrated that ZnO polymer core–shell nanoparticles with size of 3–5 nm were highly compatible as bioimaging agents [124]. Luminescent ZnO QDs were successfully used for the imaging of the Gram and bacteria *Staphylococcus aureus*. Cytotoxicity experiments showed that these QDs could be used with concentrations up to 1 mM without altering cell growth [125]. The micelle encapsulated ZnO nanocrystals are highly stable and retain their non-linear optical properties in aqueous dispersions. Folic acid coated ZnO nanocrystals when targeted into human nasopharyngeal epidermal carcinoma cells under in vitro conditions generated robust intracellular signals thus providing effective bioimaging. This study also showed no cytotoxicity at experimental dosages [126].

8.3.4 Newer Generation Nanoprobes

8.3.4.1 III–V semiconductor nanoprobes

Quantum dots are very effective bioimaging nanoprobes. Bioimaging with II–VI semiconductor Quantum dots like CdS, CdSe, CdTe or core shell particles like CdSe/ZnS has been extensively investigated in the past decade especially because of their excellent photochemical stability, high quantum yield and resistance to photochemical degradation and blinking. They also have excellent size tunable properties which exhibit high photoluminescence with near infra red bioimaging. But due to the intrinsic Cytotoxicity of elements like cadmium, tellurium, these quantum dots cannot be considered biologically safe. Even with the addition of a shell like ZnS that usually prevents the core detachment and increases the photoluminescence, it is still not an advisable therapeutic route for drug delivery and imaging.

To find a less toxic, environment-friendly replacement, extensive research has been focused on the III–V semiconductor Quantum dots. This positive shift mainly lies on the fact that III–V semiconductor quantum dots like InP are less toxic than their II–VI group counterparts (CdS, CdSe, CdTe). This has been experimentally proved with Syrian golden hamsters that survived over a two year observation period following regular intake of InP quantum dots even with considerable large size of 1.2 μm. This non-toxicity can be attributed to the formation of covalent bonds by the InP quantum dots that renders them biocompatible when compared with cadmium-based quantum dots that bond ionically. Despite being biocompatible, InP quantum dots still suffer from the disadvantage of low photoluminescence. This can be overcome by coating a shell of ZnS over the InP Quantum dots that can tune the photoluminescent efficiency up to 10–15% with ~680 nm emission. In addition to the above factors, site specific targeting with appropriate biofunctionalization is highly essential for the development of an effective nanoprobe. Usually folic acid, anticlaudin 4 and anti-PSCA (antibodies) are used in biofunctionalizing InP Quantum dots. They are very specific to their respective receptors present in the targeted cell. In case of folic acid biofunctionalization, the folate receptors are over-expressed in cancerous tissue more than normal cells. In addition, folic acid interacts via receptor mediated endocytosis. Folic acid is biocompatible and does not incite any immune reaction and is stable in most biological environments. Using a surfactant like mercaptoacetic acid helps in better attachment with folic acid [53].

8.3.4.2 Lanthanide-based nanoprobes

Lanthanide-dopant ions are entrenched in an insulating host lattice of nanoprobes [54–60]. The intra-configurational 4fn electron transitions in the dopant ions give rise to the luminescence properties of these lanthanide-based nanoprobes. Interestingly, quantum confinement effects are negligible in these nanoprobes because of the small exciton Bohr radius and weak interaction between the 4fn electron of lanthanide ions and host material. Conversely, since there is no quantum confine effect, the emission properties are similar with that of the bulk and can be effectively tuned by changing the composition of the dopant ions [61,62–72]. Due to the inherent toxicity of cadmium-based quantum dots, researchers are shifting their focus towards lanthanide-doped nanoparticles which are non-toxic, biocompatible, and optically stable with insignificant photobleaching effects and good luminescent properties.

Because of the above properties they have been suggested as effective replacements for Cadmium-based quantum dots. The research in this particular area is rapidly escalating but one important quandary that needs to be overcome is that these lanthanide doped ions needs deep UV excitation which is harmful to biological cells as UV damages and destabilizes the DNA. In order to trounce this, many researchers are now studying a process called upconversion (UC) by which the lanthanide ions are subjected to NIR excitation and visible emission. Photon UC processes are distinctive non-linear optical phenomenon that convert two or more low-energy pump photons into a higher-energy output photon through chronological photon inclusion.

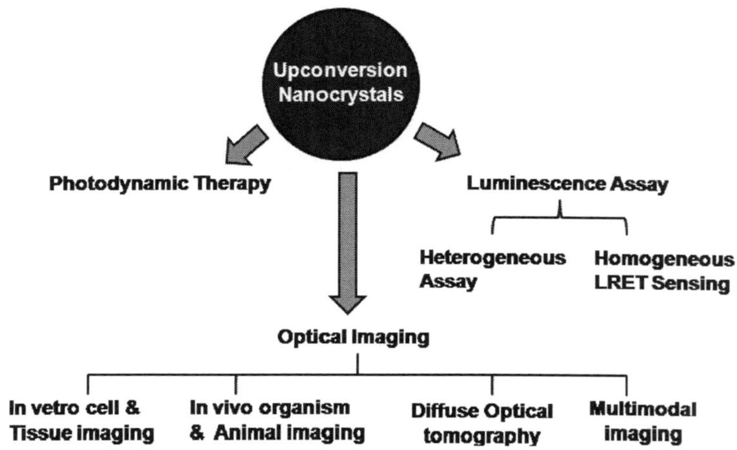

Figure 8.5 Flowchart of biological application of lanthanide doped upconversion nanocrystals.

8.3.4.2.1 *Origin of up conversion*

Excited-state excitation, energy transfer up conversion, and photon avalanche, ladder-like arrangement of energy levels of lanthanide dopant ions are mainly accountable for UC emission. Chronological inclusion of two or more photons within energy states of a dopant ion in the crystalline particle are mainly responsible for up conversion emission. UC luminescence is a by-product of one or more metastable level with *f* and *d* ions which help in the characteristic luminescence. Extremely long-lived intermediate energy states of trivalent lanthanide ions make them highly luminescent upconversion nanomaterials where the lanthanide dopant ions are entrenched in the host material lattice. Equally spaced energy levels facilitate energy transfer steps in upconversion as shown in Fig. 8.6.

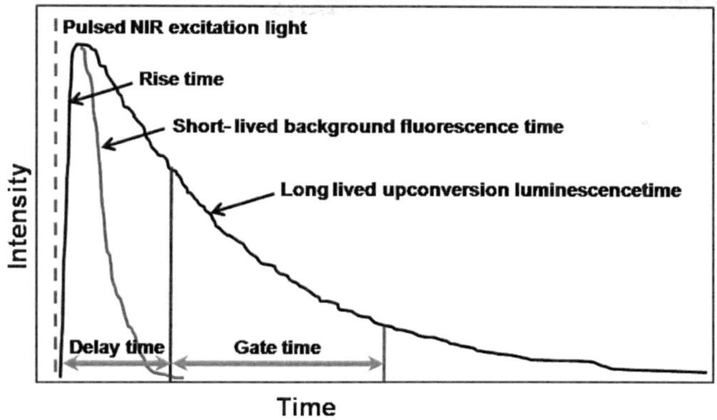

Figure 8.6 The measurement principle of time-resolved luminescence detection technique based on UC nanoparticles. The short-lived background fluorescence can be effectively eliminated by setting an appropriate delay time and gate time. The long-lived UC luminescence of trivalent lanthanide ions is collected during the gate time.

For example, to enhance UC efficiency, Yb^{3+} with a larger absorption cross section in the NIR spectral region is frequently doped as a sensitizer in combination with the activators. The doping percentage of the activator is always kept below 2% to prevent the loss of excitation activator by cross relaxation process. Also excited energies of dopant ions are absorbed by the host lattice by lattice vibration. Furthermore, by varying the crystal structure we can vary the optical properties and level of doping. The host material can be a number of transition metal ions (e.g., Zr^{4+}, Ti^{4+}, and Mn^{2+}) or alkaline earth materials, etc. The Conventional synthesis method of lanthanide-based nanoprobes aims to achieve highly crystalline and negligibly defective with small size and high dispersity with good biocompatibility. However, these techniques suffer from particle aggregation and large particle size. Yu et al. have suggested a facile route for the formation of $NaYF_4$ UC nanocrystals with Gd^{3+} ions doping which yielded effective UC nanoprobes bioimaging tools as demonstrate in Fig. 8.7 [73].

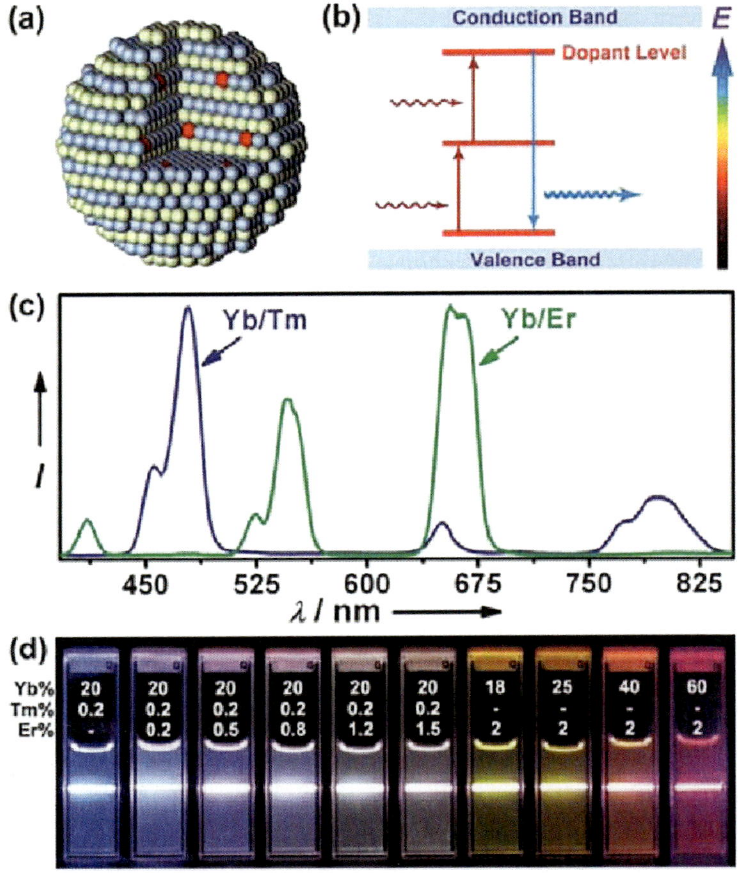

Figure 8.7 Structure and optical properties of UC nanoparticles. (a) Schematic illustration of UC nanoparticles composed of a crystalline host and lanthanide dopant ions embedded in the host lattice. (b) Schematic energy level diagram showing that UC luminescence primarily originates from electron transitions between energy levels of localized dopant ions. (c) Typical emission spectra showing multiple narrow and well-separated emissions produced by cubic $NaYF_4$:Yb/Tm (20/0.2 mol%) and $NaYF_4$:Yb/Er (18/2 mol%) nanoparticles. (d) UC multi-colour fine-tuning through the use of lanthanide-doped $NaYF_4$ nanoparticles with varied dopant ratios. Note that the emission spectra and colours are associated with the host composition, particle size, and particle surface properties. (Emission spectra and luminescent photos are reprinted with permission from [108], *J. Am. Chem. Soc.*, 2008, **130**, 5642–5643, Copyright 2008, American Chemical Society.)

The luminescent property of UC lanthanide-based probes is due to the 4f electrons, which are wholly protected by the $5s^2$ and $5p^6$ sub-shells; hence, they do not have any direct interaction with the host lattice. The narrow absorption profile of lanthanide doped nanoprobes warrants a specific excitation source. InGaAs diode laser systems which function at a wavelength of ca. 980 nm coincides with the absorption profile of Yb^{3+}, providing an ideal excitation source for UC nanoparticles. Lanthanide ions characteristically show a different set of sharp emission peaks, thus providing discernible spectroscopic profiles for accurate understanding of the emission spectra in the event of overlapping emission spectra. No chemical bonds are broken due to the UC emission of the intra 4f electron transitions. This is a highly desirable trait as this renders the nanomaterial safe against photo bleaching and degradation. Even after continuous exposure for hours with UV or NIR irradiation, the nanomaterials remain highly stable. Since the f–f electron transitions are Laporte-forbidden, the UC process has high luminescent times. This makes bioimaging more accurate as this detection technique diminishes the intrusion of undesired short-lived background fluorescence that occasionally originates from biological tissues under multi-photon excitation. Moreover, this UC process has a high signal to noise ratio that makes it more sensitive for imaging biological samples.

8.3.4.2.2 *Surface chemistry*

Surface chemistry plays important role for producing biocompatible nanoprobes. Capping ligands are very essential while preparing UC nanoparticle as they stabilize these nanomaterials against aggregation and control the particle size. Immobilization can also be done for effective biofunctionalization [74]. In addition, surface passivation or treatment is done before bioconjugation to impart aqueous solubility. A ligand exchange reaction or an oxidation of the native ligand (6-aminohexanoic acid (AHA), poly-(ethylene glycol) (PEG) diacid, hexanedioic acid (HDA), citrate, polyacrylic acid (PAA) and phosphate derived molecules) provides the base for ligand engineering [75–79]. Surface functionalization can be done by ligand attraction, Electrostatic layer-by-layer assembly and surface polymerization. Surface coating for better biocompatibility is taken care of by a silica coating by reverse-micro-emulsion method that is applicable to both hydrophilic and hydrophobic nanoparticles.

8.3.4.2.3 *Toxicity of lanthanide-based nanoparticles*

Lanthanide-based nanoprobes have excellent scope as bioimaging tools but even then the Cytotoxicity issues have to be taken into consideration while designing such nanoprobes. Cytotoxicity tests through valuation of cellular morphology and mitochondrial function (MTT and MTS assays) prove that lanthanide-doped nanoparticles are non-cytotoxic to a broad variety of cell lines. Different concentrations of nanomaterials are incubated with varying times to determine Cytotoxicity. Control and exposed data are then scored followed by the various assays. Li and co-workers [80] have tested the biocompatibility of Yb/Er-doped rare earth fluorides coated with a silica shell or a layer of azelaic acid molecules. They have concluded that these nanoprobes did not affect the cell viability of human nasopharyngeal epidermal carcinoma KB cells after incubation for 20 h at a particle concentration of 800 mg/mL. Further biodistribution studies are essential to confirm the long term toxicity of the lanthanide doped nanomaterials.

8.3.4.2.4 *Applications*

High-quality UC nanoparticles have excellent properties like high photon conversion efficiency, non-blinking emission behaviour, negligible fading effect over time, no auto-fluorescence and good biocompatibility. Recently, Jiang et al. [81] have demonstrated that UC nanoparticle (silica-coated $NaYF_4$:Yb/Er nanoparticles biofunctionalized with anti-Her2 antibody) can be used for the delivery and tracking of small interference RNA (siRNA). A luciferase assay was used to systematically study the gene silencing effect of the siRNA while a confocal microscope clearly visualizes the cellular uptake and penetration of the nanomaterials. The results from the assay were conclusive enough to suggest that target specific binding had taken place that brought about the gene silencing. Recently, Wang et al. have proposed a sandwich-hybridization assay for the accurate DNA detection using sub-50 nm $NaYF_4$: Yb/Er nanoparticles. The main advantage of this method is DNA separation based on magnetic nanoparticles separation without PCR amplification. A detection limit of ca. 10 nM can be achieved by this method as shown in Fig. 8.8.

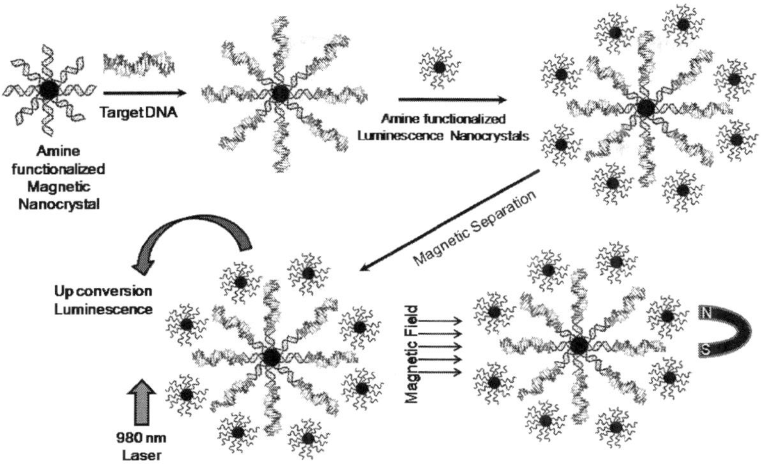

Figure 8.8 Schematic representation of the upconversion nanoparticle-based DNA detection. Magnetic nanoparticles are used and modified with capture DNA strands. Up conversion nanoparticles are modified with probe DNA strands. Upon incubation with target DNA stands, the upconversion and magnetic nanoparticles from binary nanoparticles aggregates. The aggregates can be purified with magnetic separation and examine through upconversion luminescence assays.

8.3.4.2.5 *In vivo organism and animal imaging*

Lim et al. [82] have previously demonstrated initiated the use of UC particles in live organism imaging. They used live nematode *Caenorhabditis elegans* (*C. elegans*) worms as test specimens for the sub 150 nm Y_2O_3:Yb/Er nanoparticles inoculation. Upon NIR irradiation at 980 nm, the digestive tract of the nematode clearly showed the distribution of the nanomaterials. They nanomaterials have significantly shown good biocompatibility to the lanthanide doped nanomaterials. The same research group has further improved their procedure by preparing a sub 10 nm Y_2O_3:Yb/Er nanoparticles. But they found loss of photo intensity due to surface quenching. In recent times, the facile synthesis of small-sized UC nanoparticles with strong emission intensity and high aqueous dispersity has assisted the in vivo imaging of small animals by fast intravenous or intradermal injection of nanoparticles. NIR-to-NIR UC nanoparticles has been developed and demonstrated by Nyk et al. involving the whole body imaging of Balb-c mouse

[83]. This technique involves deep penetration and good imaging. While using PEG coating, the lanthanide-based nanomaterials are highly specific with better circulation times and in situ imaging. Importantly, region of interest (ROI) analysis of the UC luminescence signal in vivo proved that UC imaging achieved a high signal-to-noise ratio (SNR) of ca. 24 between the tumour and the background which normally cannot be achieved in single-photon or two-photon fluorescence imaging. Hydrophilization involving the coating of hydrophobic OA-capped UCNPs, antibody bioconjugation, and the incubation of anti-CEA8-UCNPs with HeLa cancer cells has been shown in Fig. 8.9.

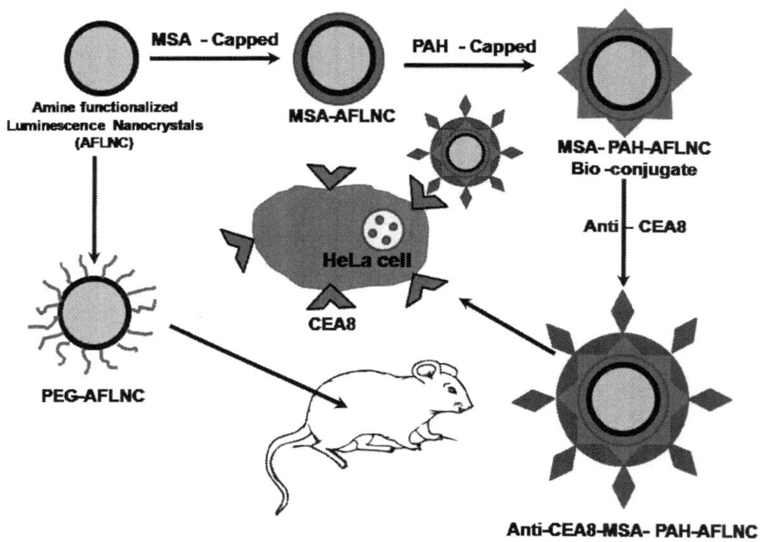

Figure 8.9 Schematic representation of the hydrophilization and coating of hydrophobic Amine-capped luminescence nanocrystals (AFLNC) the bioconjugation between MSA-PAH-AFLNCs and Antibody anti-CEA8, and the incubation of antibody-conjugated anti-CEA8-AFLNCs with HeLa cancer cells. DSPE-mPEG-5000-AFLNCs were intravenously injected into a nude mouse for in vivo animal imaging.

Multi-modal imaging has also been done using lanthanide-based nanoprobes. The basic notion of multi-modal imaging lies in the integration of two imaging modalities within the setting of a single examination. Two strategies have been used for this application. The first approach involves inclusion of gadolinium (Gd^{3+}), in the crystal host lattice to develop nanoprobes that have both optical and magnetic contrast agent properties [84,85]. Since the luminescent and magnetic cores are combined together, no special synthesis step is required to combine two different optical and magnetic agents. Fabrication can be diligently controlled with this approach. The second approach uses the silica shell encapsulation of UC nanoparticles followed by reporter molecule impregnation. Li and co-workers have fabricated folic acid biofunctionalized organic dye impregnated silica shell UC nanoparticles [86]. Folic acid is very specific to folate receptors that are commonly expressed more in the tumour tissues. They then verified receptor-mediated delivery of FA-conjugated nanocomposites targeting KB cells by using UC luminescence microscopy and down conversion flow cytometry. Besides amorphous shells, dual imaging capabilities can be achieved by crystalline materials comprising Gd^{3+} ions. The spatial confinement of the second staining agent in the shell can minimize its interaction with the UC core and thus maximize the functionalities of both the core and the shell, despite causing an increase in the particle size.

8.3.4.2.6 *As light transducers for cancer therapy*

Photodynamic therapy can be made into a facile, cost effective anticancer therapy with the help of lanthanide-based nanoparticles as illustrated in Fig. 8.10 [87, 88]. In theory, PDT involves three basic steps:

(i) selective uptake and localization of a photo sensitizer into specific tumour cell/tissue type
(ii) irradiation of the photo sensitizer with predetermined intense doses of light to activate the photo sensitizer
(iii) generation of reactive oxygen species (ROS) that kill nearby abnormal cells with little or no effect on the surrounding tissues

On the whole, photodynamic therapy acts in the following way. Initially the reactive oxygen species generate by the PDT destabilizes the DNA and thus kills the cancer cells. Also this reactive species of oxygen damages blood vessels which prevents the access of nutrients to the cancer tissue. Moreover, this therapy may stimulate the immune system to effectively control damage to normal tissues and destroy cancer cells.

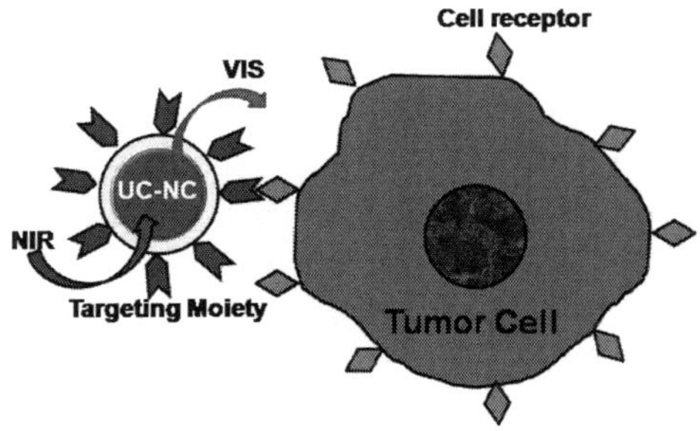

Figure 8.10 Schematic design of upconversion nanocrystal-based photodynamic therapy (PDT) for the treatment of a tumour cell. The design is composed of a nanoparticle core and a porous silica or polymer shell impregnated with photosensitizers. The shell is also modified with functional groups for targeting a specific tumour cell.

PDT is most likely to treat tumours that are not deep enough or lie just beneath the skin. Moreover, PDT cannot treat secondary tumours or metastases. Hence, scientists are focusing their research on this area to improve the depth of penetration of this therapy. In order to overcome this researchers are now intensely exploring the up conversion nanomaterials in PDT that make use of NIR radiation which penetrated deep into the biological tissue. The upconverted visible emission from the nanoparticles can excite photo sensitizers and subsequently generate reactive oxygen species. This comes with the added advantage of tunable emission spectra with efficient drug delivery systems. Zhang et al. demonstrated the application of UC nanoparticle-based PDT in MCF-7/AZ bladder cancer cells [89]. Mouse monoclonal antibodies biofunctionalized photo sensitizer-impregnated nanoparticles were irradiated with NIR rays that efficiently brought about the death of the cancer cells after one hour. To improve the therapeutic efficiency, the concentration of the coated photo sensitizers should be increased outside the UC nanomaterials. 9,10-anthracene dipropionic acid (ADPA) generally used to detect the generation of singlet oxygen from the photo sensitizer impregnated nanoparticles by fluorescence quenching. Zhang and co-workers have studied the effect of UC nanoparticles in murine bladder cancer cells [89]. The PDT activity was observed through cell viability tests, protein release and expression assays, and gel electrophoresis analysis. The tests concluded that DNA was perpetually damaged with fragmentation of nucleosomal DNA and mitochondrial protein cytochrome C release into the cytoplasm. PSA protein functioned as the carcinoma marker in transfectant cell line MB49-PSA. When the PSA protein expression is substantially reduced it showed cell damage. Irrespective of the above positive impacts, the testing of lanthanide-based ions in live systems is yet to be carried out.

8.3.4.3 Carbon-based nanomaterials as nanoprobes

Carbon nanomaterials have tremendous potential in the field of biomedical research since they are inherently biocompatible. Carbon derivatives such as fullerenes, graphene, carbon nanotubes and nanodiamonds, as shown in Fig. 8.11, have been investigated extensively for other electronic applications but their applicability towards biological systems has been studied only recently. They are known to have excellent drug delivery, imaging properties and gene regulation properties. These nanomaterials have an overwhelming variety of physical and chemical properties which have to be tuned by surface engineering for varied biological application. Biofunctionalization is an important aspect of surface engineering which includes attachment with hydrophilic moieties and cell receptor-binding ligands. In this way they can be effectively used as bionanoprobes.

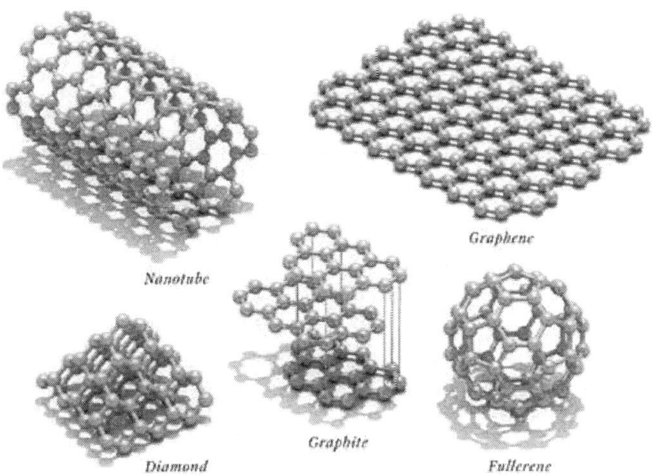

Figure 8.11 Different carbon-based nanostructures such as graphene, fullerenes, carbon nanotubes and nanodiamonds as nanoprobes for bioimaging.

8.3.4.3.1 *Graphene*

Each derivative of carbon has its own individual property and application. Graphene is a two-dimensional carbon nanomaterial with a single layer of sp^2-bonded carbon atoms. Its derivatives, namely graphene oxide and reduced graphene oxide have recently garnered much attention due to their ability to act as nanocargos which effectively deliver insoluble drugs, biomolecules like nucleic acid and other therapeutic agents into the biological systems for imaging purposes. Graphene and graphene oxide when introduced into the biological systems exhibited dose-dependent cytotoxicity due to the generation of reactive oxygen species and cell membrane disruption when non-functionalized. Consequently when these nanoparticles were functionalized with chitosan, peroxidase, etc., they exhibited excellent biocompatibility. Hence, by effectively controlling the dosage we can effectively use these graphene nanomaterials as antibacterial as well as therapeutic agents. Wenbing Hu et al. systematically studied the dosage-dependent toxicity of graphene sheets with foetal bovine serum (FBS), a common component in growth culture media [8]. They found that the cytotoxicity of GO was greatly mitigated at 10% FBS, the concentration usually employed in cell medium. Generally the graphene oxide damages the cell membrane that can be countered by using FBS which can lessen the cytotoxicty of the graphene oxide since GO has high protein absorbability.

8.3.4.3.2 *Carbon fullerenes*

Fullerenes are hollow cage like structures and made entirely of surface carbon atoms that have shown astounding neuroscience applications. They can act as potent neuroprotective agents that search for the free radicals associated with neurological disorders. Oxidative stress induced cell death is one of the main factors that affect nerve tissue in neurotic disorders. Carboxylic acid-functionalized fullerene nanomaterials can go through lipid membranes and prevent such type of cell death mechanism. Parkinson's disease, brain tumours are some of the diseases that can be treated with the help of biofunctionalized fullerenes. In addition, Liang et al. have shown that hydroxylated metallofullerene can function as effective anti cancer drug carrier for cisplatin with controlled and effective drug release both in vitro and in vivo [91].

8.3.4.3.3 *Carbon nanotubes*

Carbon nanotubes have been extensively studied because of their unique optoelectronic, mechanical, conducting and thermal properties that have proven their utility in many different areas of science [92,93]. Carbon nanotubes have high drug-loading capacities with excellent cell membrane penetration. They can also be easily functionalized either covalently or non-covalently depending on the application. To make the carbon nanotubes water soluble, amphiphilic molecule coating can be done by cycloaddition. Covalent functionalization makes them biologically stable while easily transports them across the cytoplasm. Drugs like cisplatin, palcitaxel and gene silencing siRNA molecules have been extensively studied with carbon nanotubes. Bhirde et al. and Dhar et al. have demonstrated the destruction of head and neck squamous cancer and choiocarcinoma and nasopharyngeal carcinoma respectively based on biofunctionalized carbon nanotubes drug delivery [94,95]. Thermal ablation is the inherent ability of carbon nanomaterials by which they absorb NIR irradiation (700–1100 nm) and convert it into heat. This capability of carbon nanotubes can be used to treat to human B lymphoma cells as demonstrated by Marches et al. whose group biofunctionalized the carbon nanotubes with CD22 antibodies that are very specific to the B lymphoma cells [96]. In a new study, Choi and co-workers have further confirmed that PEG-modified single-walled carbon nanotubes when injected intratumourally can effectively destroy tumour cells with NIR irradiation with no recurrence [97]. This study also proved that the carbon nanotubes were cleared completely from the body within 2 months.

6.3.3.3.4 *Nanodiamonds*

Nanodiamonds have a size range of approximately 2–8 nm in diameter with excellent biocompatibility, stability and luminescent properties. Ho et al. have demonstrated polyethylene-imine functionalized nanodiamonds effectively bind with extra-chromosomal DNA or plasmids (PEI800) and deliver them efficiently into mammalian cells [98–108]. This way they can function as non-viral gene carrying vectors with better transfection efficiency than pure plasmids (PEI800) alone. The same concepts when applied for chemotherapy with doxorubicin brought about effective cell death and shrinkage of tumour size for liver and breast cancer. Thus this carbon derivative is fast becoming a burgeoning area for therapeutic nanomaterial drug delivery.

8.4 Conclusion

In conclusion, nanoprobes derived from perfect blend of nanoscale science and biotechnology offer great promise for obtaining unprecedented bioimaging tools due to their exceptional properties, non-invasive biocompatibility, ease of preparation and their flexible dimension control. Both conventional and new generation nanomaterials have been studied extensively to bring about a breakthrough in effective drug delivery. Although conventional nanoprobes have effective photoluminescence, the toxicity issues make them unattractive for bioimaging. But the newer generation nanoprobes like III–V semiconductor and lanthanide-based quantum dots are non-toxic and can be effectively tuned to achieve better luminescence. One of the important strategies to overcome these limitations of lower photoluminescence in lanthanide-based nanoprobes is doping or core–shell strategies. Consequently, many of these limitations of these former nanoprobes can be alleviated by correctly choosing the core element of the nanoprobes and thus using these engineered nanostructured materials effectively. Several nanostructured biofunctionalized materials have shown clear potential for their use as promising bioimaging tools though further testing is needed to ensure commercial marketability of these probes. The main challenge in the coming years is to ensure the development of low-cost, non-invasive, high-photoluminescent, non-toxic nanoprobe that will change the outlook of the current medical, therapeutic and healthcare industry.

References

1. Y. Piao, J. Kim, H. B. Na, D. Kim, J. S. Baek, M. K. Ko, J. H. Lee, M. Shokouhimehr, T. Hyeon, *Nat. Mater.*, 2008, **7**, 242–247.
2. M. De, P. S. Ghosh, V. M. Rotello, *Adv. Mater.*, 2008, **20**, 1–17.
3. F. Wang, D. Banerjee, Y. Liu, X. Chen, X. Liu, *Analyst*, 2010, **135**, 1839–1854.

4. X. Xue, F. Wang, X. Liu, *J. Mater. Chem.*, 2011, **21**, 13107.
5. K.-T. Yong, H. Ding, I. Roy, W.-C. Law, E. J. Bergey, A. Maitra, P. N. Prasad, *ACS Nano*, 2009, **3**, 502–510.
6. D. J. Bharali, D. W. Lucey, H. Jayakumar, H. E. Pudavar, P. N. Prasad, *J. Am. Chem. Soc.*, 2005, **127**, 11364–11371.
7. J. Jeon, D.-K. Lim, J.-M. Nam, *J. Mater. Chem.*, 2009, **19**, 2107–2117.
8. W. Hu, C. Peng, M. Lv, X. Li, Y. Zhang, N. Chen, C. Fan, Q. Huang, *ACS Nano,* 2011, **5**, 3693–3700.
9. M. R. Dreher, W. Liu, C. R. Michelich, M. W. Dewhirst, F. Yuan, A. Chilkoti, *J. Natl. Cancer Inst.*, 2006, **98**, 335–344.
10. P. Jani, G. W. Halbert, J. Langridge, A. T. Florence, *J. Pharm. Pharmacol.,* 1990, **42**, 21–826.
11. F. M. Henretig, A. R. Temple, *Clin. Lab. Med.,* 1984, **4**, 575–586.
12. H. S. Choi, W. Liu, P. Misra, E. Tanaka, J. P. Zimmer, B. I. Ipe, M. G. Bawendi, J. V. Frangioni, *Nat. Biotechnol.,* 2007, **25**, 1165–1170.
13. N. L. Rosi, C. A. Mirkin, *Chem. Rev.*, 2005, **105**, 1547–1562.
14. F. Wang, X. G. Liu, *Chem. Soc. Rev.*, 2009, **38**, 976–989.
15. W. C. W. Chan, S. M. Nie, *Science,* 1998, **281**, 2016–2018.
16. E. Tasciotti, X. W. Liu, R. Bhavane, K. Plant, A. D. Leonard, B. K. Price, M. M. C. Cheng, P. Decuzzi, J. M. Tour, F. Robertson, M. Ferrari, *Nat. Nanotechnol.,* 2008, **3**, 151–157.
17. F. M. Kievit, O. Veiseh, N. Bhattarai, C. Fang, J. W. Gunn, D. Lee, R. G. Ellenbogen, J. M. Olson, M. Q. Zhang, *Adv. Funct. Mater.,* 2009, **19**, 2244–2251.
18. J. M. Rosenholm, A. Meinander, E. Peulhu, R. Niemi, J. E. Eriksson, C. Sahlgren, M. Linden, *ACS Nano*, 2009, **3**, 197–206.
19. J. H. Park, L. Gu, G. von Maltzahn, E. Ruoslahti, S. N. Bhatia, M. J. Sailor, *Nat. Mater.*, 2009, **8**, 331–336.
20. M. J. Hostetler, J. E. Wingate, C. J. Zhong, J. E. Harris, R. W. Vachet, M. R. Clark, J. D. Londono, S. J. Green, J. J. Stokes, G. D. Wignall, G. L. Glish, M. D. Porter, N. D. Evans, R. W. Murray, *Langmuir,* 1998, **14**, 17.
21. C. M. Niemeyer, *Angew. Chem. Int. Ed.*, 2001, **40**, 4128.
22. M. Sastry, M. Rao, K. N. Ganesh, *Acc. Chem. Res.*, 2002, **35**, 847.
23. E. Katz, I. Willner, *Angew. Chem. Int. Ed.*, 2004, **43**, 6042.
24. T. Pellegrino, S. Kudera, T. Liedl, A. M. Javier, L. Manna, W. J. Parak, *Small*, 2005, **1**, 48.
25. A. Verma, V. M. Rotello, *Chem. Commun.*, 2005, **303**.

26. N. L. Rosi, C. A. Mirkin, *Chem. Rev.*, 2005, **105**, 1547.
27. F. Caruso, *Adv. Mater.*, 2001, **13**, 11.
28. K. Naka, H. Itoh, Y. Tampo, Y. Chujo, *Langmuir,* 2003, **19**, 5546.
29. B. A. Armitage, in *DNA Binders and Related Subjects*, *Springer*, 2005, **253**, 55.
30. M. Zheng, X. Y. Huang, *J. Am. Chem. Soc.*, 2004, **126**, 12047.
31. A. Sood, R. Panchagnula, *Chem. Rev.,* 2001, **101**, 3275–3304.
32. J. S. Patton, P. R. Byron, *Nat. Rev. Drug Discovery*, 2007, **6**, 67–74.
33. N. A. Monteiro-Riviere, J. E. Riviere, *Nanotoxicology,* 2009, **3**, 188–193.
34. D. Pissuwan, S. M. Valenzuela, M. B. Cortie, *Trends Biotechnol.*, 2006, **24**, 62–67.
35. C. K. Kim, P. Ghosh, V. M. Rotello, *Nanoscale,* 2009, **1**, 61–67.
36. D. Pissuwan, S. M. Valenzuela, M. B. Cortie, *Trends Biotechnol.*, 2006, **24**, 62–67.
37. S. S. Agasti, A. Chompoosor, C. C. You, P. Ghosh, C. K. Kim, V. M. Rotello, *J. Am. Chem. Soc.,* 2009, **131**, 5728–5729.
38. R. Jin, Y. Cao, C. A. Mirkin, K. L. Kelly, G. C. Schatz, J. G. Zheng, *Science,* 2001, **294**, 1901–1903.
39. H. J. Zhang, W. Xu, X. G. Liu, F. Stellacci, J. T. L. Thong, *Appl. Phys. Lett.,* 2010, **97**, 163702.
40. R. Weissleder, *Nat. Biotechnol.*, 2001, **19**, 316–317.
41. X. H. Huang, I. H. El-Sayed, W. Qian, M. A. El-Sayed, *J. Am. Chem. Soc.,* 2006, **128**, 2115–2120.
42. L. R. Hirsch, R. J. Stafford, J. A. Bankson, S. R. Sershen, B. Rivera, R. E. Price, J. D. Hazle, N. J. Halas, J. L. West, *Proc. Natl. Acad. Sci. U. S. A.*, 2003, **100**, 13549–13554.
43. T. S. Troutman, S. J. Leung, M. Romanowski, *Adv. Mater.,* 2009, **21**, 2334–2338.
44. Y. Lu, Y. Zhao, L. Yu, L. Dong, C. Shi, M. J. Hu, Y. J. Xu, L. P. Wen, S. H. Yu, *Adv. Mater.,* 2010, **22**, 1407–1411.
45. F. Y. Cheng, C. T. Chen, C. S. Yeh, *Nanotechnology*, 2009, **20**, 425104.
46. M. S. Yavuz, Y. Cheng, J. Chen, C. M. Cobley, Q. Zhang, M. Rycenga, J. Xie, C. Kim, K. H. Song, A. G. Schwartz, L. V. Wang, Y. Xia, *Nat. Mater.,* 2009, **8**, 935–939.
47. J. Chen, C. Glaus, R. Laforest, Q. Zhang, M. Yang, M. Gidding, M. J. Welch, Y. Xia, *Small,* 2010, **6**, 811–817.

48. S.-W. Choi, Y. Zhang, Y. Xia, *Angew. Chem. Int. Ed.*, 2010, **49**, 7904–7908.

49. A. P. Alivisatos, *Science*, 1996, **271**, 933–937.

50. Q. Zhang, T. Sun, F. Cao, M. Li, M. Hong, J. Yuan, Q. Yan, H. H. Hng, N. Wu, X. Liu, *Nanoscale*, 2010, **2**, 1256–1259.

51. M. Bruchez Jr, M. Moronne, P. Gin, S. Weiss, A. P. Alivisatos, *Science*, 1998, **281**, 2013–2016.

52. A. C. S. Samia, X. B. Chen, C. Burda, *J. Am. Chem. Soc.*, 2003, **125**, 15736–15737.

53. D. J. Bharali, D. W. Lucey, H. Jayakumar, H. E. Pudavar, P. N. Prasad, *J. Am. Chem. Soc.*, 2005, **127**, 11364–11371.

54. F. Wang, X. G. Liu, *Chem. Soc. Rev.*, 2009, **38**, 976–989.

55. F. Vetrone, J. A. Capobianco, *Int. J. Nanotechnol.*, 2008, **5**, 1306–1339.

56. C. X. Li, J. Lin, *J. Mater. Chem.*, 2010, **20**, 6831–6847.

57. K. Binnemans, *Chem. Rev.*, 2009, **109**, 4283–4374.

58. S. V. Eliseeva, J. C. G. Bunzli, *Chem. Soc. Rev.*, 2010, **39**, 189–227.

59. J. Shen, L. D. Sun, C. H. Yan, *Dalton Trans.*, 2008, 5687–5697.

60. W. Q. Luo, R. F. Li, X. Y. Chen, *J. Phys. Chem. C*, 2009, **113**, 8772–8777.

61. F. Wang, X. G. Liu, *J. Am. Chem. Soc.*, 2008, **130**, 5642–5643.

62. F. Wang, X. P. Fan, M. Q. Wang, Y. Zhang, *Nanotechnology*, 2007, **18**, 025701.

63. F. Wang, J. Wang, J. Xu, X. J. Xue, X. G. Liu, *Spectrosc. Lett.*, 2010, **43**, 400–405.

64. F. Wang, X. J. Xue, X. G. Liu, *Angew. Chem. Int. Ed.*, 2008, **47**, 906–909.

65. O. Ehlert, R. Thomann, M. Darbandi, T. Nann, *ACS Nano*, 2008, **2**, 120–124.

66. J. Wang, F. Wang, J. Xu, Y. Wang, Y. Liu, X. Chen, H. Chen, X. Liu, *C. R. Chim.*, 2010, **13**, 731–736.

67. C. Jiang, F. Wang, N. Wu, X. Liu, *Adv. Mater.*, 2008, **20**, 4826–4829.

68. R. Sivakumar, F. van Veggel, M. Raudsepp, *J. Am. Chem. Soc.*, 2005, **127**, 12464–12465.

69. Y. Zhang, J. H. Hao, C. L. Mak, X. H. Wei, *Opt. Express*, 2011, **19**, 1824–1829.

70. H. T. Wong, H. L. W. Chan, J. H. Hao, *Opt. Express*, 2010, **18**, 6123–6130.

71. H. Zheng, X. J. Wang, M. J. Dejneka, W. M. Yen, R. S. Meltzer, *J. Lumin.*, 2004, **108**, 395–399.

72. H. Zhang, Y. J. Li, Y. C. Lin, Y. Huang, X. F. Duan, *Nanoscale,* 2011, **3**, 963–966.

73. X. F. Yu, M. Li, M. Y. Xie, L. D. Chen, Y. Li, Q. Q. Wang, *Nano Res.*, 2010, **3**, 51–60.

74. J. Q. Gu, J. Shen, L. D. Sun, C. H. Yan, *J. Phys. Chem. C*, 2008, **112**, 6589–6593.

75. J. Q. Gu, J. Shen, L. D. Sun, C. H. Yan, *J. Phys. Chem. C*, 2008, **112**, 6589–6593.

76. F. Wang, D. K. Chatterjee, Z. Q. Li, Y. Zhang, X. P. Fan, M. Q. Wang, *Nanotechnology*, 2006, **17**, 5786–5791.

77. F. Wang, X. P. Fan, M. Q. Wang, Y. Zhang, *Nanotechnology*, 2007, **18**, 025701.

78. F. Wang, X. G. Liu, *J. Am. Chem. Soc.,* 2008, **130**, 5642–5643.

79. P. R. Diamente, F. C. J. M. van Veggel, *J. Fluoresc.,* 2005, **15**, 543–551.

80. F. Wang, Y. Zhang, X. P. Fan, M. Q. Wang, *Nanotechnology*, 2006, **17**, 1527–1532.

81. L. Q. Xiong, Z. G. Chen, M. X. Yu, F. Y. Li, C. Liu, C. H. Huang, *Biomaterials*, 2009, **30**, 5592–5600.

82. J. Zhou, Y. Sun, X. Du, L. Xiong, H. Hu, F. Li, *Biomaterials*, 2010, **31**, 3287–3295.

83. H. Hu, L. Q. Xiong, J. Zhou, F. Y. Li, T. Y. Cao, C. H. Huang, *Chem. Eur. J.*, 2009, **15**, 3577–3584.

84. S. Jiang, Y. Zhang, K. M. Lim, E. K. W. Sim, L. Ye, *Nanotechnology*, 2009, **20**, 155101.

85. S. F. Lim, R. Riehn, C. K. Tung, W. S. Ryu, R. Zhuo, J. Dalland, R. H. Austin, *Nanotechnology*, 2009, **20**, 405701.

86. M. Nyk, R. Kumar, T. Y. Ohulchanskyy, E. J. Bergey, P. N. Prasad, *Nano Lett.,* 2008, **8**, 3834–3838.

87. P. H. Kuo, E. Kanal, A. K. Abu-Alfa, S. E. Cowper, *Radiology*, 2007, **242**, 647–649.

88. P. Caravan, *Chem. Soc. Rev.,* 2006, **35**, 512–523.

89. J. Zhou, Y. Sun, X. Du, L. Xiong, H. Hu, F. Li, *Biomaterials*, 2010, **31**, 3287–3295.

90. D. Dolmans, D. Fukumura, R. K. Jain, *Nat. Rev. Cancer*, 2003, **3**, 380–387.

91. X. J. Liang, H. Meng, Y. Z. Wang, H. Y. He, J. Meng, J. Lu, P. C. Wang, Y. L. Zhao, X. Y. Gao, B. Y. Sun, C. Y. Chen, G. M. Xing, D. W. Shen, M. M. Gottesman, Y. Wu, J. J. Yin, L. Jia, *Proc. Natl. Acad. Sci. U. S. A.*, 2010, **107**, 7449–7454.

92. L. Lacerda, A. Bianco, M. Prato, K. Kostarelos, *Adv. Drug Delivery Rev.*, 2006, **58**, 1460–1470.

93. T. A. Hilder, J. M. Hill, *Small*, 2009, **5**, 300–308.

94. A. A. Bhirde, V. Patel, J. Gavard, G. F. Zhang, A. A. Sousa, A. Masedunskas, R. D. Leapman, R. Weigert, J. S. Gutkind, J. F. Rusling, *ACS Nano*, 2009, **3**, 307–316.

95. S. Dhar, Z. Liu, J. Thomale, H. J. Dai, S. J. Lippard, *J. Am. Chem. Soc.*, 2008, **130**, 11467–11476.

96. R. Marches, P. Chakravarty, I. H. Musselman, P. Bajaj, R. N. Azad, P. Pantano, R. K. Draper, E. S. Vitetta, *Int. J. Cancer*, 2009, **125**, 2970–2977.

97. R. Marches, P. Chakravarty, I. H. Musselman, P. Bajaj, R. N. Azad, P. Pantano, R. K. Draper E. S. Vitetta, *Int. J. Cancer*, 2009, **125**, 2970–2977.

98. E. K. Chow, X.-Q. Zhang, M. Chen, R. Lam, E. Robinson, H. Huang, D. Schaffer, E. Osawa, A. Goga, D. Ho, *Sci. Transl. Med.*, 2011, **3**, 73ra21.

99. M. Chen, E. D. Pierstorff, R. Lam, S. Y. Li, H. Huang, E. Osawa, D. Ho, *ACS Nano*, 2009, **3**, 2016–2022.

100. J. Zheng, C. Zhang, R. M. Dickson, *Phys. Rev. Lett.*, 2004, **93**, 077402.

101. J. W. Kim, E. I. Galanzha, E. V. Shashkov, H. M. Moon, V. P. Zharov, *Nat. Nanotechnol.*, 2009, **4**, 688–694.

102. D. H. Kim, E. A. Rozhkova, I. V. Ulasov, S. D. Bader, T. Rajh, M. S. Lesniak, V. Novosad, *Nat. Mater.*, 2009, **9**, 165–171.

103. D. Pissuwan, S. M. Valenzuela, C. M. Miller, M. C. Killingsworth, M. B. Cortie, *Small*, 2009, **5**, 1030–1034.

104. N. L. Rosi, C. A. Mirkin, *Chem. Rev.*, 2005, **105**, 1547–1562.

105. A. P. Castano, P. Mroz, M. R. Hamblin, *Nat. Rev. Cancer*, 2006, **6**, 535–545.

106. P. Zhang, W. Steelant, M. Kumar, M. Scholfield, *J. Am. Chem. Soc.*, 2007, **129**, 4526–4527.

107. F. Wang, X. G. Liu, *J. Am. Chem. Soc.*, 2008, **130**, 5642–5643.

108. Q. Zhan, J. Qian, H. Liang, G. Somesfalean, D. Wang, S. He, Z. Zhang, S. A. Engels, *ACS Nano,* 2011, **5**, 3744–3757.

109. R. V. Kumar, Y. Koltypin, Y. S. Cohen, Y. Cohen, D. Aurbach, O. Palchik, I. Felner, Gedanken, *J. Mater. Chem.*, 2000, **10**, 1125.

110. J. W. Bulte, D. L. Kraitchman, *NMR Biomed.*, 2004, **17**, 484–499.

111. M. Saeed, M. F. Wendland, M. Engelbrecht, H. Sakuma, C. B. Higgins, *Eur. Radiol.,* 1998, **8**, 1047–1053.

112. L. A. Harris, J. D. Goff, A. Y. Carmichael, J. S. Riffle, J. J. Garburn, T. G. St. Pierre, *Chem. Mater.*, 2003, **15**, 1367–1377.

113. L. Wang, R. Yan, Z. Huo, L. Wang, J. Zeng, J. Bao, X. Wang, Q. Peng, Y. Li, *Angew. Chem. Int.,* 2005, **44**, 6054–6057.

114. H. Wang, M. Uehara, H. Nakamura, M. Miyazaki, H. Maeda, *Adv. Mater.*, 2005, **17**, 2506–2509.

115. H. Wang, M. Yu, C. Lin, X. Liu, J. Lin, *J. Phys. Chem. C,* 2007, **111**, 11223–11230.

116. J.-L. Yuan, X.-Y. Zeng, J.-T. Zhao, Z.-J. Zhang, H.-H. Chen, X.-X. Yang, *J. Phys. D Appl. Phys.*, 2008, **41**, 105406.

117. Y. Li, Y. Zhang, G. Hong, Y. Yu, *J. Rare Earths*, 2008, **26**, 450–454.

118. J. K. Han, G. A. Hirata, J. B. Talbot, Mc Kittrick, *Mater. Sci. Eng. B,* 2011, **176**, 436–441.

119. F. Zhang, S.S. Wong, *ACS Nano*, 2010, **4**, 99–112.

120. T. S. Atabaev, J. H. Lee, D.-W. Han, Y.-H. Hwang, H.-K. Kim, *J. Biomed. Mater. Res.*, 2012, **3**, 1–8.

121. F. Erogbogbo, K.-T. Yong, R. Hu, W.-C. Law, H. Ding, C.-W. Chang, P. N. Prasad, M. T. Swihart, *ACS Nano,* 2010, **4**, 5131–5138.

122. H.-H. Yu, M. K. F. Wong, E. M. Ali, J. Y. Ying, *Chem. Commun.*, 2008, 4912.

123. N. R. Jana, H. Yu, E. M. Ali, Y. Zheng, J. Y. Ying, *Chem. Commun.*, 2007, 1406–1408.

124. H.-M. Xiong, Y. Xu, Q.-G. Ren, Y.-Y. Xia, *J. Am. Chem. Soc.,* 2008, **130**, 7522.

125. R.-O. Moussodia, L. Balan, C. Merlin, C. M. R. Schneider, *J. Mater. Chem.,* 2010, **20**, 1147–1155.

126. D. J. Bharali, D. W. Lucey, H. Jayakumar, H. E. Pudavar, P. N. Prasad, *J. Am. Chem. Soc.* 2005, **127**, 11364–11371.

127. J.-H. Kim, K. Park, H. Y. Nam, S. Lee, K. Kim, I. C. Kwon, *Prog. Polym. Sci.,* 2007, **32**, 1031–1053.

128. S. A. Corr, A Yury, P. Rakovich, A Yurii, K. Gun'ko, *Nanoscale. Res. Lett.*, 2008, **3**, 87–104.

129. D. J. Czewski, Y. Zhang, G. K. Das, D. K. Yi, P. Padmanabhan, K. K. Bhakoo, T. T. Y. Tan, S. T. Selvan, *Microsc. Res. Tech.*, 2011, **74**, 563–576.

130. T. Sh. Atabaev, J. H. Lee, D.-W. Han, Y.-H. Hwang, H.-K. Kim, *J. Biomed. Mater. Res.* A, 2012, 1–8.

131. P. K. Sharma, R. K. Dutta, M. Kumar, P. K. Singh, A. C. Pandey, *J. Luminescence,* 2009, **129**, 605–610.

Chapter 9

Band Energy and Crystal Structure Employing Density Functional Theory

Piyush Dua[a] and Avanish Kumar Srivastava[b]

[a]*Vidya College of Engineering, Vidya Knowledge Park, Baghpat Road, Meerut, UP, India*
[b]*National Physical Laboratory, Council of Scientific and Industrial Research, Dr. K. S. Krishnan Road, New Delhi 110012, India*

aks@nplindia.org

In this chapter, we address the development in the field of oxide nanostructures, especially ZnO and TiO_2, using density functional theory (DFT). Density functional theory has been playing a very important role to understand the fundamental aspects like electronic structure, conducting properties, binding energies as well as absorption properties for the electronic as well as sensing applications.

9.1 Importance of Oxide Nanostructures

From the forms of materials, nanomaterials are classified as zero- (0D), one- (1D), two- (2D) and three-dimensional (3D) nanostructures.

Oxide Nanostructures: Growth, Microstructures, and Properties
Edited by Avanish Kumar Srivastava

ISBN 978-981-4411-35-6 (Hardcover), 978-981-4411-36-3 (eBook)
www.panstanford.com

(i) In general, the particles with all of the three dimensions below 100 nm are called nanoparticles or zero-dimensional nanostructures: 0D nanostructures, also named as nanoparticles, include single crystal, polycrystalline and amorphous particles with all possible morphologies, such as spheres, cubes and platelets. If the nanoparticles are single crystalline, they are often referred to as nanocrystals. When the characteristic dimension of the nanoparticles is sufficiently small and quantum effects are observed, quantum dots are the common term used to describe such nanoparticles.

(ii) One-dimensional (1D) nanostructures are called by a variety of names, including whiskers, fibres, nanowires, nanorods and nanotubes. 1D nanostructures are a class of low dimensional nanomaterials which have one of their dimensions in the nanometre range, but with a much larger length scale (Vayssieres and Manthiram, 2004), which goes from hundreds of nanometres to hundreds of microns and up to millimetres (Hu et al., 1999; Yang et al., 2002; Zhang et al., 2002; Louie, 2001). In such cases, dimensionality can give aspect ratios (length over diameter) of several thousands: For example, 10,000 for a 1D object of 10 nm in diameter and 0.1 mm in length. The shortest 1D nanomaterials are called nanorods or nanowires. In literature other names can be noted such as quantum wires, nanofibres, nanopillars, nanocables and nanolines.

(iii) Thin films are two-dimensional nanostructures, another important nanostructure, and have been a subject of intensive study for almost a century, and many methods have been developed and improved.

(iv) Bulk materials of alloys and compounds constituted of nanograins may be classified as three-dimensional nanostructures. These materials are normally possible in the field of metals & alloys and ceramic. However, polymers are also prepared as bulk nanostructures depending on the processes employed for synthesis.

Vast variety of nanostructures have been synthesized and studied by researches. Out of a big list of nanostructures, oxide nanostructures are a class of nanomaterials which is useful for the industry in various applications as well as being studied by researchers from the fundamental point of view. Many oxide nanostructures do exist such as aluminium oxide (Zhou et al. (2002)), antimony oxide (Zhang (2002)), bismuth oxide (Chen et al. (2002)), cobalt oxide (Shi et al. (2002)), copper oxide (Xu et al. (2002); Wang et al. (2002)), europium oxide (Pol et al. (2002)), gadolinium oxide (Decca et al. (1998)), gallium oxide (Liang et al. (2001)), indium oxide (Peng et al. (2002)), iron oxide (Fu et al. (2001)), magnesium oxide (Yin et al. (2002)), manganese oxide (Wang and Li. (2002)), molybdenum oxide (Niederberger (2001)), nickel oxide (Jeevanandam et al. (2001); Matsui et al. (2002)), rhenium oxide (Ryan et al. (2000)), silicon oxide (Zheng et al. (2002)), tin oxide (Hu et al. (2002); Liu et al. (2003)), titanium oxide (Yi et al. (2002); Kasuga et al. (1998), Lin et al. (2002)), tungsten oxide (Gu et al. (2002); Liao et al. (2000)), vanadium oxide (Lopez et al. (2002); Muster et al. (2000)), yttrium oxide (Wu et al. (2003)) zinc oxide (Kong et al. (2001); Pacholski et al. (2002); Vayssieres (2003); Wu et al. (2002)), zirconium oxide (Liu et al. (2002)). It is really a difficult task to consider all of them together. Various attempts have been made to fundamentally understand the nanoscale system, for which theoretical methods do exist. The advantage of using nanostructures can be as simple as their large surface to volume ratio, such as for catalytic nanoparticles. But it can also be because of their intrinsic changes in electronic and optical properties.

Computational modelling is one of the most important techniques to simulate the nanostructures and nanomaterial. It can give the probable direction of research and can give a direction for the development across the entire scientific scope by reducing the cost in the development and commercialization of technologies and materials. Various computational approaches have been developed and used to predict the materials properties (e.g., electronic, magnetic, optical) at different length and time scales. For example, at an atomic scale, DFT is widely used for binding energy calculation, while at a microscopic scale, molecular dynamics (MD) are able to provide insights into atomic/molecular systems. This chapter is restricted to DFT studies only. Various approximations have been proposed, such as local density approximation (LDA), generalized gradient approximation (GGA), LDA + U and hybrid functional. LDA takes into account the average density. The term GGA denotes a variety of ways proposed for functions that modify the behaviour at large gradients in such a way as to preserve desired properties.

Modern GGA functionals appear formally like the LDA, but with the local function including not just the density or spin densities in the local spin density approximation (LSDA), but also the local gradient (or spin density gradients). LDA + U stands for the methods that involve LDA or GGA type calculations coupled with an additional orbital-dependent interaction. U stands for the intra-site Coulomb interaction (similar as Hubbard model). The additional interaction U is usually considered only for highly localized atomic like orbitals on the same atomic site. The form of the coupling constant integration for the exchange-correlation energy is the basis for constructing a class of functionals called hybrid because they are a combination of orbital-dependent Hartree–Fock and explicit density functional (Martin (2004)). To understand electronic structures, surface and impurity effects, atomic relaxations, interior electric fields, carrier dynamics, and transports of nanostructures, it is often necessary to carry out atomistic simulations. For that purpose various DFT codes are available like WIEN2K (Blaha et al. (2001)), VASP (Kresse et al. (1996)), CASTEP (Segall et al.) and GAUSSIAN (Frisch et al. (2009)). Due to the advance of recent algorithm developments and improved supercomputer powers, it is now possible to calculate such nanostructures based on ab initio methods. Most of software (codes) can work in parallelization. In nanoscience research, numerical simulation plays an important role (Hafner (2008); Marzani (2006); Hoshi and Fujiwara (2006); Wang (2004)). As some of the well-established bulk analytical models may not be directly applied. An additional challenge is the size of the system as nanosystems present a very large scale problem in computation, because a large unit cell has to be considered for simulation. In the bulk system, one only needs to calculate one unit cell of the infinite periodic system. For molecules, the typical number of atoms is smaller than 100. However, a realistic nanosystem can contain a few thousand to hundreds of thousands of atoms, yet atomistic ab initio calculations are needed in many cases (Zunger (1998)). Therefore, large-scale simulations need to be carried out on supercomputers or big clusters. To understand oxide nanostructures from the fundamental point of view, we have discussed ZnO and TiO_2.

9.2 Zinc Oxide Nanostructures

ZnO is a good candidate in electronic industry for low-cost devices and it is being utilized as fundamental building blocks for various applications such as: field effect transistors, gas sensors, resonators, transducers, actuators, cantilevers, and field emitters (Özgür et al. (2005)).

To calculate the bond length of ZnO, we have used the full-potential linear augmented plane-wave (FLAPW) method as implemented in the WIEN2K code (Blaha et al. (2001)). The calculations are based on DFT with the exchange and correlations effects being described by the GGA (Wu and Cohen (2006)), which includes improvements over other GGA methods. Other simulation parameters can be seen in (Srivastava et al. (2010)). There have been various experimental measurements and theoretical calculations to determine the lattice parameters of ZnO. In the available literature, lattice parameters a and c vary from 3.2475 to 3.2501 Å and 5.2042 to 5.2075 Å, respectively. Experimentally, for wurtzite ZnO, u and c/a varies from 0.3817 to 0.3856 and 1.593 to 1.6035, respectively (Desgreniers et al. (1998); Jaffe and Hess (1993)). To determine the equilibrium geometry of the wurtzite (hexagonal unit-cell), we optimize the independent parameters V (volume of the unit cell), c/a ratio, and u (internal parameter or structure factor). Having found c/a and V', we vary the internal parameter u to minimize the total energy. The values obtained after calculations are assigned as: a (Bohr) = 6.140665, c (Bohr) = 9.833757, c/a = 1.601415, V (Bohr^3) = 321.13003 and u = 0.380057. Using these results, we achieve the Zn–O bond length of 0.19772 nm, which is close to that of lattice scale information inferred from electron microscopy (Srivastava et al. (2010)).

9.2.1 1D ZnO Nanostructures

Due to surface and quantum confinement effects, 1D ZnO nanostructures (Kong et al.; (2004); Duan et al. (2001); Huang et al. (2001); Gao et al. (2004); Roy et al. (2003); Li et al. (2004); Djurisǐc et al. (2004); Zhang et al. (2004), Zhou et al. (2005); Zhang et al. (2007); Stan et al. (2007); Song et al. (2008), Liu et al. (2008);

Rühle et al. (2008); Tang et al. (2007)) have unique physical and chemical properties. Zhen et al. (2008) did simulations of ZnO 1D nanostructures by relaxing without any symmetry constraints by retaining the interatomic topology and the wurtzite structure, whereas the bond lengths and angles change in the lateral surface. As discussed for the nanowire, the bond lengths of Zn–O in the surface parallel to the wire axis are 1.880 Å, and the zigzag Zn–O bond lengths are about 1.890 and 1.965 Å, respectively. ZnO nanowires usually grow along the (0001) direction and have a hexagonal or triangular cross section (Wang et al. (2006)). Various theoretical and computational investigations have focused on ZnO nanostructures (Xu et al. (2007); Tu et al. (2006; Fan et al. (1995), Li et al. (2007)) such as nanowire, nanofilm, nanoplate, graphitic planar sheet and single-walled nanotubes (SWNTs).

ZnO usually exhibits a typical wurtzite structure (space group *P*63*mc*) with two zinc and two oxygen atoms in a tetrahedral coordination per unit cell. The experimentally obtained lattice parameters are as follows: $a = b = 3.2465$ Å, $c = 5.203$ Å, and $u = 0.3812.48$. To create the models, a large 10 × 10 × 1 ZnO supercell was first constructed based on the above experimental wurtzite structure (Abrahams and Bernstein (1969)). They have compared ZnO crystalline nanowires and faceted nanotubes with conventional SWNTs. In comparison, the corresponding Zn–O bond lengths in bulk ZnO are 1.974, 1.968, and 1.968 Å. The Zn–O–Zn and O–Zn–O bond angles are 103.7 and 118.5°, respectively, in the nanowire, whereas in bulk ZnO they are 108.3° and 110.6°. The surface Zn and O atoms are all 3-fold coordinated, and the large surface strain induces structural distortions. The surface atoms exhibit apparent shrinkage, but the atoms in the core region are accordingly expanded, so the lengths of the Zn–O bonds parallel to the wire axis in the inner core are elongated to 2.017 Å. In addition, the optimized lattice constant c of the nanowire unit cell is longer than that of the bulk ZnO unit cell (5.27 vs. 5.182 Å).

9.2.2 Stability of Various ZnO 1D Nanostructures

The binding energy measures the stability of different structures; those with larger binding energies are more stable. The binding energies of nanowires with hexagonal cross sections clearly increase with wire diameter (the wire diameter is defined as the largest lateral distance between atoms on the surface of the wire). This increase in binding energy indicates high stability for large nanowires.

Electronic band structures were computed based on the equilibrium ZnO 1D nanostructures. The bandgaps for the ZnO nanowires with hexagonal cross section, triangular cross section, the faceted nanotube, and the (12,0) SWNT. A direct bandgap of 0.72 eV is computed for the ZnO wurtzite solid at the Γ point along the Γ–Z direction in reciprocal space, which is close to previous GGA (Tu et al. (2006)) and DA (Xiang et al. (2006)) results but is much lower than the experimental bandgap which is 3.437 eV at 2 K. This is due to the well-known fact that GGA and LDA approaches underestimate bandgaps (Fillppi et al. (1994)). The bandgaps of all the ZnO 1D nanostructures are larger than that of bulk ZnO. It has been shown that as surface atom ratio increases the bandgap increases and also as diameter of the nanostructure increases, the bandgap decreases. Generally the quantum size effect enlarges the bandgaps; however, the bandgap change due to quantum confinement effects is also related to the surface effect (surface atom ratio) and diameter.

9.2.3 Geometric and Electronic Structures of Pristine ZnO (6,0) SWNT

Various computational studies of pristine ZnO nanotube, nanorod and nanowire have been reported in the literature (Erkoc and Kökten (2005); Tu et al. (2006); Shen et al. (2007)). A single (ZnO)*n* sheet has been shown to be more flexible than a graphene layer (Tu et al. (2006)), i.e. less energy is required to roll up a planar (ZnO)*n* sheet to form tube-like nanostructure than rolling up a graphene layer to form nanotube. Recently, Shen et al. showed that ZnO SWNT is more stable than ZnO nanorod or nanowire if the diameter is small but becomes less stable than ZnO nanorod or nanowire if the diameter is larger than that of ZnO (24,0) SWNT (Shen et al. (2007)). Hence, the small-diameter ZnO (6,0) SWNT is expected to be more stable than ZnO nanorod or nanowire with the same diameter.

9.3 TiO_2 Nanostructures

TiO_2 rutile is commonly used in electronic devices such as thin film capacitors (Wu et al. (1990)) for the fabrication of antireflection coatings, interference filters (Griffin et al. (1990)), as well as optical waveguides (Griffin et al. (1990)) or gas sensors (Varghese et al. (2003)). Furthermore, it is used as white pigment and opacifier for its high reflection across the visible spectrum. On the other hand, the anatase polymorph is commonly used for photocatalysis applications due to its high photoreactivity (Sclafani et al. (1991)) or, more recently, in both homogeneous (Simons (2004)) and heterogeneous catalyses. TiO_2 has been, also, found to be well suited as a semiconductor in electrodes for photoelectrochemical processes such as (i) conversion of sunlight to chemical fuel (H2) through the photocatalytic splitting of water (Fujishima and Honda (1972)) and (ii) conversion of sunlight into electricity in the so-called promising "dye-sensitized solar cells" (Hagfeldt and Grätzel (1995); Kalyanasundaram and Grätzel (1998)).

TiO_2 is being utilized in the inactivation of bacteria, and in self-cleaning coatings (Diebold (2003); Carp et al. (2004); Fernández-García et al. (2004); Chen and Mao (2007)). TiO_2 is a wide bandgap, cheap, and biocompatible n-type semiconductor able to photocatalyse several heterogeneous reactions. Actually, zero-dimensional (0D) TiO_2 nanoparticles and one-dimensional (1D) nanowires share the same benefits from having larger surface area, more active sites, and quantum confinement related properties that eventually lift the rates of their activity (Chen and Mao (2007); Adachi et al. (2004); Mao et al. (2006); Yu et al. (2006); Wang et al. (2006); Koo et al. (2006); Kitano et al. (2007); Feng et al. (2008)). 1D TiO_2 nanostructures from their 0D counterpart is a better ability to transport (photo) generated charges away from possible (recombination) reaction sites, thus revealing the heavy attention that they are receiving for applications in photovoltaic cells, sensors, electrodes, and optoelectronics (Dürr et al. (2005); Tan and Wu (2006); Kang et al. (2008); Centi and Perathoner (2007); Zhu et al. (2007); Wang et al. (2008)).

Recently, TiO_2 in the anatase and rutile forms is used to make high-temperature ferromagnetic semiconductors, and Co-doped anatase TiO_2 is used to realize electric manipulation of ferromagnetism in ferromagnetic semiconductors above room temperature (Matsumoto et al. 2001; Hong et al. (2004); Griffin et al. (2005); Yamada et al. (2011)). These are of huge potential for spintronic and memristor applications in the future. In fact, TiO_2 (rutile) has been most widely used for a long time as white pigment and sunscreen because of its whiteness, high refractive index, strong ultraviolet light absorbing capabilities, and resistance to discolouration to ultraviolet light (Fujishima and Honda (1972)). In addition, it is often used to increase palatability of skimmed milk. Therefore, researchers have studied TiO_2 intensively to understand its physical and chemical properties and to explore more important applications in various fields. Recently HP labs (Strukov et al. (2008)) has shown that it is capable to show memristance which is a milestone discovery from the electronic industry as well as for artificial intelligence.

TiO_2 exhibit tetragonal structure in both the phases. In rutile phase TiO_2 the lattice parameters are a = 4.586 Å and c = 2.954 Å with $P42/mnm$ space group and for the anatase phase the lattice parameters are a = 3.782 Å and c = 9.502 Å with space group $I41/amd$. Authors (Sai and Liu (2012)) adopted the internal atomic position parameters measured at 15 K (Burdett et al. (1987)). The electronic structures was studied by using the three exchange correlation potentials: LDA, GGA (PBE) and TB-mBJ (Tran and Blaha's modified Becke-Johnson plus LDA correlation). The values of the bandgaps calculated are as follows: for rutile LDA, GGA, mBJ and GW calculations produces bandgap 1.79, 1.89, 2.60 and 3.34 eV, respectively; however, the experimental value is 3.0 eV, whereas for anatase structure LDA, GGA, mBJ and GW calculations produces bandgap 1.98, 2.12, 3.01 and 3.56 eV; however, the experimental value is 3.2 eV (Burdett et al. (1987); Cardona and Harbeke (1965); Tang et al. (1994); Amtout and Leonelli (1995); Tang et al. (1995)), and compared with previous calculated results (van Schilfgaarde et al. 2006); Kotani et al. (2007); Thulin and Guerra (2008); Chiodo et al. (2010); Kang and Hybertsen (2010)). It is quite obvious that the LDA and GGA underestimates the gap values in for both of the rutile and the anatase, but the mBJ approximation gives better agreement with the experimental values. The mBJ results are in good agreement with the best of the gap values calculated with the GW approximation (van Schilfgaarde et al. 2006); Kotani et al. (2007); Thulin and Guerra (2008); Chiodo et al. (2010); Kang and Hybertsen (2010)) with respect to the experimental results, but the mBJ values and GW ones are on different sides of the experimental ones. For rutile, DFT produces the direct gap as other DFT calculations, (Liu et al. (2011; Chiodo et al. 2010), but experimentally it exhibits indirect gap (Burdett et al. (1987)), which can be reproduced using GW approximation between Γ and R points (Kang and Hybertsen (2010)), whereas, DFT produces indirect bandgap for the anatase (Liu et al. (2011)).

9.3.1 TiO_2 Nanosheets

To do the structure optimization of lepidocrocite-type titanium dioxides unit cell is defined by a lattice vector (a, b and c) where the ab-plane is a two-dimensional plane and layers are stacked along the c-axis, which comes out to be a C-centred orthorhombic with a space group of *Cmcm* but *Bmmb* space group of stacked TiO_2 was considered to adopt the ab-plane as a two-dimensional plane. Ultrasoft pseudo potentials have been used for electronic structure calculation to describe electron–core interactions by small-core for both oxygen and titanium (Vanderbilt (1990)) as valence states include 2s and 2p states for O, and 3s, 3p, 3d, and 4s states for Ti. Kinetic energy cut-off of 340 eV have been used. The results of total-energy calculations revealed that stacked TiO_2 was a little less stable than anatase and rutile. The difference between stacked two-dimensional TiO_2 and rutile is similar to that between rutile and anatase, indicating that stacked two-dimensional TiO_2 can exist as a metastable phase of TiO_2 polymorphs (Sato et al. (2003)).

The structure of a single layer of titanium dioxides (single-layered TiO_2) was examined. Single-layered TiO_2 was modelled as a supercell in which a vacuum slab and two-dimensional TiO_2 alternate. The vacuum slab was inserted to isolate two-dimensional TiO_2 by diminishing interactions between layers. The inter-layer distance was optimized to avoid the interaction between the layers by putting a vacuum layer between them and found that the distance 1 nm is enough to avoid interaction between the layers. Therefore, in the calculations on single-layered TiO_2, all of the results were conducted with a 1 nm-thick vacuum slab. The symmetry of a structure is reduced to *Pmmn*. The unit cell contains two TiO_2 units and the ab-plane is a two-dimensional plane.

The calculated bandgap comes out to be 3.15 eV for both stacked and single-layered TiO_2, which is larger than those of rutile (2.28 eV) and anatase (2.67 eV). In the band structures of stacked and single-layered TiO_2, no band dispersion is observed along the c^*-axis in the Brillouin zone. In contrast, a large band dispersion is observed parallel to the sheet-plane (ab-plane) in the reciprocal space, indicating highly anisotropic two-dimensional electronic structures in stacked and single layered TiO_2. The larger bandgap is considered to be ascribed to a quantum size effect or to the lower dimensionality. On lowering the dimension the bandwidth decreases because the transfer integrals are smaller in a two-dimensional sheet than in a three-dimensional bulk material. Although it is well-known that the DFT calculation in LDA or GGA underestimates the bandgap, the trend of a larger bandgap in two-dimensional TiO_2, both stacked and single-layered, compared with rutile and anatase is essential (Perdew and Zunger (1981)).

However, the large optical bandgap of bulk TiO_2 (~3 eV) means that only high energy UV light may excite its electrons. As far as the photoelectrochemical water splitting is concerned, this effectively blocks most of the photons which pierce the atmosphere, typically in the visible range, from participating in any bulk TiO_2 based photocatalytic reaction. On the other hand, the difference in energy between excited electrons and holes, i.e. the bandgap, must be large enough (1.23 eV) to dissociate water into hydrogen and oxygen. For these reasons it is of great interest to adjust the bandgap "gap of TiO_2 into the range 1.23 to 2.5 eV," while maintaining the useful properties mentioned above. For low dimensional nanostructured materials, electrons and holes have to travel shorter distances to reach the surface, allowing for a shorter quasi-particle lifetime. However, due to quantum confinement effects, lower dimensional TiO_2 nanostructures tend to have *larger* bandgaps (Bavykin et al. 2006)) The experimental synthesis and characterization of nanostructured materials is in general a costly and difficult task. However, predictions of a dye or nanostructure's properties from simulations can prove a great boon to experimentalists. Modern large-scale electronic structure calculations have become important tools by providing realistic descriptions and predictions of structure and electronic properties for systems of technological interest. As mentioned earlier, standard DFT calculations tend to underestimate electronic bandgaps for bulk TiO_2 by approximately 2 eV, due in part to self-interaction errors (Cai et al. (2002); Salzner et al. (1997)).

These errors arise from an incomplete cancellation of the electron's Coulomb potential in the exchange-correlation (xc)-functional. This may be partially addressed by the use of hybrid functionals such as B3LYP (Becke (1993)), which generally seem to improve bandgaps for bulk systems (Valentin et al. (2006); Muscat et al. (2001); Angelis et al. (2004)). However, such calculations are computationally more expensive, due to the added dependence of the xc-functional on the electron's wavefunction. Moreover, B3LYP calculations for TiO_2 clusters largely underestimate the gap relative to the more reliable difference between standard DFT calculated ionization potential I_p and electron affinity E_a. For isolated systems such as clusters, the needed energetics of charged species are quantitatively described by standard DFT. Also, B3LYP and RPBE (Hammer (1999)) calculations provide the same qualitative description of the trends in the energy gaps for TiO_2. Another methodology is thus needed to describe the bandgaps of periodic systems.

9.4 Summary

Quantum confinement effects increases the energy gap significantly for both 0D and 1D systems while to a lesser extent in 2D systems. This suggests that 2D laminar structures are viable candidates for reducing the minimal quasi-particle, while leaving the bandgap nearly unchanged. The change in the width of the band has been clearly shown when we go down from 3D to 1D in case of transition elements. Earlier, it has been shown by one of the authors (PD) that on reducing the dimension i.e., going from 3D to 1D, the bandwidth decreases (Dua et al. (2011)). Understanding the adsorption behaviour of various molecules on 1D ZnO nanotubes is thus of both practical and fundamental interest. In general, adsorption behaviour can be classified as physisorption, chemisorption and molecular chemisorption. DFT is being playing its role to understand the behaviour of absorption of molecules on nanostructures. For example, adsorption of O_2, H_2, CO, NO_2, and NH_3 has been studied and found good results to explore the tuning of the band structure as well as the conducting properties which can be applied to both in electronic components as well as in sensing applications (An et al. (2008)).

References

Abrahams SC, Bernstein JL. (1969). Remeasurement of the Structure of Hexagonal ZnO. *Acta Crystallogr, B*, **25:** 1233–1236.

Adachi M, Murata Y, Takao J, Jiu J, Sakamoto M, Wang F. (2004). Highly Efficient Dye-Sensitized Solar Cells with a Titania Thin-Film Electrode Composed of a Network Structure of Single-Crystal-like TiO_2 Nanowires Made by the Oriented Attachment Mechanism. *J. Am. Chem. Soc.*, **126:** 14943–14949.

Amtout A, Leonelli R. (1995). Optical Properties of Rutile Near its Fundamental Band Gap. *Phys. Rev. B*, **51:** 6842–6851.

An W, Wu X, Zeng XC. (2008). Adsorption of O_2, H_2, CO, NH_3, and NO_2 on ZnO Nanotube: A Density Functional Theory Study. *J. Phys. Chem. C*, **112:** 5747–5755.

Angelis FD, Tilocca A, Selloni A. (2004). Time-dependent DFT Study of $[Fe(CN)_6]_4$–Sensitization of TiO_2 Nanoparticles, *J. Am. Chem. Soc.*, **126:** 15024–15025.

Bavykin DV, Friedrich JM, Walsh FC. (2006). Protonates Titanates and TiO_2 Nanostructured Materials: Synthesis, Properties, and Applications. *Adv. Mater.*, **18:** 2807–2824.

Becke AD. (1993). Density-Functional Thermochemistry. III. the Role of Exact Exchange. *J. Chem. Phys.*, **98:** 5648–5652.

Blaha P, Schwarz K, Madsen GKH, Kvasnicka D, Luitz J, Karlheinz S. (2001). WIEN2K. Techn. Universität Wien, Austria.

Burdett JK, Hughbanks T, Miller GJ, Richardson Jr JW, Smith JV. (1987). Structural-Electronic Relationships in Inorganic Solids: Powder Neutron Diffraction Studies of the Rutile and Anatase Polymorphs of Titanium Dioxide at 15 and 295 K, *J. Am. Chem. Soc.*, **109:** 3639–3646.

Cai Z-L, Sendt K, Reimers JR. (2002). Failure of Density-Functional Theory and Time-Dependent Density Functional Theory for Large Extended Systems. *J. Chem. Phys.*, **117:** 5543–5549.

Cardona M, Harbeke G. (1965). Optical Properties and Band Structure of Wurtzite-Type Crystals and Rutile. *Phys. Rev.*, **137:** A1467–A1476.

Carp O, Huisman CL, Reller A. (2004). Photoinduced Reactivity of Titanium Dioxide. *Prog. Solid State Chem.*, **32:** 33–177.

Centi G, Perathoner S. (2007). Nano-Architecture and Reactivity of Titania Catalytic Materials. Quasi-1D Nanostructures. *Catalysis.*, **20:** 367–402.

Chen X, Mao SS. (2007). Titanium Dioxide Nanomaterials: Synthesis, Properties, Modifications, and Applications. *Chem. Rev.*, **107:** 2891–2959.

Chen YJ, Li JB, Han YS, Wei QM, Dai JH. (2002). A Novel Morphology of SiOx Nanowires with a Modified Diameter. *Appl. Phys. A*, **74:** 433–435.

Chiodo L, Garcia-Lastra JM, Iacomino A, Ossicini S, Zhao J, Petek H, Rubio A. (2010). Self-Energy and Excitonic Effects in the Electronic and Optical Properties of TiO_2 Crystalline Phases. *Phys. Rev. B*, **82:** 045207 (1–12).

Decca RS, Drew HD, Maiorov B, Guimpel J, Osquiguil E. (1998). Photoinduced Superconducting Nanowires in $GdBa_2Cu_3O_{6.5}$ Films. *Appl. Phys. Lett.*, **73:** 120–122.

Desgreniers S. (1998). Structural and Compressive Parameters High-Density Phases of ZnO. *Phys. Rev. B*. **58:** 14102–14105.

Diebold U. (2003). The Surface Science of Titanium Dioxide. *Surf. Sci. Rep.*, **48:** 53–229.

Djurisǐc AB. Leung YH, Choy WCH, Cheah KW, Chan WK. (2004). Visible Photoluminescence in ZnO Tetrapod and Multipod Structures. *Appl. Phys. Lett.*, **84:** 2635 (1–3).

Dua P, Sharma S, Kumar P. (2011). Density Functional Study of 1D Nanowire for Biomedical Applications. *J. Biomed. Nanotech.*, **7:** 93–94.

Duan X, Huang Y, Cui Y, Wang J, Lieber CM. (2001). Indium Phosphide Nanowires as Building Blocks for Nanoscale Electronic and Optoelectronic Devices. *Nature*, **409:** 66–69.

Dürr M, Schmid A, Obermaier M, Rosselli S, Yasuda A, Nelles G. (2005). Low-Temperature Fabrication of Dye-Sensitized Solar Cells by Transfer of Composite Porous Layers. *Nat. Mater.*, **4:** 607–611.

Erkoc S, Kökten H. (2005). Structural and Electronic Properties of Single-Wall ZnO Nanotubes. *Phys. E*, **28:** 162–170.

Fan W, Xu H, Rosa AL, Frauenheim T, Zhang RQ. (2007). First-Principles Calculations of Reconstructed [0001] ZnO Nanowires. *Phys. Rev. B*, **76:** 073302 (1–4).

Feng X, Shankar K, Varghese OK, Paulose M, Latempa TJ, Grimes CA. (2008). Vertically Aligned Single Crystal TiO_2 Nanowire Arrays Grown Directly on Transparent Conducting Oxide Coated Glass: Synthesis Details and Applications. *Nano Lett.*, **8:** 3781–3786.

Fernández-García M, Martínez-Arias A, Hanson JC, Rodriguez JA. (2004). Nanostructured Oxides in Chemistry: Characterization and Properties. *Chem. Rev.*, **104:** 4063–4104.

Filippi C, Singh DJ, Umrigar CJ. (1994). All-Electron Local-Density and Generalized-Gradient Calculations of the Structural Properties of Semiconductors. *Phys. Rev. B*, **50:** 14947–14951.

Frisch MJ, Trucks GW, Schlegel HB, Scuseria GE, Robb MA, Cheeseman JR, Scalmani G, Barone V, Mennucci B, Petersson GA, Nakatsuji H, Caricato M, Li X, Hratchian HP, Izmaylov AF, Bloino J, Zheng G, Sonnenberg JL,

Hada M, Ehara M, Toyota K, Fukuda R, Hasegawa J, Ishida M, Nakajima T, Honda Y, Kitao O, Nakai H, Vreven T, Montgomery Jr. JA, Peralta JE, Ogliaro F, Bearpark M, Heyd JJ, Brothers E, Kudin KN, Staroverov VN, Kobayashi R, Normand J, Raghavachari K, Rendell A, Burant JC, Iyengar SS, Tomasi J, Cossi M, Rega N, Millam JM, Klene M, Knox JE, Cross JB, Bakken V, Adamo C, Jaramillo J, Gomperts R, Stratmann RE, Yazyev O, Austin AJ, Cammi R, Pomelli C, Ochterski JW, Martin RL, Morokuma K, Zakrzewski VG. Voth GA, Salvador P, Dannenberg JJ, Dapprich S, Daniels AD, Farkas Ö, Foresman JB, Ortiz JV, Cioslowski J, Fox DJ. (2009) Gaussian 09, Revision A.1. Gaussian, Inc., Wallingford CT.

Fu Y, Chen J, Zhang H. (2001). Synthesis of Fe_2O_3 Nanowires by Oxidation of Iron. *Chem. Phys. Lett.*, **350:** 491–494.

Fujishima A, Honda K. (1972). Electrochemical Photolysis of Water at a Semiconductor Electrode. *Nature (London)*, **238:** 37–38.

Gao XP, Zheng ZF, Zhu HY, Pan GL, Bao JL, Wu F, Song DY. (2004). Rotor-like ZnO by Epitaxial Growth Under Hydrothermal Conditions. *Chem. Commun.*, **12:** 1428–1429.

Griffin GL, Siefering KL. (1990). Growth Kinetics of CVD TiO_2: Influence of Carrier Gas. *J. Electrochem. Soc.*, **137:** 1206–1208.

Griffin KA, Pakhomov AB, Wang CM, Heald SM, Krishnan KM. (2005). Intrinsic Ferromagnetism in Insulating Cobalt Doped Anatase TiO_2. *Phys. Rev. Lett.*, **94:** 157204 (1–4).

Gu WG, Zheng B, Han WQ, Roth S, Liu J. (2002). Tungsten Oxide Nanowires on Tungsten Substrates. *Nano Lett.*, **2:** 849–851.

Hafner J. (2008). Ab-Initio Simulations of Materials Using VASP: Density-Functional Theory and Beyond. *J. Comput. Chem.*, **29:** 2044–2078.

Hagfeldt A, Grätzel M. (1995). Light-Induced Redox Reactions in Nanocrystalline Systems. *Chem. Rev. (Washington, D.C.)*, **95:** 49–68.

Hammer B, Hansen LB, Norskov JK. (1999). Improved Adsorption Energetics with Density Functional Theory Using Revised Perdew–Burke–Ernzerhof Functional. *Phys. Rev. B*, **59:** 7413–7421.

Hong NH, Ruyter A, Prellier W, Sakai J. (2004). Room Temperature Ferromagnetism in Anatase $Ti_{0.95}Cr_{0.05}O_2$ Thin Films: Clusters or Not? *Appl. Phys. Lett.*, **85:** 6212 (1–3).

Hoshi T, Fujiwara T. (2006). Large-Scale Electronic Structure Theory for Simulating Nanostructure Processes. *J. Phys. Condens. Matter.*, **18:** 10787–10802.

Hu J, Odom TW, Lieber CM. (1999). Chemistry and Physics in One Dimension: Synthesis and Properties of Nanowires and Nanotubes. *Acc. Chem. Res.*, **32:** 435–445.

Hu JQ, Ma XL, Shang NG, Xie ZY, Wong NB, Lee CS, Lee ST. (2002). Large-Scale Rapid Oxidation Synthesis of SnO_2 Nanoribbons. *J. Phys. Chem. B*, **106:** 3823–3826.

Huang MH, Mao S, Feick H, Yan HQ, Wu YY, Kind H, Weber E, Russo R, Yang PD. (2001). Room-Temperature Ultraviolet Nanowire Nanolasers. *Science*, **292:** 1897–1899.

Jaffe JE, Hess AC. (1993). Hartree-Fock Study of Phase Changes in ZnO at High Pressure. *Phys. Rev. B.*, **48:** 7903–7909.

Jeevanandam P, Koltypin Yu, Gedanken A. (2001). Synthesis of Nanosized α-Nickel Hydroxide by a Sonochemical Method. *Nano Lett.*, **1:** 263–266.

Kalyanasundaram K, Grätzel M. (1998). Applications of Functionalized Transition Metal Complexes in Photonic and Optoelectronic Devices. *Coord. Chem. Rev.*, **177:** 347–414.

Kang SH, Choi S-H, Kang M-S, Kim J-Y, Kim H-S, Hyeon T, Sung Y-E. (2008). Nanorod-Based Dye-Sensitized Solar Cells with Improved Charge Collection Efficiency. *Adv. Mater.*, **20:** 54–58.

Kang W, Hybertsen MS. (2010). Quasiparticle and Optical Properties of Rutile and Anatase TiO_2. *Phys. Rev. B*, **82:** 085203 (1–11).

Kasuga T, Hiramatsu M, Hoson A, Sekino T, Niihara K. (1998). Formation of Titanium Oxide Nanotube. *Langmuir*, **14:** 3160–3163.

Kitano M, Mitsui R, Eddy DR, El-Bahy ZMA, Matsuoka M, Ueshima M, Anpo M. (2007). Synthesis of Nanowire TiO_2 Thin Films by Hydrothermal Treatment and their Photoelectrochemical Properties. *Catal. Lett.*, **119:** 217–221.

Kong XY, Ding Y, Yang R, Wang ZL. (2004). Single-Crystal Nanorings Formed by Epitaxial Self-Coiling of Polar Nanobelts. *Science*, **303:** 1348–1351.

Kong YC, Yu DP, Zhang B, Fang W, Feng SQ. (2001). Ultraviolet-Emitting ZnO Nanowires Synthesized by a Physical Vapor Deposition Approach. *Appl. Phys. Lett.*, **78:** 407–409.

Koo B, Park J, Kim Y, Choi S-H, Sung Y-E, Hyeon T. (2006). Simultaneous Phase- and Size-Controlled Synthesis of TiO_2 Nanorods via Non-Hydrolytic Sol–Gel Reaction of Syringe Pump Delivered Precursors. *J. Phys. Chem. B*, **110:** 24318–24323.

Kotani T, van Schilfgaarde M, Faleev SV, Chantis A. (2007). Quasiparticle Self-Consistent *GW* Method: a Short Summary. *J. Phys.: Condens. Matter.*, **19:** 365236 (1–4).

Kresse G, Furthmüller J. (1996). Efficiency of Ab-Initio Total Energy Calculations for Metals and Semiconductors Using a Plane-Wave Basis Set. *Comput. Mat. Sci.*, **6:** 15–50.

Li C, Guo WL, Kong Y, Gao HJ. (2007). First-Principles Study of the Dependence of Ground-State Structural Properties on the Dimensionality and Size of ZnO Nanostructures. *Phys. Rev. B*, **76:** 035322 (1–4).

Li YB, Bando Y, Golberg D. (2004). ZnO Nanoneedles with Tip Surface Perturbations: Excellent Field Emitters. *Appl. Phys. Lett.*, **84:** 3603–3605.

Liang CH, Meng GW, Wang GZ, Wang YW, Zhang LD, Zhang SY. (2001). Catalytic Synthesis and Photoluminescence of β-Ga_2O_3 Nanowires. *Appl. Phys. Lett.*, **78:** 3202 (1–3).

Liao H-W, Wang Y-F, Liu X-M, Li Y-D, Qian Y-T. (2000). Hydrothermal Preparation and Characterization of Luminescent $CdWO_4$ Nanorods. *Chem. Mater.*, **12:** 2819–2821.

Lin C-H, Chien S-H, Chao J-H, Sheu C-Y, Cheng Y-C, Huang Y-J, Tsai C-H. (2002). The Synthesis of Sulfated Titanium Oxide Nanotubes. *Catal. Lett.*, **80:** 153–159.

Liu H-L, Hsueh H-C, Lin I-N, Yang M-T, Lee W-C, Chen Y-C, Chia C-T, Cheng H-C. (2011). The Dielectric Properties of $La(Mg_{0.5}Ti_{0.5})O_3$ Ceramics Studied by Raman-Scattering, Infrared Reflectivity Spectroscopy, and First-Principles Calculations, *J. Phys.: Condens. Matter.*, **23:** 025901 (1–4).

Liu J, Fei P, Song J, Wang X, Lao C, Tummala R, Wang ZL. (2008). Carrier Density and Schottky Barrier on the Performance of DC Nanogenerator. *Nano Lett.*, **8:** 328–332.

Liu Y, Dong Y, Wang G. (2003). Far-Infrared Absorption Spectra and Properties of SnO_2 Nanorods. *Appl. Phys. Lett.*, **82:** 260–262.

Liu Y, Zheng C, Wang W, Zhan Y, Wang G. (2002). Synthesis and Characterization of Zirconia Nanorods. *J. Am. Ceram. Soc.*, **85:** 3120–3122.

Lopez R, Haynes TE, Boatner LA, Feldman LC, Haglund RF Jr. (2002). Temperature-Controlled Surface Plasmon Resonance in VO_2 Nanorods. *Opt. Lett.*, **27:** 1327–1329.

Louie SG. (2001). Electronic Properties, Junctions, and Defects of Carbon Nanotubes. eds. M.S. Dresselhaus, G. Dresselhaus, and P. Avouris (Springer-Verlag, Heidelberg) *Topics Appl. Phys.*, **80:** 113–145.

Mao Y, Kanungo M, Hemraj-Benny T, Wong SS. (2006). Synthesis and Growth Mechanism of Titanate and Titania One-Dimensional Nanostructures Self-Assembled into Hollow Micrometer-Scale Spherical Aggregates. *J. Phys. Chem. B*, **110:** 702–710.

Martin RM. (2004). *Electronic Structure: Basic Theory and Practical Methods.* Cambridge University Press.

Marzari N. (2006). Realistic Modeling of Nanostructures Using Density Functional Theory. *Mater. Res. Bull.,* **31:** 681–687.

Matsui K, Kyotani T, Tomita A. (2002). Hydrothermal Synthesis of Single-Crystal $Ni(OH)_2$ Nanorods in a Carbon-Coated Anodic Alumina Film. *Adv. Mater.,* **14:** 1216–1219.

Matsumoto Y, Murakami M, Shono T, Hasegawa T, Fukumura T, Kawasaki M, Ahmet P, Chikyow T, Koshihara SY, Koinuma H. (2001). Room-Temperature Ferromagnetism in Transparent Transition Metal-Doped Titanium Dioxide. *Science,* **291:** 854–856.

Muscat J, Wander A, Harrison NM. (2001). On the Prediction of Band Gaps from Hybrid Functional Theory. *Chem. Phys. Lett.,* **342:** 397–401.

Muster J, Kim GT, Krstic V, Park JG, Park YW, Roth S, Burghard M. (2000). Electrical Transport Through Individual Vanadium Pentoxide Nanowires. *Adv. Mater.,* **12:** 420–424.

Niederberger M, Krumeich F, Muhr H-J, Mller M, Nesper R. (2001). Synthesis and Characterization of Novel Nanoscopic Molybdenum Oxide Fibers. *J. Mater. Chem.,* **11:** 1941–1945.

Özgür Ü, Alivov YI, Liu C, Teke A, Reshchikov MA, Doðan S, Avrutin V, Cho S-J, Morkoç H. (2005). A Comprehensive Review of ZnO Materials and Devices. *Appl. Phys.,* **98:** 041301 (1–103).

Pacholski C, Kornowski A, Weller H. (2002). Self-Assembly of ZnO: From Nanodots to Nanorods, *Angew. Chem. Int. Ed. Engl.,* **41:** 1188–1191.

Peng XS, Meng GW, Zhang J, Wang XF, Wang YW, Wang CZ, Zhang LD. (2002). Synthesis and Photoluminescence of Single-Crystalline In_2O_3 Nanowires. *J. Mater. Chem.,* **12:** 1602–1605.

Perdew JP, Zunger A. (1981). Self-Interaction Correction to Density-Functional Approximations for Many-Electron Systems. *Phys. Rev. B,* **23:** 5048–5079.

Pol VG, Palchik O, Gedanken A, Felner I. (2002). Synthesis of Europium Oxide Nanorods by Ultrasound Irradiation. *J. Phys. Chem. B,* **106:** 9737–9743.

Roy VAL, Djurisic AB, Chan WK, Gao J, Lui HF, Surya C. (2003). Luminescent and Structural Properties of ZnO Nanorods Prepared Under Different Conditions. *Appl. Phys. Lett.,* **83:** 141–143.

Rühle S, van Vugt LK, Li H-Y, Keizer NA, Kuipers L, Vanmaekelbergh D. (2008). Nature of Sub-Band Gap Luminescent Eigenmodes in a ZnO Nanowire. *Nano Lett.,* **8:** 119–123.

Ryan JV, Berry AD, Anderson ML, Long JW, Stroud RM, Cepak VM, Browning VM, Rolison DR, Merzbacher CI. (2000). Electronic Connection to the Interior of a Mesoporous Insulator with Nanowires of Crystalline RuO_2. *Nature (London)*, **406:** 169–172.

Sai G, Liu B-G. (2012) Electronic Structures and Optical Properties of TiO_2: Improved Density-Functional-Theory Investigation. *Chin. Phys. B*, **21:** 057104 (1–7).

Salzner U, Lagowski JB, Pickup PG, Poirier RA. (1997). Design of Low Band Gap Polymers Employing Density Functional Theory — Hybrid Functionals Ameliorate Band Gap Problem. *J. Comput. Chem.*, **18:** 1943–1953.

Sato H, Ono K, Sasaki T, Yamagishi A. (2003). First-Principles Study of Two-Dimensional Titanium Dioxides. *J. Phys. Chem. B*, **107:** 9824–9828.

Sclafani A, Palmisano L, Davi E. (1991). Photocatalytic Degradaton of Phenol in Aqueous Polycrystalline TiO_2 Dispersions: the Influence of Fe^{3+}, Fe^{2+} and Ag^+ on the Reaction Rate. *J. Photochem. Photobiol. A*, **56:** 113–123.

Segall M, Probert M, Pickard C, Hasnip P, Clark S, Refson K, Payne M. CASTEP. http://www.tcm.phy.cam.ac.uk/castep/index.html.

Shen X, Allen PB, Muckerman JT, Davenport JW, Zheng J-C. (2007). Wire Versus Tube: Stability of Small One-Dimensional ZnO Nanostructures. *Nano Lett.*, **7:** 2267–2271.

Shi X, Han S, Sanedrin RJ, Galvez C, Ho DG, Hernandez B, Zhou F, Selke M. (2002). Formation of Cobalt Oxide Nanotubes: Effect of Intermolecular Hydrogen Bonding between Co(III) Complex Precursors Incorporated onto Colloidal Templates. *Nano Lett.*, **2:** 289–293.

Simons C. (2004). Noncovalent Anchoring of Asymmetric Hydrogenation Catalysts on a New Mesoporous Aluminosilicate: Application and Solvent Effects. *Chem. Eur. J.*, **10:** 5829–5835.

Song J, Wang X, Liu J, Liu H, Li Y, Wang ZL. (2008). Piezoelectric Potential Output from ZnO Nanowire Functionalized with p-Type Oligomer. *Nano Lett.*, **8:** 203–207.

Srivastava AK, Gakhar R, Dua P, Senthil K, Sood KN, Yong K. (2010). Structural Determination of Zn–O Dumbbells in Facetted Nano-Particles. Book chapter: *Microscopy, Science, Technology, Applications and Education*, **4:** 1820–1823.S

Stan G, Ciobanu CV, Parthangal PM, Cook RF. (2007). Diameter-Dependent Radial and Tangential Elastic Moduli of ZnO Nanowires. *Nano Lett.*, **7:** 3691–3697.

Strukov DB, Snider GS, Stewart DR, Williams RS. (2008). The Missing Memristor Found. *Nature*, **453:** 80–83.

Tan B, Wu Y. (2006). Dye-Sensitized Solar Cells Based on Anatase TiO_2 Nanoparticle/Nanowire Composites. *J. Phys. Chem. B*, **110:** 15932–15938.

Tang H, Levy F, Berger H, Schmid PE. (1995). Urbach Tail of Anatase TiO_2, *Phys. Rev. B*, **52:** 7771–7774.

Tang H, Prasad K, Sanjines R, Schmid PE, Levy F. (1994). Electrical and Optical Properties of TiO_2 Anatase Thin Films. *J. Appl. Phys.*, **75:** 2042 (1–6).

Tang H, Zhu L, Ye Z, He H, Zhang Y, Zhi M, Yang F, Yang Z, Zhao B. (2007). Synthesis of Two Kinds of ZnO Nanostructures by Vapor Phase Method, *Mater. Lett.*, **61:** 1170–1173.

Thulin L, Guerra J. (2008). Calculations of Strain-Modified Anatase TiO_2 Band Structures. *Phys. Rev. B*, **77:** 195112 (1–5).

Tu ZC, Hu X. (2006). Elasticity and Piezoelectricity of Zinc Oxide Crystals, Single Layers, and Possible Single-Walled Nanotubes. *Phys. Rev. B*, **74:** 035434 (1–6).

Valentin CD, Pacchioni G, Selloni A. (2006). Electronic Structure of Defect States in Hydroxylated and Reduced Rutile TiO_2 (110) Surfaces. *Phys. Rev. Lett.*, **97:** 166803 (1–4).

Van Schilfgaarde M, Kotani T, Faleev S. (2006). Quasiparticle Self-Consistent GW Theory. *Phys. Rev. Lett.*, **96:** 226402 (1–4).

Vanderbilt D. (1990). Soft Self-Consistent Pseudopotentials in a Generalized Eigenvalue Formalism. *Phys. Rev. B.*, **41:** 7892–7895.

Varghese OK, Grimes CA. (2003). Metal Oxide Nanoarchitectures for Environmental Sensing. *J. Nanosci. Nanotechnol.*, **3:** 277–293.

Vayssieres L, Manthiram A. (2004). One-Dimensional Metal Oxide Nanostructures. *Encyclopedia Nanosci. Nanotechnol.*, **8:** 147–166.

Vayssieres L. (2003). Growth of Arrayed Nanorods and Nanowires of ZnO from Aqueous Solutions. *Adv. Mater.*, **15:** 464–466.

Wang LW. (2004). Theoretical Investigations of Silicon Quantum Dots. *Handbook of Theoretical and Computational Nanotechnology*, ed. M Rieth, W Schommers, **10:** 635–82. New York: Am. Sci.

Wang Q, Wen Z, Li J. (2006). Solvent-Controlled Synthesis and Electrochemical Lithium Storage of One-Dimensional TiO_2 Nanostructures. *Inorg. Chem.*, **45:** 6944–6949.

Wang R, Ruan C, Kanayeva D, Lassiter K, Li Y. (2008). TiO_2 Nanowire Bundle Microelectrode Based Impedance Immunosensor for Rapid and Sensitive Detection of Listeria Monocytogenes. *Nano Lett.,* **8:** 2625–2631.

Wang W, Zhan Y, Wang X, Liu Y, Zheng C, Wang G. (2002). Synthesis and Characterization of CuO Nanowhiskers by a Novel One-Step, Solid-State Reaction in the Presence of a Nonionic Surfactant. *Mater. Res. Bull.,* **37:** 1093–1100.

Wang X, Li Y. (2002). Selected-Control Hydrothermal Synthesis of α- and β-MnO_2 Single Crystal Nanowires. *J. Am. Chem. Soc.,* **124:** 2880–2881.

Wang X, Song J, Wang ZL. (2006). Single-Crystal Nanocastles of ZnO. *Chem. Phys. Lett.,* **424:** 86–90.

Wu C, Qin W, Qin G, Zhao D, Zhang J, Huang S, Lu S, Liu H, Lin H. (2003). Photoluminescence from Surfactant-Assembled Y_2O_3:Eu Nanotubes. *Appl. Phys. Lett.,* **82:** 520–522.

Wu J-J, Liu S-C, Wu C-T, Chen K-H, Chen L-C. (2002). Heterostructures of ZnO–Zn Coaxial Nanocables and ZnO Nanotubes. *Appl. Phys. Lett.,* **81:** 1312–1314.

Wu JM, Chen CJ. (1990). Effect of Powder Characteristics on Microstructures and Dielectric Properties of (Ba, Nb)-Doped Titania Ceramics. *J. Am. Ceram. Soc.,* **73:** 420–424.

Wu Z, Cohen RE. (2006). More Accurate Generalized Gradient Approximation for Solids. *Phys. Rev. B.,* **73:** 235116 (1–6).

Xu C, Liu Y, Xu G, Wang G. (2002). Preparation and Characterization of CuO Nanorods by Thermal Decomposition of CuC_2O_4 Precursor. *Mater. Res. Bull.,* **14:** 2365–2372.

Xu H, Rosa AL, Frauenheim T, Zhang RQ, Lee ST. (2007). Density-Functional Theory Calculations of Bare and Passivated Triangular-Shaped ZnO Nanowires. *Appl. Phys. Lett.,* **91:** 031914 (1–3).

Yamada Y, Ueno K, Fukumura T, Yuan HT, Shimotani H, Iwasa Y, Gu L, Tsukimoto S, Ikuhara Y, Kawasaki M. (2011). Electrically Induced Ferromagnetism at Room Temperature in Cobalt-Doped Titanium Dioxide. *Science,* **332:** 1065–1067.

Yang P, Wu Y, Fan R. (2002). Inorganic Semiconductor Nanowires. *Int. J. Nanosci.* **1:** 1–40.

Yi DK, Yoo SJ, Kim D-Y. (2002). Spin-on-Based Fabrication of Titania Nanowires Using a Sol–Gel Process. *Nano Lett.,* **2:** 1101–1104.

Yin Y, Zhang G, Xia Y. (2002). Synthesis and Characterization of MgO Nanowires Through a Vapor-Phase Precursor Method. *Adv. Funct. Mater.*, **12:** 293–298.

Yu H, Yu J, Cheng B. (2006). Preparation, Characterization and Photocatalytic Activity of novel TiO_2 Nanoparticle-Coated Titanate Nanorods. *J. Mol. Catal. A: Chem.*, **253:** 99–106.

Zhang BP, Binh NT, Segawa Y, Kashiwaba Y, Haga K. (2004). Photoluminescence Study of ZnO Nanorods Epitaxially Grown on Sapphire $(11\hat{2}0)$ Substrates. *Appl. Phys. Lett.*, **84:** 586–588.

Zhang Y, Zhu J. (2002). Synthesis and Characterization of Several One-Dimensional Nanomaterials. *Micron*, **33:** 523–534.

Zhang Z, Wang SJ, Yu T, Wu T. (2007). Controlling the Growth Mechanism of ZnO Nanowires by Selecting Catalysts. *J. Phys. Chem. C.*, **111:** 17500–17505.

Zhang Z. (2002). Synthesis and Microstructure of Antimony Oxide Nanorods. *J. Mater. Res.*, **17:** 1698–1701.

Zhen Z, Yafei L, Lu L, Yongsheng C, Zhang SB, Zhongfang C. (2008). Size- and Surface-dependent Stability, Electronic Properties, and Potential as Chemical Sensors: Computational Studies on One-dimensional ZnO Nanostructures, *J. Phys. Chem. C*, **112:** 13926–13931.

Zheng B, Wu Y, Yang P, Liu J. (2002). Synthesis of Ultra-Long and Highly Oriented Silicon Oxide Nanowires from Liquid Alloys. *Adv. Mater.*, **14:** 122–124.

Zhou J, Deng SZ, Chen J, She JC, Xu NS. (2002). Synthesis of Crystalline Alumina Nanowires and Nanotrees. *Chem. Phys. Lett.*, **365:** 505–508.

Zhou Z, Komori T, Ayukawa T, Morinaga M, Koizumi A, Takeda A. (2005). Li- and Er-codoped ZnO with enhanced 1.54 μm photoemission. Y. *Appl. Phys. Lett.*, **87:** 091109 (1–3).

Zhu K, Vinzant TB, Neale NR, Frank AJ. (2007). Removing Structural Disorder from Oriented TiO_2 Nanotube Arrays: Reducing the Dimensionality of Transport and Recombination in Dye-Sensitized Solar Cells. *Nano Lett.*, **7:** 3739–3746.

Zunger A. (1998). Electronic-Structure Theory of Semiconductor Quantum Dots. *Mater. Res. Bull.*, **23:** 35–42.

Chapter 10

Paramagnetic Lattice Defects in Natural Crystalline Quartz

Shin Toyoda

Department of Applied Physics, Okayama University of Science, 1-1 Ridai, Okayama, 700-0005, Japan

toyoda@dap.ous.ac.jp

Quartz is one of the most abundant minerals on the surface of the earth, paramagnetic defects in which have been the subject of scientific interest for a long time. It was found that paramagnetic defects in quartz make "chemical reactions" in its crystal lattice so that the formation of the E_1' center (unpaired electron at an oxygen vacancy) is controlled by the concentration of aluminum impurity. Impurity centers of hole type and electron type seem also to react each other which affects the decay of the centers on heating. Formation of the oxygen vacancies in quartz is one of the topics further to be investigated where temperature dependence of formation efficiency will be a key feature to discuss the issue. Dating of natural quartz based on the measurement by ESR (electron

Oxide Nanostructures: Growth, Microstructures, and Properties
Edited by Avanish Kumar Srivastava

ISBN 978-981-4411-35-6 (Hardcover), 978-981-4411-36-3 (eBook)
www.panstanford.com

spin resonance) of natural accumulated dose is one of the important applications, which have been applied to determine the ages of fault movements, volcanic eruption, heating of implements, and sedimentation in river. The E_1' center in quartz was also found to be a useful proxy in investigating material transport on the surface of the earth, based on which the environmental changes in the past are discussed.

10.1 Introduction

Point defects in quartz have been studied by ESR (electron spin resonance, equivalent to EPR, electron paramagnetic resonance) extensively since about a decade after the discovery of ESR by Zavoisky (1945). Aluminum hole centers (Griffiths et al., 1954), germanium electron centers (Anderson and Weil, 1959), and E′ centers (Weeks, 1956) were subject of early discovery of the paramagnetic point defects in crystalline quartz. Since then, many paramagnetic point defects have been found as summarized by Weil (1984).

Crystalline quartz occurs commonly on the surface of the earth. Most sediments contain small grains of crystalline quartz as a major component, which is usually the result of weathering of granites and of transportation by rivers. Large land surface area of the earth is covered by granites in which quartz is also a major component. Ikeya et al. (1982) first started investigations to read out geological information recorded as lattice defects in natural quartz, which was dating of fault movements. Subsequently, dating with quartz was found to be possible also for tephra (Imai et al., 1985), archaeological flints (e.g., Porat et al., 1994) and for sediments (e.g., Yokoyama et al., 1985). More recently, lattice defects in quartz can also be used as proxy of transportation of sediments on the surface of the earth, which gives information on the environmental change (e.g., Toyoda and Naruse, 2002).

For these latter application studies, formation and decay properties of paramagnetic defects are the essential parameters while atomic configuration and electronic states are the major interest in the former studies. The present article will first briefly summarize the models of paramagnetic centers, which are observed in natural crystalline quartz, and secondary describe the formation and decay characteristics of these centers, then, geologic and geomorphologic applications.

10.2 Paramagnetic Centers Observable in Natural Crystalline Quartz

10.2.1 Aluminum Hole Center

The ESR signal of a paramagnetic defect attributed to aluminum was first observed by Griffith et al. (1954). Its electronic structure is $[AlO_4]^0$ where a hole is trapped in a nonbonding p-orbital of an oxygen anion linking a substitutional aluminum (O'Brien, 1955). By an ELDOR experiment, Taylor and Farnell (1964) showed that the hole associated with a substitutional aluminum atom moves back and forth between the two oxygen atoms linking the aluminum with long bonds. Schnadt and Räuber (1971) analyzed the broadening with increasing temperature of the line widths of the signal to obtain an activation energy of 70 ± 10 meV for this hopping process. They pointed out that the dielectric loss is due to this hopping process. Precise angular dependence of the ESR spectrum was studied by Nuttal and Weil (1981a) to determine the spin Hamiltonian parameters including ^{17}O and ^{29}Si hyperfine matrices where a spectrum calculated from the obtained parameters coincided with the observed one. Its single crystal ESR spectrum shows many complex lines where simple hyperfine lines due to ^{27}Al (I = 5/2, 100% abundance) are not prominent because the contributions of nuclear Zeeman and quadruple terms to the Zeeman energy levels.

A typical powder spectrum of the aluminum hole center (Al center) observed in natural quartz extracted from a tephra layer is shown in Fig. 10.1 where only principal g factors are indicated as it is not possible to indicate the hyperfine lines due to the above reason.

When an aluminum atom replaces a silicon, it is a precursor with a formula, $[AlO_4/M^+]^0$, which is diamagnetic, accompanying an alkali ion as a charge compensator. It turns $[AlO_4/M^+]^+$, after irradiation at 77 K (Nuttal and Weil, 1981b), and then $[AlO_4]^0$, when warming to room temperature, after an alkali ion diffuses away as charge compensation is not necessary any more. When the sample is irradiated at room temperature, above two processes occur simultaneously. In natural quartz, this aluminum hole center, $[AlO_4]^0$, created by natural radiation is observed.

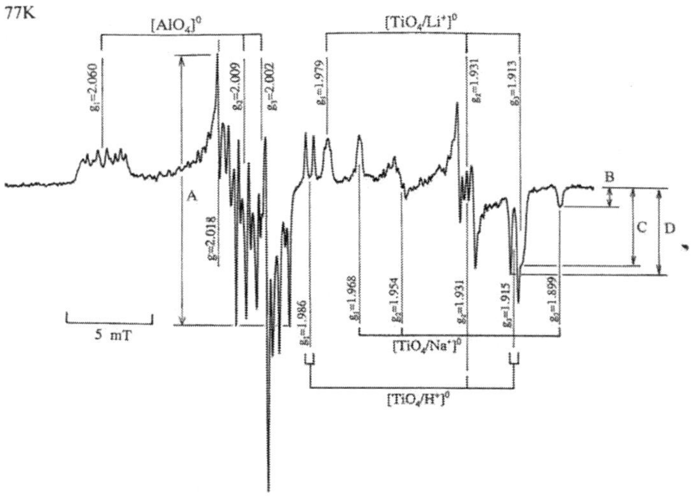

Figure 10.1 A typical powder ESR spectrum of natural quartz (Toyoda and Goff, 1996). Signals due to the aluminum hole center, Ti–Li, Ti–H, and T–Na centers are observed where the g factors are consistent with the ones obtained by single-crystal analysis.

10.2.2 Germanium Centers

Many germanium centers were found after Anderson and Weil (1959) had first reported germanium centers in quartz. They found two types with different set of g factors, designated as A and C. Weil and Anderson (1961) presumed that an unpaired electron is localized predominantly on a germanium ion replacing a silicon atom. A germanium atom replaces a silicon with a formula $[GeO_4]^0$ as it has an electron orbital configuration similar with silicon. When quartz is irradiated at 15 K, it traps an electron to form $[GeO_4]^-$ (Mackey, 1963). Alkali ions, possibly released by aluminum hole center, are mobile on heating to be trapped by this center for charge compensation to form $[GeO_4/M^+]^0$ (Weil, 1984). Accompanying alkali ions are detected by a set of four hyperfine lines, i.e., ^7Li or ^{23}Na with a nuclear spin of 3/2. Weil (1984) pointed out that the difference between Type A (equivalent to $[GeO_4/Li^+]^0_1$) and Type C (equivalent to $[GeO_4/Li^+]^0_2$) are presumably due to the different positions of alkali ion in the crystal. The spin Hamiltonian parameters for $[GeO_4/Li^+]_1^0$) and $[GeO_4/Li^+]_2^0$ were obtained by Weil (1971) at room temperature.

Irradiated natural quartz samples commonly exhibit the signal of germanium centers as composite of the two signals after gamma ray irradiation, but not without irradiation. It is because that the signal is not thermally stable in geological time scale (Ikeya, 1993), although no systematic detailed investigations on thermal stabilities on this signal have been made.

10.2.3 Titanium Centers

A substitutional titanium atom traps an unpaired electron and accompanies an alkali ion as does a germanium center. There are also two types designated by 1 and 2 or B and A. In a piece of rose quartz, Wright et al. (1963) first reported titanium centers with Li^+ and/or Na^+, and H^+, which are $[TiO_4/M^+]_2^0$. The spin Hamiltonian parameters for $[TiO_4/Li^+]_1^0$ (B–Ti–Li), $[TiO_4/Li^+]_2^0$ (A–Ti–Li), $[TiO_4/H^+]_1^0$ (B–Ti–H), and $[TiO_4/H^+]_2^0$ (A–Ti–H) were obtained by Rinneberg and Weil (1972). Okada et al. (1971) found $[TiO_4/Na^+]_2^0$ (A–Ti–Na) and obtained the parameters for this center. They also reported that $[TiO_4/Li^+]_1^0$ (B–Ti–Li) and $[TiO_4/H^+]_1^0$ (B–Ti–H) decay rapidly at room temperature.

In natural quartz, only thermally stable $[TiO_4/M^+]_2^0$ (M=H, Li, Na) are observable where they are denoted as Ti–H, Ti–Li, and Ti–Na centers in some literature. The g factors shown in a typical ESR powder spectrum of titanium centers (Fig. 10.1) are consistent with these obtained principal g factors. The Ti–Na center is not so common as Ti–H and Ti–Li centers in natural quartz.

10.2.4 E_1' Center

E_1' centers are paramagnetic defects associated with oxygen vacancies. Three species, E_1', E_2', and E_4' centers were known until recently new E_1' centers (E_1' to E_9') were found by Mashkovtsev and Pan (2012). While the E_2' and E_4' centers were found to be an oxygen vacancy with an unpaired electron associated with a hydrogen atom (Weeks and Nelson, 1960a; Isoya et al., 1981), the E_1' center is a fundamental defect discovered at the earliest stage of investigation (Weeks, 1956). According to Feigl et al. (1974), it is an oxygen vacancy with an unpaired electron in a single silicon sp^3 orbital oriented along a bond direction where neighboring atoms have relaxed asymmetrically. Rudra and Fowler (1987) made theoretical calculation to show that above asymmetric relaxation are probable as shown in Fig. 10.3, and also to obtained the energy levels of each charged state of oxygen vacancy, where that of the positively charged oxygen vacancy (the E_1' center) is 5.5 eV above the neutral oxygen vacancy with two electrons (Si–Si bond). However, there is a debate that it may be a divacancy to explain the weak hyperfine splitting of 8 G and 9 G (Jani et al., 1983) while strong hyperfine splitting (400 G) is explained by the one due to ^{29}Si (Silsbee, 1961).

A powder ESR spectrum of the E_1' center is shown in Fig. 10.2 (Wieser and Regulla, 1989) where the g factors are consistent with those obtained for single crystal analysis by Jani et al. (1983). The E_1' is commonly observed in natural quartz extracted from granites and from sediments but not in the one from young volcanic tephra or of hydrothermal origin, the reason of which will be discussed in a later section.

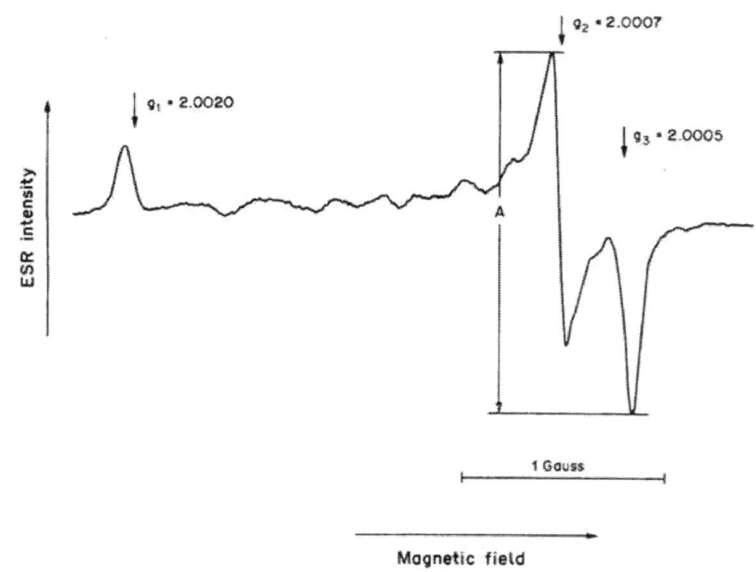

Figure 10.2 A powder ESR spectrum of the E_1' center (Wieser et al., 1989) where the g factors are consistent with the ones obtained by single crystal analysis.

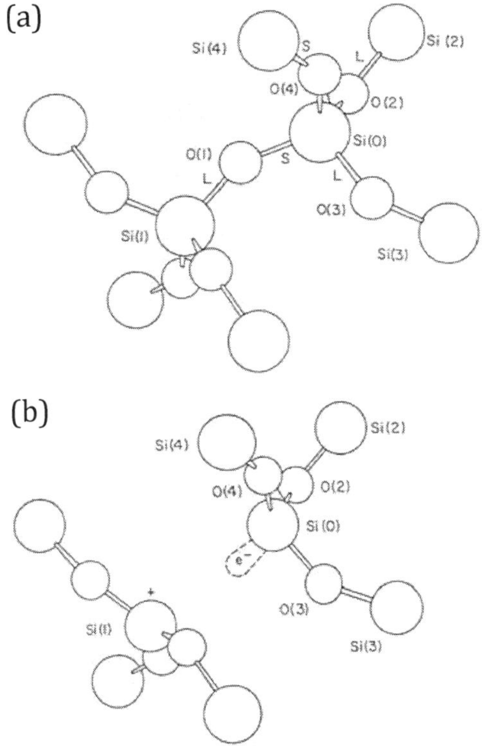

Figure 10.3 Atomic configuration models for (a) quartz lattice and (b) the E_1' center (partially modified after Rudra and Fowler, 1987).

10.3 Formation of the E_1' Center

While the model and the atomic configuration of the E_1' center have been the subject and primary interest of the previous studies, its formation and decay processes have not well been investigated although some studies mentioned its importance.

Usually paramagnetic lattice defects are less stable than their precursors, which are indicated by the observation that the signals decay on heating. However, it has been well known that the E_1' center increases on heating since the early stage of the investigations (Weeks and Nelson, 1960b) as typically shown in Fig. 10.4 (Toyoda and Ikeya, 1991) where the intensity is at the maximum at 300°C in a stepwise heating experiment with a heating duration of 15 min. Jani et al. (1983) found a correlation between the decay of the aluminum hole center and the increase of the E_1' center in electron irradiated quartz to propose that electronic holes released on heating from aluminum hole centers are trapped by neutral oxygen vacancies and recombine one of the electrons to form the paramagnetic E_1' center.

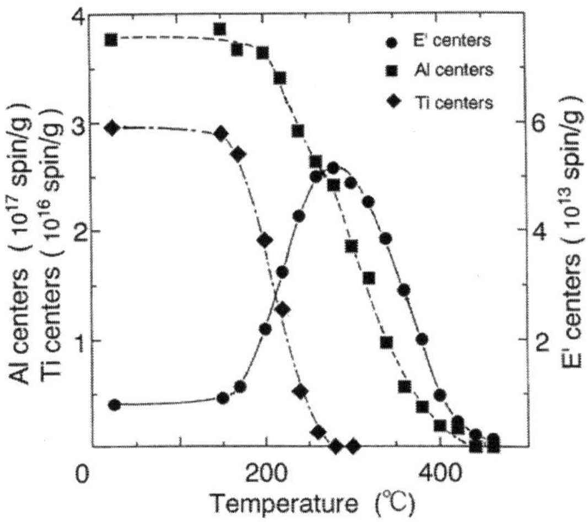

Figure 10.4 Results of a stepwise heating experiment with a duration of 15 min (Toyoda and Ikeya, 1991).

Toyoda and Ikeya (1991) pointed out that two processes are necessary for the E_1' center to be formed, which are the formation of oxygen vacancies, part of which can be diamagnetic neutral oxygen vacancies, and the transfer of the holes from the aluminum hole center to neutral oxygen vacancies. Based on the electronic process proposed by Jani et al. (1983), the authors proposed a protocol to estimate the relative total number of the oxygen vacancies, by measuring the ESR intensity of the E_1' center after gamma ray irradiation to more than 200 Gy followed by heating at 300°C for 15 min (the heat-treated E_1' center). Toyoda and Hattori (2000) supported this idea by studying the dose responses of the heat-treated E_1' center and of the aluminum hole center. They observed in granitic quartz heated at 450°C, at which the E_1' center is erased, that the intensity of the heat-treated E_1' center saturates above 200 Gy while that of the aluminum hole center still increases with dose (Fig. 10.5). They attributed this increase below 200 Gy of the heat-treated E_1' center with dose to the increase of the number of the transferred holes while it saturates in the higher dose region because of the limited number of oxygen vacancies in quartz.

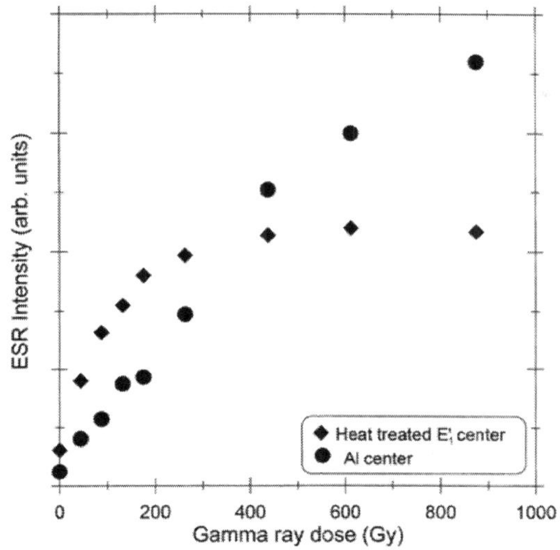

Figure 10.5 Dose response of the Al center and the heat-treated E_1' center in granitic quartz (Toyoda and Hattori, 2000). The intensity of the Al center increases with dose while that of the heat-treated (regenerated) E_1' center saturates above 200 Gy.

Recently, Usami et al. (2009) made a detailed study on the same topic to show that the concentration of aluminum in quartz also controls the formation of the E_1' center. They investigated natural and synthetic crystalline quartz samples with various aluminum concentrations after electron irradiation at room temperature. When the aluminum concentration is low, the number of aluminum hole center created by irradiation is also low to result in a smaller number of the heat-treated E_1' center as the number of holes to be transferred is smaller (Fig. 10.6). Furthermore, with increasing the number of aluminum hole center, the signal intensity of the heat-treated E_1' center saturates at certain levels depending on the electron dose. Then this saturating level has a dose response shown in Fig. 10.7. As discussed above (Fig. 10.5), this value indicates the amount of oxygen vacancies, the precursor of the E_1' center, created by electron irradiation.

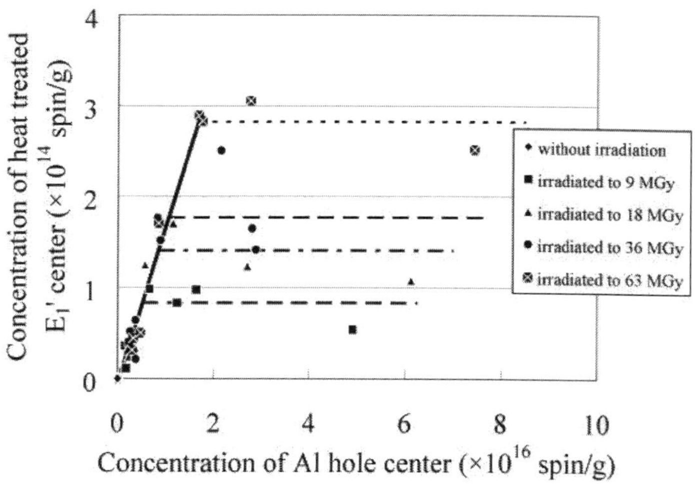

Figure 10.6 The correlation between the concentrations of the $[AlO_4]^0$ center and of the heat-treated E_1' center (Usami et al., 2009, partially modified).

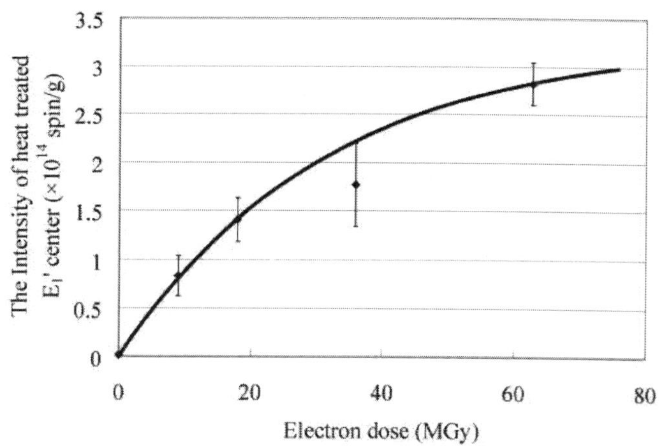

Figure 10.7 The dose response of the "saturating" level in Fig. 10.6 of the heat-treated E_1' center to electron dose.

10.4 Decay of Oxygen Vacancies

The E_1' center is annihilated above 430°C (Fig. 10.4) but is regenerated by gamma ray irradiation and subsequent heat treatment (Fig. 10.5). Toyoda and Ikeya (1991) showed that this regenerated heat-treated E_1' center decreases with increasing the temperature of first heating as shown in Fig. 10.8. This can be interpreted that with increasing the temperature, the oxygen vacancies, the precursors of the E_1' center, decay on heating. The authors found that it decays with second order decay kinetics and obtained the activation energy to be 1.9 ± 0.1 eV, pointing out that its lifetime at room temperature is 1.9×10^{15} years.

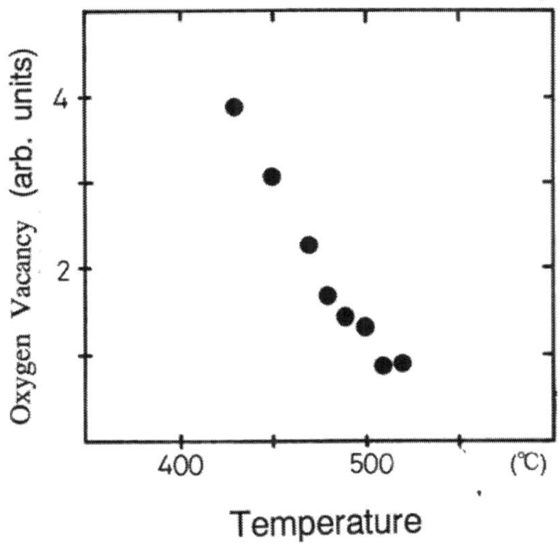

Figure 10.8 Decay of the intensity of the regenerated E_1' center after heat treatment (heat-treated E_1' center) with heating temperature (Toyoda and Ikeya, 1991).

10.5 Formation of Oxygen Vacancies

The quartz samples were irradiated by electrons (e.g., Jani et al., 1983) or by fast neutrons (e.g., Garrison et al., 1981) in the studies to investigate the models of the E_1' center. The process in nature which creates oxygen vacancies was in debate in the aspect of possible dating tool for quartz.

A correlation was found between the natural (without any treatment) intensities of the E_1' center in granitic quartz and the ages of the host rocks (Shimoyama, 1986; Odom and Rink, 1989) while the hydrothermal quartz is out of the correlation. This result implied that the ESR dating of quartz may be applicable for the whole history of the earth (Grün, 1989) if the formation process in nature of the oxygen vacancies is clarified. Their subsequent paper

(Rink and Odom, 1991) suggested that alpha recoil nuclei are the source to create the E_1' center in quartz as high LET (linear energy transfer) radiation would be necessary to provide the energy enough for the displacement of oxygen atoms. The authors measured uranium and thorium concentrations in quartz matrix in ppb order. They did not examine the heat-treated E_1' center, therefore, they could not tell the increase of the E_1' center with age is due whether to increase of oxygen vacancies, the precursor, or to increase with age of oxygen vacancies, the precursor, or to the increase with age of the fraction of the paramagnetic states (the E_1' center) of oxygen vacancies.

Our later works revealed that the number of oxygen vacancies is a measure more essential than the superficial "natural" (no treatment) intensity of the E_1' center (Toyoda, 1992; Toyoda and Hattori, 2000) by showing the correlation between the ages of the host granites and the relative number of oxygen vacancies in quartz in the age range of 10 Ma to 1 Ga (Fig. 10.9).

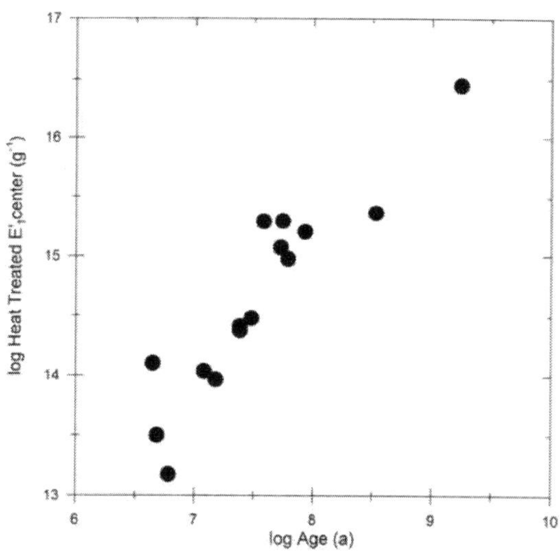

Figure 10.9 The numbers of oxygen vacancies (heat-treated E_1' center) per gram in quartz are plotted against the ages of the host granites to find the correlation between them (Toyoda, 1992).

Wieser and Regulla (1989) showed that gamma ray irradiation creates the heat-treated E_1' center in synthetic quartz although they did not describe their results in the context of oxygen vacancies.

Referring to this study, Toyoda et al. (1996) found that the oxygen vacancies (heat-treated E_1' center) are created in natural quartz by gamma ray irradiation in the range of 1×10^4 to 5×10^6 Gy (Fig. 10.10). The number of oxygen vacancies is consistent with the number in natural granitic quartz samples for the external beta and gamma ray doses calculated from the ages of the granites and its U, Th, and K concentrations.

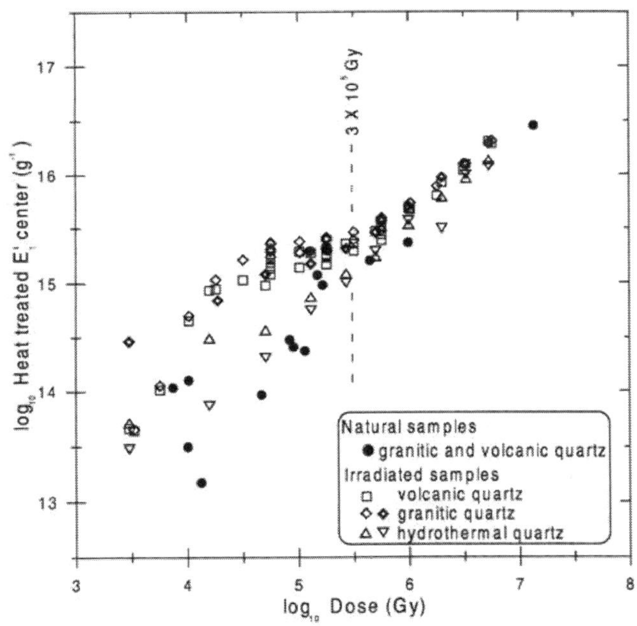

Figure 10.10 The number of oxygen vacancies per gram in quartz plotted against the irradiated gamma ray doses together with those observed in natural quartz (Toyoda et al., 1996). The horizontal axis for the natural quartz (closed circles) denotes the beta and gamma ray doses calculated from the ages of host granites and the dose rates obtained from U, Th, and K concentrations. Irradiated samples consist of five volcanic, granitic, and hydrothermal quartz samples, individually indicated by different symbols. The samples were heated at 600°C first to erase out the oxygen vacancies. Subsequently, the samples were given the gamma ray doses indicated by the horizontal axis.

The formation efficiency of oxygen vacancies by internal particles were also examined by Toyoda et al. (2001) by looking at neutron irradiated boron doped quartz where internal alpha particles are created by a nuclear reaction. The number of the oxygen vacancies created by internal alpha and alpha recoil particles were roughly in agreement with those calculated. This result indicates that the alpha and alpha recoil particles from U and Th contained in quartz matrix of ppb order can also create observed number of oxygen vacancies.

Toyoda et al. (2005) examined the spin–spin relaxation times of the heat-treated E_1' center by pulsed ESR technique. The T_2, the spin–spin relaxation time, obtained by Hahn's spin echo method, corresponds to the local concentration of the spins, i.e., the T_2 is shorter when the distances between local spins are smaller (indicating the defects created by high LET radiation such as alpha particles) while it is longer when the distance is larger (indicating low LET). As results, the T_2 values for granitic quartz and for the ones from uranium ore are between 3.7 μs and 15 μs, being similar as those of electron irradiated quartz, between 7 and 16 μs, but larger than those, less than 2 μs, for the samples with helium ions implantation which simulated alpha particles. These results indicate that the oxygen vacancies in granitic quartz as well as in quartz of uranium ore have been created by low LET radiation such as beta particles (electron beam) but not by high LET radiation such as alpha or alpha recoil particles. These results supported the idea that the oxygen vacancies in natural quartz are created by external beta and gamma rays.

In summary, any radiations, i.e., heavy ions (such as He ions), fast neutrons, electrons (equivalent to beta particles), and photons (gamma rays), can create oxygen vacancies in crystalline quartz, which are the precursor of the E_1' centers. Among those, beta and gamma rays are the main sources that produce oxygen vacancies in nature, where the formation efficiency denote as the G value (the number of vacancies created per 100 eV of absorbed dose) was obtained to be as low as 2.5×10^{-5} for gamma rays (Toyoda et al., 1996). However, a question how such low LET radiation induces atomic displacements would still remain. The temperature dependence of the formation efficiencies of oxygen vacancies would be a clue for this question. Toyoda et al. (1996) irradiated quartz grains at room and elevated temperatures by gamma rays after erasing the oxygen vacancies by heating at 600°C for 1 h. They found a shoulder for sample aliquots irradiated at room temperature

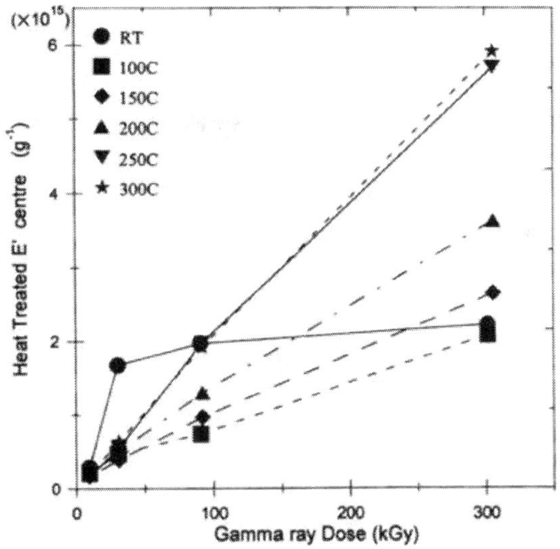

Figure 10.11 Formation of oxygen vacancies (heat-treated E_1' center) by gamma ray irradiation with varying the temperature during irradiation (Toyoda et al., 1996).

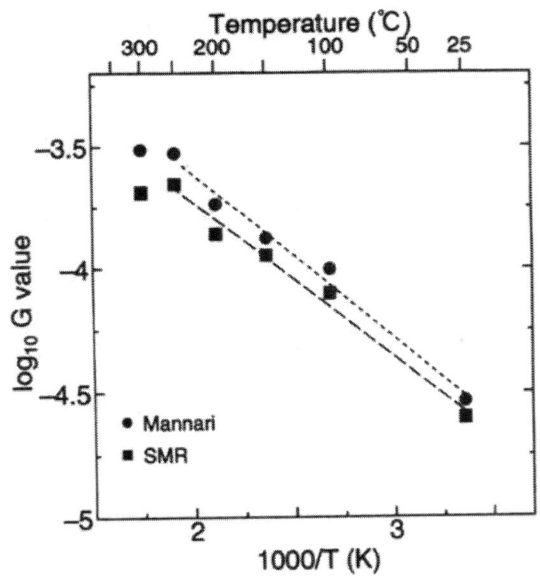

Figure 10.12 The Arrhenius diagram for formation efficiency (G-value) for oxygen vacancies (heat-treated E_1' center) by gamma ray irradiation with varying the temperature during irradiation (Toyoda et al., 1996).

in the dose responses in the dose region between 5 × 10^4 and 5 × 10^5 Gy (Fig. 10.10) which disappear at elevated temperature (above 100°C) during irradiation (Fig. 10.11) They interpreted the results with two components for created oxygen vacancies, Sites A with larger formation efficiencies saturating above 5 × 10^4 Gy, and Site B with smaller formation efficiencies showing no saturation. The formation efficiencies for Site B have temperature dependence as shown in Fig. 10.10, deducing an Arrhenius diagrams (Fig. 10.12) to obtain an activation energy of 0.13 ± 0.01 eV. They suggested that, in Site A, oxygen vacancies are formed through a thermally unstable excited state, which are not made at elevated temperature while, in Site B, they are formed through an energy barrier defined the activation energy, the process of which more easily happens at elevated temperatures.

10.6 Applications to Provenance Research

The provenance of surface sediments is a subject of interest as it indicates the history of transportation of the sediments, which is directly related with climatic changes of that region. Based on the correlation between the number of oxygen vacancies in quartz and the age of the host granites (Fig. 10.9), it was proposed to use the heat-treated E_1' center as a indicator for provenance studies. It works when the number of the oxygen vacancies in quartz of a dust reservoir differs from that of another, which is a function of the original crystallization age of the basement rocks from which dust is generated by weathering. When the ages of the basement rocks are different from location to location of dust sources, we can distinguish the locations by measuring the number of oxygen vacancies.

The numbers of the oxygen vacancies in quartz of loess samples in Japanese Islands were systematically measured based on this idea. It was shown that the number in Japanese loess is about the same with that in Chinese loess plateau in MIS 1. In MIS 2, the case also applies to the loess samples in southern Japan while the numbers are systematically larger in those in northern Japan, implying that aeolian dust from sources which have derived from older basement rocks, possibly Precambrian rocks in Siberia, has contributed the loess deposited in northern Japan in MIS 2 (Naruse et al, 1997; Ono et al., 1998; Toyoda and Naruse, 2002).

Nagashima et al. (2007a) proposed that the Crystallinity Index (CI) measured by X ray diffraction is another proxy to reveal the source of the aeolian dust. From the number of oxygen vacancies and the crystallinity index, Nagashima et al. (2007b) calculated the ratios of the contributions from three sources, Taklamakan, Siberia, and Japanese river, to the sediments taken from a core in Japan Sea down to 150 ka. It was found that the contribution from Taklamakan dominates in warmer period while that from Siberia is larger in colder period (Fig. 10.13), and that the Taklamakan contribution correlates with the isolation at 30°N in June.

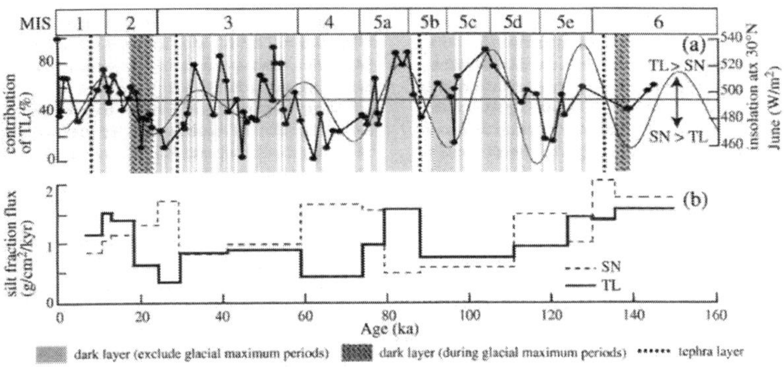

Figure 10.13 The contributions from eolian dusts from Taklamakan desert and Siberia to sediment in Japan sea as a function of the deposition age. The change of the contribution from Taklamakan desert is consistent with insolation at 30°N in June (solid curve) (Nagashima et al., 2007b).

Yamamoto et al. (2010) investigated, based on the same idea, the number of oxygen vacancies in quartz of atmospheric deposition to Japanese cities in recent past. They found that the number of oxygen vacancies in quartz is larger in finer fractions, where the highest values in those fractions are consistent with the value of the samples from Taklamakan desert. Comparing with the grain size distributions, they concluded that the number of oxygen vacancies in quartz in the atmospheric depositions varies due to the variation of the value in the source rather than due to the contamination of local components.

The same idea was also successful using impurity centers in addition to the oxygen vacancies for a series of river sediments. Shimada (2008) studied the ESR signals of the impurity centers, aluminum hole and Ti–Li centers as well as the E_1' center in sediments in Nara basin, central Japan. She succeeded to specify one of the rivers which could have brought sediments to an alluvial fan.

10.7 Impurity Centers

The ESR signals of the aluminum hole and Ti–Li centers have been used to determine the ages of geological materials such as volcanic tephra (e.g., Imai et al., 1985; Toyoda et al., 1995), fault movements (e.g., Lee and Schwarcz, 1994), and sediments (e.g., Yokoyama et al., 1985; Voichet et al., 2003). The method is based on the idea that paramagnetic defects with unpaired electrons are created by natural radiation and have been accumulated through geological time scale. From the signal intensity of the paramagnetic defects, the amount of the accumulated natural radiation dose is estimated, usually by the additive dose method. A dose response is obtained for a sample by measuring the intensity of the signals as a function of given dose of gamma ray irradiation, and is then extrapolated to the zero ordinate to obtain the accumulated dose. To deduce the age, this dose is subsequently divided by the dose rate, which is estimated separately usually from the concentration of U, Th, and K concentrations of surrounding sediment with appropriate corrections such as the water content in the sediment and the beta particle attenuation depending on the grain size of the mineral. The method assumes that the geological events to be dated (volcanic eruption, fault movement, and sedimentation) erase the signal intensities to zero before starting the "clock."

In some papers (e.g., Lee and Schwarcz, 1994; Toyoda et al., 2000), the authors insist that obtaining consistent accumulated doses with two or more paramagnetic centers will assure that the dating method works in the aspect of complete bleaching at the time of the geological events to be dated. The idea is based on the assumption that the species are independent from one another. However, it would be possible that they exchange electrons and/or some alkali ions, as pointed out by Weil (1984) that alkali ions which come to germanium centers may originate from those released by aluminum hole centers. We have actually observed the "reaction" between the aluminum hole centers and the oxygen vacancies in the previous sections.

The author of this chapter has an idea that such reactions between impurity centers would be one of the important factors that are the reason for "unsuccessful" dating results on quartz. For example, Fig. 10.14 shows the results of dating of Japanese volcanic tephra where older samples show scattered ages, being inconsistent with age references (Toyoda et al., 2006). Figure 10.15 shows a result of stepwise heating experiment for a tephra sample where the Ti–Li center decays at higher temperature than the one at which the aluminum hole center decays (Toyoda and Ikeya, 1994) which is contrary to the result for granitic quartz shown in Fig. 10.4. The Ti–Li center increase on heating in some dose range while it decreases in the other range as shown in Fig. 10.16 (Toyoda et al., 2006). It is possible that these resulted from exchange of electrons between impurity centers where the relative amount of each species are different from sample to sample.

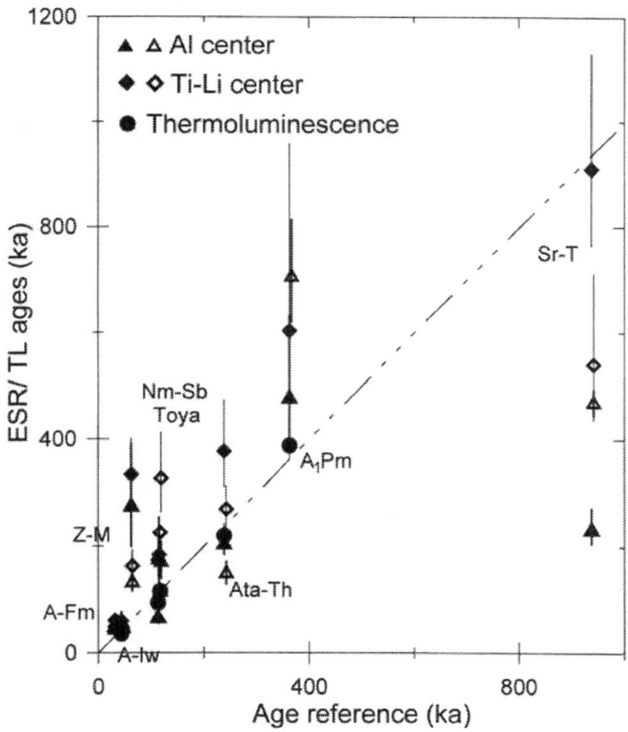

Figure 10.14 Results of ESR dating of Japanese tephra in comparison with thermoluminescence method as a function of reference ages (Toyoda et al. 2006).

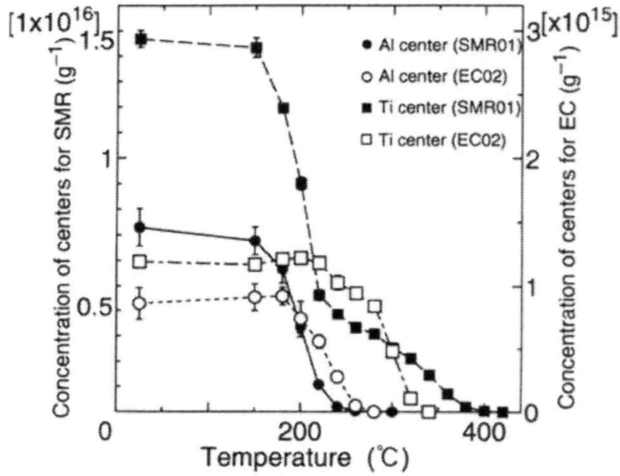

Figure 10.15 Results of stepwise heating experiments for samples of tephra of Valles Rhyolite, USA (Toyoda and Ikeya, 1994).

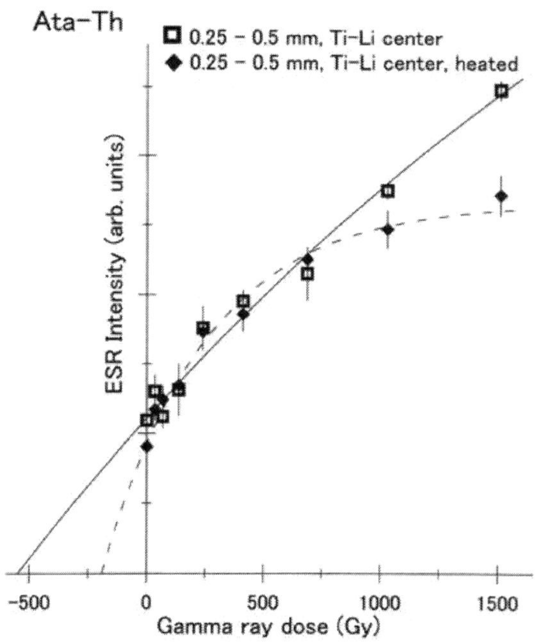

Figure 10.16 Change of dose response on heating for a Japanese tephra sample. The signal intensity of Ti–Li center increases slightly on heating in the lower dose region while it decreases in higher dose region (Toyoda et al., 2006).

10.8 Summary

The "chemical reactions," i.e., transfer of electrons and holes, also alkali ions, within crystalline matrix would be key issues to have comprehensive understanding on formation and decay of paramagnetic defect centers in crystalline quartz. Formation of the E_1' center is controlled by the decay of the aluminum hole center. Its precursor, the oxygen vacancy, is created by various radiations while beta and gamma rays are the main sources in nature. Using the correlation between the ages of host rocks and the number of oxygen vacancies in the age range of 10 Ma to 1 Ga, provenance of quartz on the surface of the earth have been investigated to discuss the aeolian dust source. Electron and hole reactions between impurity centers have not well been studied, but can be an important issue further to be investigated at least in the aspect to establish the method of ESR dating of quartz.

References

Anderson JH, Weil JA, (1959). Paramagnetic resonance of color centers in germanium-doped quartz. *J. Chem. Phys.*, **31**, 427–434.

Feigl FJ, Fowler WB, Yip KL, (1974). Oxygen vacancy model for the E_1' centre in SiO_2. *Solid State Commun.*, **14**, 225–229.

Garrison EG, Rowlett RM, Cowan DL, Holroyd LV, (1981). ESR dating of ancient flints. *Nature*, **290**, 44–45.

Griffiths JHE, Owen J, Ward IM, (1954). Paramagnetic resonance in neutron-irradiated diamond and smoky quartz. *Nature*, **173**, 439–442.

Grün R, (1989). ESR dating for the early Earth. *Nature*, **338**, 543–544.

Ikeya M, (1993). *New Applications of Electron Spin Resonance, Dating, Dosimetry, and Microscopy*. World Scientific, Singapore, 500p.

Ikeya M, Miki T, Tanaka K, (1982). Dating of fault by electron spin resonance on intrafault materials, *Science*, **215**, 1392–1393.

Imai N, Shimokawa K, Hirota M, (1985). ESR dating of volcanic ash, *Nature*, **314**, 81–85.

Isoya J, Weil JA, Halliburton LE, (1981). EPR and ab initio SCF-MO Studies of the Si·H–Si system in the E_4' center of alpha-quartz, *J. Chem. Phys.*, **74**, 5436–5448.

Jani MG, Bossoli RB, Halliburton LE, (1983). Further characterization of the E_1' center in crystalline SiO_2, *Phys. Rev. B*, **27**, 2285–2293.

Lee H–K, Schwarcz HP, (1994). Criteria for complete zeroing of ESR signals during faulting of the San Gabriel fault zone, southern California. *Tectonophysics*, **35**, 317–337.

Mackey JH Jr., (1963). EPR study of impurity-related color centers in Germanium-doped quartz. *J. Chem. Phys.*, **39**, 74–83.

Mashkovtsev RI, Pan Y, (2012). Five new E′ centers and their ^{29}Si hyperfine structures in electron-irradiated α-quartz. *Phys. Chem. Miner.*, **39**, 79–85.

Nagashima K, Tada R, Tani A, Toyoda S, Sun Y, Isozaki Y, (2007a). Contribution of Aeolian dust in Japan Sea sediments estimated from ESR signal intensity and crystallinity of quartz. *Geochem. Geophys. Geosyst.*, **8**, Q02Q04, doi:10.1029/2006GC001364.

Nagashima K, Tada R, Matsui H, Irino T, Tani A, Toyoda S, (2007b). Orbital- and Millennial-scale variations in Asian dust transport path to the Japan Sea. *Palaeogeography, Palaeoclimatol., Palaeoecol.*, **247**, 144–161.

Naruse T, Ono Y, Hirakawa K, Okashita M, Ikeya M, (1997). Source areas of eolian dust quartz in East Asia: a tentative reconstruction of prevailing winds in Isotope Stage 2 using electron spin resonance. *Geographical Rev. Japan*, 70A-1, 15–27 (in Japanese with English abstract).

Nuttal RHD, Weil JA, (1981a). The magnetic properties of the oxygen-hole aluminum centers in crystalline SiO_2. I. $[AlO_4]^0$. *Can. J. Phys.*, **59**, 1696–1708.

Nuttal RHD, Weil JA, (1981b). The magnetic properties of the oxygen-hole aluminum centers in crystalline SiO_2. II. $[AlO_4/H^+]^+$ and $[AlO_4/Li^+]^+$. *Can. Jour. Phys.* **59**, 1709–1718.

O'Brien MCM (1955). The structure of the color centers in smoky quartz. *Proc. Royal. Soc. London A*, **231**, 404–414.

Okada M, Rinneberg J, Weil JA, Wright PM, (1971). EPR of Ti^{3+} centers in alpha quartz. *Chem. Phys. Lett.*, **11**, 275–276.

Odom AL, Rink WJ, (1989). Natural accumulation of Shottky–Frenkel defects: implications for a quartz geochronometer. *Geology*, **17**, 55–58.

Ono Y, Naruse T, Ikeya M, Kohno H, Toyoda S, (1998). Origin and derived courses of eolian dust quartz deposited during marine isotope stage 2 in East Asia, suggested by ESR signal intensity. *Global Planet. Change*, **18**, 129–135.

Porat N, Schwarcz HP, Valladas H, Bar-Yosef O, Vandermeersch B, (1994). Electron spin resonance dating of burned flint from Kebara cave, Israel. *Geoarchaeology*, **9**, 393–407.

Rink WJ, Odom AL, (1991). Natural alpha recoil particle radiation and ionizing radiation sensitivities in quartz detected with EPR: implication for geochronometery. *Nucl. Tracks Radiat. Meas.*, **18**, 163–173.

Rinneberg H, Weil JA, (1972). EPR studies of Ti^{3+}–H^{+} centers in X-irradiated alpha quartz. *J. Chem. Phys.*, **56**, 2019–2028.

Rudra JK, Fowler WB, (1987). Oxygen vacancy and the E_1' center in crystalline SiO_2. *Phys. Rev. B*, **35**, 8223–8230.

Schnadt R., Räuber A. (1971). Motional effects in the trappcd-hole center in smoky quartz. *Solid State Commun.*, **9**, 159–161.

Shimada A, 2008. Characteristics of ESR signals and thermoluminescence color images of quartz grains to study the provenance of sediments (in Japanese, Japanese title translated). Ph.D Thesis, Nara Women's University, Japan, 153p.

Shimoyama, Y. (1986). ESR dating of volcanic rocks, *Abstr. First Workshop ESR Appl. Metrol.*, 47–48, IONICS, Tokyo (in Japanese).

Silsbee RH, (1961). Electron spin resonance in neutron-irradiated quartz. *J. Appl. Phys.*, **32**, 1459–1462.

Taylor AL, Farnell GW, (1964). Spin-lattice interaction experiments on color centers in quartz. *Can. J. Phys.*, **42**, 595–607.

Toyoda S, (1992). Production and decay characteristics of paramagnetic defects in quartz: application to ESR dating. Ph.D Thesis, Osaka University, Japan, 106p.

Toyoda S, Goff F, (1996). Quartz in post-caldera volcanic rocks of Valles caldera, New Mexico: ESR finger printing and discussion in ESR ages, New Mexico Geological Society Guidebook, 47th Field Conference, 303–309.

Toyoda S, Goff F, Ikeda S, Ikeya M, (1995). ESR dating of El Cajete and Battleship Rock Member of Valles Rhyolite, Valles Caldera, New Mexico. *J. Volcanol. Geotherm. Res.*, **67**, 29–40.

Toyoda S, Hattori M, (2000). Formation and decay of the E_1' center and of its precursor. *Appl. Radiat. Isotopes*, **52**, 1351–1356.

Toyoda S, Ikeya M, (1994). ESR dating of quartz with stable component of impurity centers. *Quatern. Sci. Rev.* (*Quatern. Geochronol.*), **13**, 625–628.

Toyoda S, Ikeya M, (1991). Thermal stabilities of paramagnetic defect and impurity centers in quartz: basis for ESR dating of thermal history. *Geochem. J.*, **25**, 437–445.

Toyoda S, Naruse T, (2002). Eolian dust from the Asian deserts to the Japanese Islands since the Last Glacial Maximum; the basis for the ESR method. *Trans., Japanese Geomorphol. Union*, **23**, 811–820.

Toyoda S, Rink WJ, Schwarcz HP, Ikeya M, (1996). Formation of E_1' precursors in quartz: applications to dosimetry and dating. *Appl. Radiat. Isot.*, **47**, 1393–1398.

Toyoda S, Rink WJ, Yonezawa C, Kagami T, (2001). In-situ production of alpha particles and alpha recoil particles in quartz applied to ESR studies of oxygen vacancies. *Quatern. Sci. Rev.*, **20**, 1057–1061.

Toyoda S, Takeuchi D, Asai T, Komuro K, Horikawa Y, (2005). Spin-spin relaxation times of the E_1' center in quartz with and without irradiation: implications for the formation process of the oxygen vacancies in nature. *Radiat. Meas.*, **39**, 503–508.

Toyoda S, Tsukamoto S, Hameau S, Usui H, Suzuki T, (2006). Dating of Japanese Quaternary tephras by ESR and Luminescence methods. *Quatern. Geochronol.*, **1**, 320–326.

Toyoda S, Voinchet P, Falgueres C, Dolo JM, Laurent M, (2000). Bleaching of ESR signals by the sunlight: a laboratory experiment for establishing the ESR dating of sediments, *Appl. Radiat. Isot.*, **52**, 1357–1362.

Usami T, Toyoda S, Bahadur H, Srivastava AK, Nishido H, (2009). Characterization of the E_1' center in quartz: Role of aluminum hole centers and oxygen vacancies, *Phys. B: Condensed Matter.*, **404**, 3819–3823.

Voinchet P, Falguères C, Laurent M, Toyoda S, Bahain JJ, Dolo JM, (2003). Artificial optical bleaching of the aluminium center in quartz. Implications to ESR dating of sediments. *Quatern. Sci. Rev.* (*Quatern. Geochronol.*), **22**, 1335–1338.

Weeks RA., (1956). Paramagnetic resonance of lattice defects in irradiated quartz. *J. Appl. Phys.*, **27**, 1376–1381.

Weeks RA, Nelson CM, (1960a). Trapped electrons in irradiated quartz and silica: II. Electron spin resonance. *J. Am. Ceramic Soc.*, **43**, 399–404.

Weeks, RA, Nelson CM, (1960b). Irradiation effects and short-range order in fused silica and quartz. *J. Appl. Phys.*, **31**, 1555–1558.

Weil JA, (1971). The analysis of large hyperfine splitting in paramagnetic resonance spectroscopy. *J. Magn. Reson.*, **4**, 394–399.

Weil JA, (1984). A review of electron spin spectroscopy and its applications to the study of paramagnetic defects in crystalline quartz. *Phys. Chem. Minerals*, **10**, 149–165.

Weil JA, Anderson JH, (1961). Direct field effects in electron paramagnetic resonance hyperfine spectra. *J. Chem. Phys.*, **35**, 1410–1417.

Wieser A, Regulla D, (1989). ESR dosimetry in the "Giga-rad" range. *Appl. Radiat. Isot.*, **40**, 911–913.

Wright PM, Weil JA, Buch T, Anderson JH, (1963). Titanium colour centers in rose quartz. *Nature*, **197**, 246–248.

Yamamoto Y, Toyoda S, Nagasima K, Igarashi Y, Tada R, (2010). The grain size dependence of the E_1' center observed in quartz of atmospheric deposition at two Japanese cities. *Geochronometria.*, **37**, 9–12.

Yokoyama Y, Quaegebeur JP, Falgueres C, (1985). ESR dating of quartz from quaternary sediments: first attempt. *Nucl. Tracks*, **10**, 921–928.

Zavoisky E, (1945). Spin-magnetic resonance in paramagnetics. *J. Phys.*, **9**, 245.

Chapter 11

ZnO Nanoparticles: Defect Structure, Space-Charge Depletion Layer, and Core–Shell Model

Emre Erdem[a] **and Rüdiger-A. Eichel**[b]

[a]*Albert-Ludwigs-Universität Freiburg, Institut für Physikalische Chemie I, Freiburg, 79104, Germany*
[b]*Forschungszentrum Jülich, Institute of Energy & Climate Research (IEK-9), D-52425 Jülich, Germany*

emre.erdem@physchem.uni-freiburg.de, r.eichel@fz-juelich.de

11.1 Introduction

Current research interest in nanostructured ZnO is motivated by novel applications as electrode material for dye-sensitized solar cells [1], light-emitting diodes [2], photoluminescence [3], thermoelectrics [4], printable electronics [5] varistors [6,7], sensors [8], or nanogenerators [9]. All of these applications exploit the semiconducting properties of ZnO, which is a direct wide-

Oxide Nanostructures: Growth, Microstructures, and Properties
Edited by Avanish Kumar Srivastava

ISBN 978-981-4411-35-6 (Hardcover), 978-981-4411-36-3 (eBook)
www.panstanford.com

bandgap (3.3 eV) semiconductor [10], and take advantage of the fact that desired materials properties may be tailored by means of controlling the defect structure by aliovalent doping.

Although various ZnO compounds have been extensively investigated by many researchers over the last decades, there exists still controversy in various properties and some issues have even remained unclear. Things become more complicated and controversial when the crystallite size is confined to the nano-dimension. Although aliovalent doping and compound morphology play a decisive role to control materials properties, the synthesis method defines an additional, most crucial role that markedly impacts the nature and amount of intrinsic defects or impurities.

Over recent years, a number of reviews concerning various ZnO aspects have been reported that discussed very detailed the optical properties [10,11], role of intrinsic defects [12,13], impact of doping with transition metal ions [14], characterization of donor/acceptor impurities [15,16]. In addition to these reviews, two important books have been published summarizing recent developments of ZnO research and its technological implications [17,18].

Developing materials with novel electronic, mechanical, and optical properties and confined size to the nano-dimension define two important requirements for integrated device technology. In that respect, ZnO nanostructures such as nanowires [19–21], nanorods [22–24], nanotetrapods [25–27], nanofibers [28], nanowalls [29], nanotubes [30], nanobelts [31–33], nanoflowers [34], and nanorings/nanobows [35] most recently have attracted significant attention since they open a way for the development of new technological devices [36,37].

A feature in common for all nano-architectures is a *core-shell* structure that accounts for marked variations in materials properties at the surface or the "bulk" of the nanocompounds. Electron paramagnetic resonance (EPR) spectroscopy provides a sensitive means to investigate corresponding features. Furthermore, still controversially discussed phenomena, such as the existence of room-temperature ferromagnetism [38,39] or p-type ZnO semiconductors [40], may be investigated.

11.2 Possible Defect Centers in Undoped ZnO Nanomaterials

The band structure of ZnO has been discussed over several decades, and there are still contradictions that prevent a unique band diagram. In Fig. 11.1, the band scheme is depicted where the contradictory defect assignments are indicated as given in different experimental and theoretical studies from the literature [10,41,42].

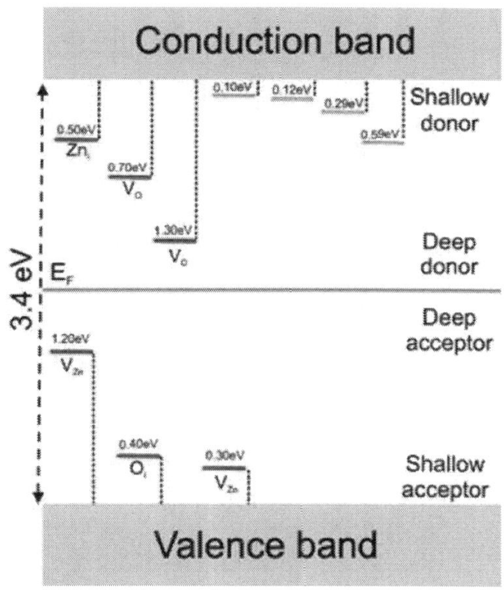

Figure 11.1 Schematic band diagram of ZnO. Contradictory assignments and the energy values for different defect centers were taken from literature. Zn_i, V_O, V_{Zn}, and O_i indicate neutral zinc interstitials, neutral oxygen vacancy, neutral zinc vacancy and oxygen interstitials, respectively.

The above-mentioned contra-verse assignments can be partly solved by analytical EPR in the framework of *core–shell* model. Indeed, there are two main kinds of EPR signal in ZnO nanomaterials. One originates from the bulk defects and the other originates from the surface defects. Due to their different environment each intrinsic defect centers give distinctly different *g*-factor. In this model, EPR signal at around $g \sim 1.96$ and around $g \sim 2.00$ directly assigned to the core and shell, respectively. Below we will discuss the contradictory assignments of not only bulk and surface defects at g-factor 1.96 and 2.00, respectively, but also we will discuss the defects in doped ZnO.

11.2.1 Bulk Defects

Bulk defects in ZnO can be identified according to a characteristic fingerprint type EPR resonance at around $g \sim 1.96$ (cf. Fig. 11.2a). It is extremely difficult to resolve the g-anisotropy by standard cw-EPR at X- and Q-band microwave frequencies. The bulk defects are located at the hexagonal core of ZnO; thus they carry typical intrinsic properties of it. However, the real origin of the signal with a g-factor at around 1.96 is still unclear. This EPR signal is in general assigned to shallow donors [43–45] owing to a singly ionized oxygen vacancy defect (V_O^+) center [46–50]. However, it is sometimes also attributed to unpaired electrons trapped on oxygen vacancies [47] or shallow donor centers such as ionized impurity atoms in the crystal lattice of ZnO [43,51,52] or to free carriers in the conduction band [53]. In contrast to common conviction [43], there are reports supposing that the EPR signal at $g \sim 1.96$ may be due to one electron being weakly bound to ionized impurity atoms [12]. This implies that the electron from a shallow donor level is not completely delocalized into the conduction band but stays weakly bound to the original atoms, and behaves much like an electron in a hydrogen-like atom with a large Bohr radius [12]. The irradiation with UV light is expected to produce excitations and, once bound, the excitation would give rise to a paramagnetic signal at $g \sim 1.96$ suggesting that the signal is not related to oxygen vacancies [12].

Furthermore, in previous works several arguments in favor of a Zn–O divacancy or even Zn–O–Zn trivacancy for the origin of this bulk defect center were presented [54]. Recently, this center was used for monitoring electron photoexcitation from the deep level centers in the ZnO gap [55] or photoexcitation of the conducting electrons from Pb centers [56] where they related this photosensitive defect center to conduction electrons located at donor centers in the bulk and near the surface of ZnO particles. Photosensitive feature of this signal, e.g., enhancement after UV light [5] (see also Fig. 11.2b), supports that this signal can also safely be attributed to the electrons in conduction-band states, since UV light can promote electrons into these states [42].

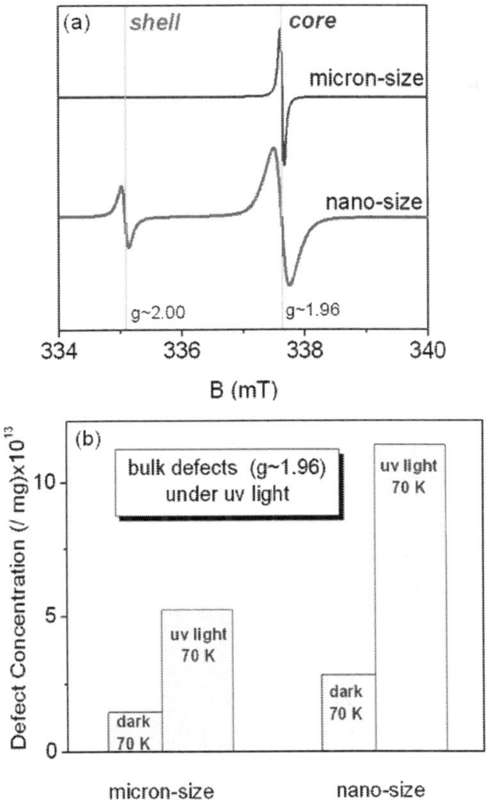

Figure 11.2 (a) Typical X-band EPR spectra for the micron- (bulk) and nano-size undoped ZnO. (b) Defect concentrations obtained from X-band cw-EPR at 70 K for micron- and nano-sized ZnO, respectively, both for dark and under UV light illumination. The concentrations were determined from the bulk defects signal at $g \sim 1.96$ (cf. the EPR spectra inset).

Figure 11.2b shows the defect concentration in ZnO both for micron- and nano-sized samples. It is clearly seen that nano-size samples have more defects than the micron-size, and defect concentrations were almost doubled under UV light illumination. Independent from the origin of the defect center this result shows that by light excitation one can induce more intrinsic defects.

11.2.2 Surface Defects

The amount of surface-related defects enormously increases due to the increase of surface-to-volume ratio by size reduction to the nano dimension. Since the surface of ZnO is an undefined structure, the defects in this region possess an electronic environment, which is difficult to describe and assign by EPR. In general, surface defects are detectable by EPR at g-factor between 1.99–2.00 (cf. Fig. 11.2a). These defect centers indeed form defect complexes consisting of magnetically inequivalent anisotropic, axial or isotropic EPR signals. To resolve such overlapped or highly anisotropic signals is an extremely difficult issue.

In literature many different origins have been pointed out for the $g \sim 1.99$–2.00 signal. Zinc interstitials, negatively charged Zn vacancies, singly positively charge oxygen vacancies [57], and an unpaired electron trapped on an oxygen vacancy site [12] were reported. In some reports, the EPR spectrum of singly positively charge oxygen vacancies (sometimes called F^+-center) is observed at g-factor $g \sim 1.99$ only in irradiated ZnO samples under or after illumination and stable up to 400–450°C annealing [58]. In very rare cases this signal has been assigned to Pb^{3+} deep donor impurities which the Pb^{3+} isotropic EPR spectrum has weak axial symmetry of g-tensor around c-axis with $g_z = 2.0133$ and $g_{x,y} = 2.0135$ [58]. Other contradictory assignments for the 1.99–2.00 signal such as di- or trivacancy centers, shallow H donor, zinc vacancy-related center, hole on an oxygen ion is summarized in EPR related review [16].

11.3 Defect Centers in Doped ZnO

Doping ZnO with transition metal elements [14], such as Fe, Co, Mn, and Cr, leads to materials with entirely different behavior toward magnetic and optical excitation. This new research direction is named as diluted magnetic semiconductors (DMS) [39,59–61]. DMS materials that are based on ZnO could not only show ferromagnetism at room temperature upon doping with transition metal ions, e.g., Fe^{3+} [62], Cr^{3+} [63], Co^{2+} [64], and Mn^{2+} [65] but also simultaneously also exhibit ferromagnetic and semiconducting properties [39]. Recently, Mn-doped ZnO revealed ferromagnetism above room temperature [66], which is important for new spintronic devices (spin LEDs, spin valve transistors) and magneto-optic components. On the contrary, there are also studies [38] that report that there is no indication of ferromagnetism or antiferromagnetism in Mn-doped ZnO. Due to lack of advanced EPR studies and insufficient characterization of the materials, there is an ongoing controversy regarding the magnetic properties of ZnO-based DMS [38,39,66].

In Fig. 11.3, various metal ion–doped (Fe^{3+}, Cr^{3+}, Co^{2+}, Mn^{2+}) ZnO materials reveal unique EPR signals at room temperature that can be analyzed and compared with complementary standard and analytical characterization methods. Moreover, in the case of externally metal-ion doped ZnO having high-spins $S > 1/2$ such as $Mn^{2+,4+}$, Cr^{3+}, and Fe^{3+}, we may have detailed information about the EPR zero field splitting parameter in the framework of crystal field.

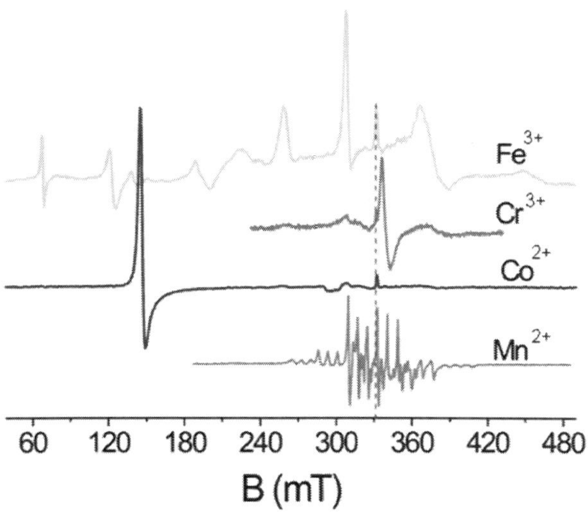

Figure 11.3 X-band (9.86 GHz) room temperature EPR spectra of ZnO materials with extrinsic defects (doped with transition metal ions such as Mn^{2+}, Co^{2+}, Cr^{3+}, and Fe^{3+}).

11.4 Defect Chemistry in ZnO

ZnO possesses a relatively open structure, with a hexagonal close packed lattice where Zn atoms occupy half of the tetrahedral sites and all of the octahedral sites are non-occupied. Hence, there are plenty of sites for ZnO to accommodate intrinsic defects, such as zinc interstitials, or extrinsic dopants. More generally, the semiconducting properties of ZnO strongly depend on defects, and the control of defect structure impacts the number of charge carriers. For that reason, controlling the defect structure of ZnO is of paramount importance for the above-mentioned applications that exploit the wide range of properties of doped ZnO.

Concerning its defect structure, there are a number of intrinsic defect states within the bandgap of ZnO; the donor defects being zinc interstitials ($\mathrm{Zn}_{\mathrm{Zn}}^{\bullet\bullet}$, $\mathrm{Zn}_{\mathrm{Zn}}^{\bullet}$), and oxygen vacancies ($V_{\mathrm{O}}^{\bullet\bullet}$, $V_{\mathrm{O}}^{\bullet}$), whereas zinc vacancies rather form acceptor states ($V_{\mathrm{Zn}}^{//}$, $V_{\mathrm{Zn}}^{/}$). The also discussed $V_{\mathrm{O}}^{\times}$, $V_{\mathrm{Zn}}^{\times}$, and $\mathrm{Zn}_{\mathrm{i}}^{\times}$ rather form isovalent defects in the bandgap. Zn interstitials and oxygen vacancies are known to be the predominant ionic defect types. However, which point and complex defect dominates in native, undoped ZnO is still a matter of great controversy [67]. The different defects are designated by using the standard Kröger–Vink notation [68], according to which the subscript indicates the lattice site that a defect occupies and the superscript corresponds to the electronic charge of the defect relative to the site that it occupies. To indicate isovalent substitution charge, × is used, • indicates a single positive charge, and the prime (′) signifies negative charge.

11.5 High-Field, High-Frequency EPR

As the majority of the relevant defects in ZnO are paramagnetic and thus accessible by EPR and because moreover different defects generally exhibit varying electronic g-values, an accurate determination of g-values is of utmost importance for a reliable characterization of defect structure. However, the corresponding g-value differences are sometimes very small such that the corresponding resonances may overlap for EPR experiments at low resonant fields. As consequence, high-field (high-frequency) is an indispensable tool to study the defect structure of nanostructured ZnO.

In order to characterize the nature of *intrinsic* defects in nano-sized ZnO particles, EPR at various Larmor frequencies between X-band (9.85 GHz) and 406.4 GHz has been performed. The corresponding X-band and high-field EPR spectra are shown in

Fig. 11.4. Obviously, the benefit of high-field EPR is demonstrated by disentangling overlapping contributions with small g-value differences. In that respect, the two isotropic resonances observed at X-band, are resolved to at least three different lines at low field and an anisotropic line at high field.

The origin of the signal at $g \approx 1.96$ is generally explained by the existence of singly charged oxygen vacancies [46,50,69] where the $V_O^\bullet$ act as shallow donor centers. Whereas the X-band spectrum exhibits a pseudo-isotropic line shape for this resonance, at high Larmor frequency a small axial g-anisotropy can be resolved (cf. Fig. 11.5b), which is characteristic for an axial site symmetry for the $V_O^\bullet$. This observation is in line with recent ab initio calculations [70].

The existence of the $V_O^\bullet$ can be ascribed to the "bulk" defect chemistry of ZnO, after which oxygen vacancies arise from the Schottky-type defect equilibrium

$$\mathrm{Zn}_{\mathrm{Zn}}^{\times} + \mathrm{O}_{\mathrm{O}}^{\times} \leftrightarrow V_{\mathrm{Zn}}^{//} + V_{\mathrm{O}}^{\bullet\bullet} + \mathrm{Zn}_i^{\times} + \frac{1}{2}\,\mathrm{O}_2(\mathrm{g})$$

which describes the fact that molecular oxygen evaporates to the atmosphere during high-temperature processing, whereas the zinc atoms rather remain in the lattice and form interstitials. Concerning the oxygen vacancies, different ionization equilibria allow various charge states, where only the singly charged $V_O^\bullet$ is paramagnetic with $S = 1/2$

$$V_{\mathrm{O}}^{\times} \rightarrow V_{\mathrm{O}}^{\bullet} + \mathrm{e}'$$

$$V_{\mathrm{O}}^{\bullet} \rightarrow V_{\mathrm{O}}^{\bullet\bullet} + \mathrm{e}'$$

On the other hand, at $g \approx 2.0$ the single resonance observed at X-band transforms into a manifold of resonance lines at high field, demonstrating that different defects are simultaneously present. The sextet hyperfine pattern owing to the hyperfine interaction with a ^{55}Mn-defect center with spin-Hamiltonian parameters $g(\mathrm{Mn}^{2+}) = 2.0019$ and $A(^{55}\mathrm{Mn}) = 260$ MHz, is indicative for a divalent oxidation state [71] and thus isovalent substitution of the Mn^{2+} for the Zn site. The corresponding resonances are marked with asterisks in Fig. 11.4.

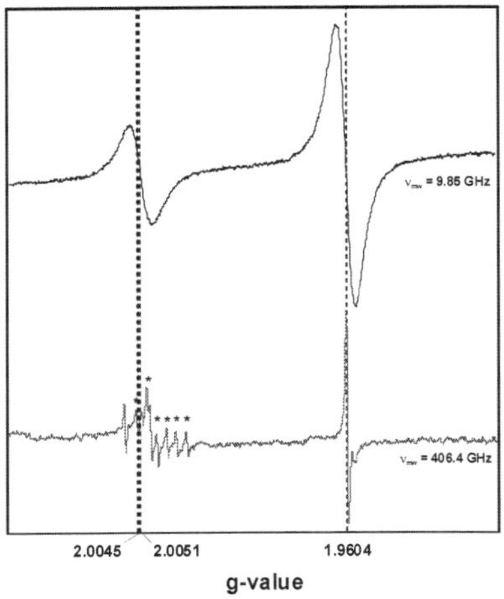

Figure 11.4 EPR spectra of nanoscale ZnO recorded at 290 K in a g-value representation. X-band spectrum at a Larmor frequency of 9.85 GHz (top) compared to high-field EPR at a Larmor frequency of 406.4 GHz. The sextet hyperfine resonances owing to an Mn^{2+}-impurity center are indicated by asterisks.

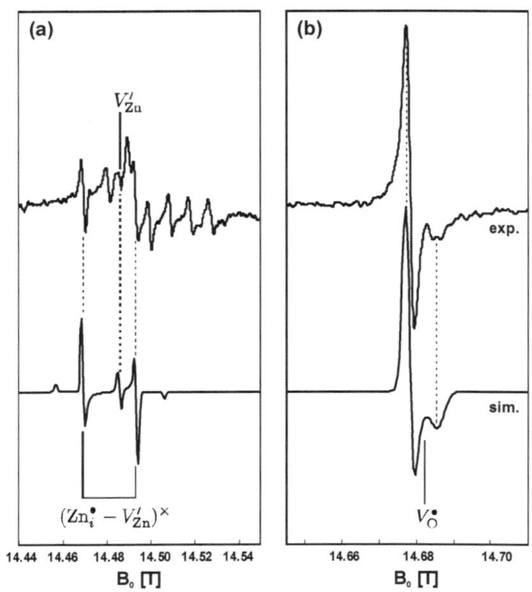

Figure 11.5 High-field EPR spectrum of nanoscale ZnO at Larmor frequency of 406.4 GHz recorded at 290 K. Experimental spectrum (top) compared to numerical spectrum simulation (bottom) invoking the spin-Hamiltonian parameters given in Table 11.1.

Besides the Mn sextet hyperfine pattern, three distinct additional resonances are resolved, where the two outer resonances show identical intensity. Principally, this observation may be explained by three different scenarios: First, this "doublet" may be due to two different paramagnetic species of different g-value. In this case, the splitting between the lines should be field dependent. Because the splitting of the doublet remained invariant upon changing the Larmor frequency from 300 to 400 GHz (not shown), this scenario can be ruled out. Accordingly, a field-independent interaction has to be the origin of the observed doublet, which could be either owing to a hyperfine interaction with a magnetically active nucleus of nuclear spin $I = 1/2$ or due to a triplet state with electron spin $S = 1$. However, as the observed splitting between the lines is orders of magnitude larger than the one reported for the hyperfine interactions of ^{1}H-impurities [44,72], this model can be discarded as well. Correspondingly, the origin of the observed doublet is due a triplet state with electron spin $S = 1$.

Concerning the defect-structural origin of the resonances about $g \approx 2.0$, it was recently demonstrated that these signals are owing to surface defects [5,49,73]. Commonly, oxygen vacancies are depleted at the surface layers [7] and defects at the surface are rather attributed to zinc interstitials or zinc vacancies [74,75]. Additional to the interstitials formed according to the Schottky-type defect equilibrium, zinc interstitials and zinc vacancies are generated owing to a Frenkel-type defect equilibrium according to

$$\mathrm{Zn}_{\mathrm{Zn}}^{\times} \rightarrow V_{\mathrm{Zn}}^{\times} + \mathrm{Zn}_{i}^{\times},$$

where subsequent ionization equilibria again allow for varying charge states

$$\mathrm{V}_{\mathrm{Zn}}^{\times} \rightarrow \mathrm{V}_{\mathrm{Zn}}^{/} + h^{\bullet}$$

$$\mathrm{V}_{\mathrm{Zn}}^{/} \rightarrow \mathrm{V}_{\mathrm{Zn}}^{/} + h^{\bullet}$$

$$\mathrm{Zn}_{i}^{\times} \rightarrow \mathrm{Zn}_{i}^{\bullet} + \mathrm{e}'$$

$$\mathrm{Zn}_{i}^{\bullet} \rightarrow \mathrm{Zn}_{i}^{\bullet\bullet} + \mathrm{e}'$$

The resonance at $g = 2.0045$ is assigned to a singly charged $V_{\mathrm{Zn}}^{/}$ [74,75] and the triplet spectrum at $g = 2.0051$ is a complexation of two defects, which results in the formation of a $(\mathrm{Zn}_{i}^{\bullet} - V_{\mathrm{Zn}}^{/})^{\times}$ defect complex. For this complex to form, a zinc ion leaves a lattice site and occupies an interstitial position in the first coordination sphere of the initial site. Because both isolated defects mutually compensate when forming the defect complex, the corresponding $(\mathrm{Zn}_i^{\bullet} - V_{\mathrm{Zn}}^{/})^{\times}$ defect complex is electrically neutral.

The assignment of defect-complex formation is supported by recent photoluminescence experiments that anticipated the existence of complex defects involving zinc interstitials to explain green emission bands in the corresponding spectra [3]. Furthermore, the existence of $(\mathrm{Zn}_i^{\bullet} - V_{\mathrm{Zn}}^{/})^{\times}$ defect complexes is in line with results reported for ZnSe [76] and may be compared to donor–acceptor pair centers recently reported in Li-doped ZnO nanocrystals [77].

Taking into account the above-derived defect structure, the obtained high-field spectrum can accurately be numerically simulated as demonstrated in Fig. 11.5. The corresponding refined set of spin-Hamiltonian parameters is summarized in Table 11.1.

Table 11.1 Refined set of spin-Hamiltonian parameters obtained after numerical spectrum simulation of the high-field EPR spectrum in Fig. 11.4

	$V_{\mathrm{O}}^{\bullet}$	$V_{\mathrm{Zn}}^{/}$	$\mathrm{Zn}_{\mathrm{i}}^{\bullet} - V_{\mathrm{O}}^{/}$	$\mathrm{Mn}_{\mathrm{Zn}}^{\times}$
g_{iso}	—	2.0045	2.0051	2.0019
g_{parallel}	1.97715	—	—	—
g_{perp}	1.97825	—	—	—
B_2^0(GHz)	—	—	1.39	—
A (MHz)	—	—	—	260

11.6 Space-Charge Depletion Layer and Core–Shell Model

Exploiting the observed EPR results, information of the formation of a space-charge depletion layer can be obtained. More generally, for the here studied nanostructured ZnO compounds the confined length scales and accordingly enlarged surface-to-volume ratio promotes surface defects to play a stronger role in controlling materials properties. This effect can be interpreted in terms of a core–shell model in which the particles consist of a core region that is electronically different from the surface of the ZnO nanoparticles.

11.6.1 Core–Shell Model

Despite intense research activities, a unique model for the description of ZnO nanocrystal regardless of its structural or geometrical shape is still lacking. As a structural model, the so-called *core–shell* model of hexagonal nanoparticles, is presented to offer a possible explanation for the size effects in ZnO and to show that the surface *shell* with a cubic structure covers the particle *core* with a hexagonal structure. The model is also important for estimating the width of the space-charge depletion layer. This model has previously been applied to explain size effects for ferroelectric Cr^{3+} [78] and Fe^{3+} [79] doped $PbTiO_3$ nanoparticles prepared from a combined polymerization and pyrolysis route [80], and semiconductor ZnO prepared by the microwave-assisted method [81]. We have reported 6 nm critical size for $PbTiO_3$ below this size there is size-driven phase transition from tetragonal-to-cubic phase where ferroelectricity disappears [78]. In case of ZnO [81] by using *core–shell* model we reported that 8 nm core was embedded inside a 1 nm thick *shell*. In case of ZnO, by the aid of EPR, thus quantitative spin counting, we distinguish two EPR signal arising from interior hexagonal *core* and outer *shell* where the *shell* signal (surface defects) increases and *core* signal (bulk defects) decreases by reducing the particle size. Since we obtain the relative ratio of *core* and *shell* signal we can easily obtain the shell thickness by using the aid of following sketch and formula in Fig. 11.6.

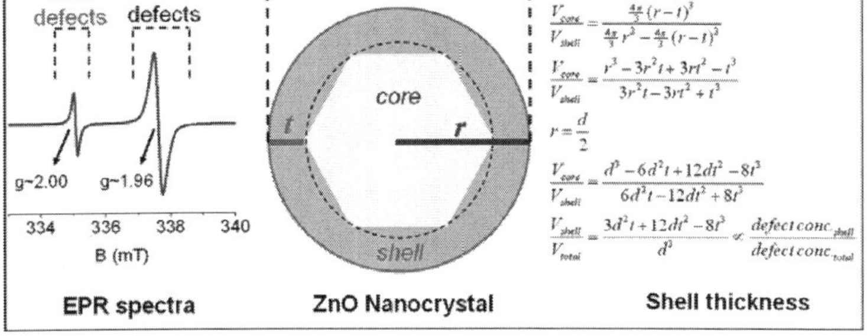

Figure 11.6 Sketch for the structural core–shell model. (Left) X-band EPR spectra for nano-sized ZnO giving two distinct signal at $g \sim 1.96$ (*core*) and $g \sim 2.00$ (*shell*). (Middle) A model for the ZnO nanocrystal with hexagonal interior *core* and cubic outer *shell*. (Right) Calculation for the *shell* thickness.

Exploiting the above-derived defect structure, the core is dominated rather by (shallow) donor-type $V_O^{\bullet}$, whereas at the surface acceptor-type $V_{Zn}^{/}$ and $(Zn_i^{\bullet}- V_{Zn}^{/})^{\times}$ prevail. As consequence, the development of a space-charge (depletion) layer as schematically illustrated in Fig. 11.7 emerges, which clarifies the defect structure anticipated for a grain-boundary defect model [82].

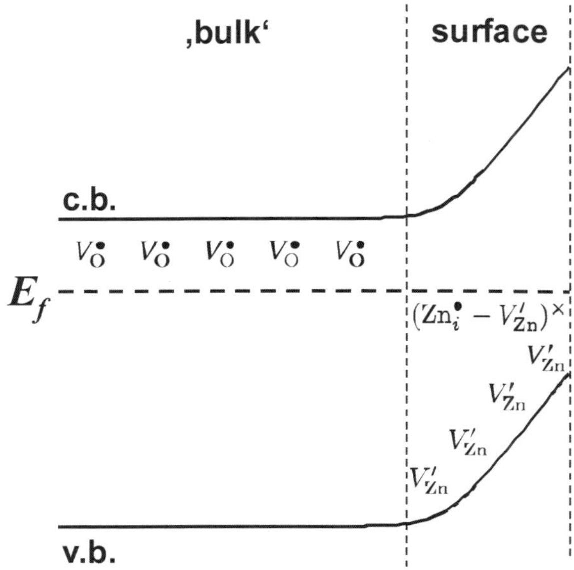

Figure 11.7 Schematic illustration of the development of a space-charge layer as function of the defect structure in nano-sized ZnO. Fermi energy (E_f), conduction (c.b.) and valence (v.b.) are indicated.

In particular, application of the core–shell model allows the determination of the thickness of the depletion layer to 1.0 nm for the compounds studied here, which defines a unique means to study the surface morphology of ZnO nanoparticles [83].

Acknowledgements

This research has been financially supported by the *Deutsche Forschungsgemeinschaft, DFG* (Grant: Er 662/1-1). Special thanks to P. Jakes, J. J. Schneider, R. Hofmann, and R. Böttcher.

References

1. K. Keis, E. Magnusson, H. Lindstrom, S. E. Lindquist, and A. Hagfeldt, *Solar Energy Materials and Solar Cells,* **73**, pp. 51–58, 2002.
2. J. Bao, M. A. Zimmler, F. Capasso, X. Wang, and Z. F. Ren, *Nano Letters,* **6**, pp. 1719–1722, 2006.
3. A. B. Djurisic, W. C. H. Choy, V. A. L. Roy, Y. H. Leung, C. Y. Kwong, K. W. Cheah, T. K. G. Rao, W. K. Chan, H. T. Lui, and C. Surya, *Advanced Functional Materials,* **14**, pp. 856–864, 2004.
4. Y. Kinemuchi, M. Mikami, K. Kobayashi, K. Watari, and Y. Hotta, *Journal of Electronic Materials,* **39**, pp. 2059–2063, 2010.
5. J. J. Schneider, R. C. Hoffmann, J. Engstler, S. Dilfer, A. Klyszcz, E. Erdem, P. Jakes, and R. A. Eichel, *Journal of Materials Chemistry,* **19**, pp. 1449–1457, 2009.
6. T. K. Gupta, *Journal of the American Ceramic Society,* **73**, pp. 1817–1840, 1990.
7. G. D. Mahan, *Journal of Applied Physics,* **54**, pp. 3825–3832, 1983.
8. M. Yang, D. Wang, L. Peng, Q. Zhao, Y. Lin, and X. Wei, *Sensors and Actuators B-Chemical,* **117**, pp. 80–85, 2006.
9. Z. L. Wang and J. H. Song, *Science,* **312**, pp. 242–246, 2006.
10. U. Ozgur, Y. I. Alivov, C. Liu, A. Teke, M. A. Reshchikov, S. Dogan, V. Avrutin, S. J. Cho, and H. Morkoc, *Journal of Applied Physics,* **98**, p. 041301, 2005.
11. A. B. Djurisic and Y. H. Leung, *Small,* **2**, pp. 944–961, 2006.
12. V. Ischenko, S. Polarz, D. Grote, V. Stavarache, K. Fink, and M. Driess, *Advanced Functional Materials,* **15**, pp. 1945–1954, 2005.
13. M. D. McCluskey and S. J. Jokela, *Journal of Applied Physics,* **106**, p. 071101, 2009.
14. S. J. Pearton, D. P. Norton, M. P. Ivill, A. F. Hebard, J. M. Zavada, W. M. Chen, and I. A. Buyanova, *Ieee Transactions on Electron Devices,* **54**, pp. 1040–1048, 2007.
15. A. Janotti and C. G. Van de Walle, *Reports on Progress in Physics,* **72**, p. 126501, 2009.
16. B. K. Meyer, H. Alves, D. M. Hofmann, W. Kriegseis, D. Forster, F. Bertram, J. Christen, A. Hoffmann, M. Strassburg, M. Dworzak, U. Haboeck, and A. V. Rodina, *Physica Status Solidi B-Basic Research,* **241**, pp. 231–260, 2004.

17. J. Jagadish and S. Pearton, *Zinc Oxide Bulk, Thin Films and Nanostructures*. Sidney: Elsevier, 2006.
18. N. H. Nickel and E. Terukov, *Zinc Oxide as a Material for Micro- and Optoelectronic Applications*. St. Petersburg: Springer, 2004.
19. P. C. Chang, Z. Fan, C. J. Chien, D. Stichtenoth, C. Ronning, and J. G. Lu, *Applied Physics Letters,* **89**, p. 3, 2006.
20. C. Czekalla, J. Guinard, C. Hanisch, B. Q. Cao, E. M. Kaidashev, N. Boukos, A. Travlos, J. Renard, B. Gayral, D. L. Dang, M. Lorenz, and M. Grundmann, *Nanotechnology,* **19**, p. 115202, 2008.
21. J. S. Reparaz, F. Guell, M. R. Wagner, A. Hoffmann, A. Cornet, and J. R. Morante, *Applied Physics Letters,* **96**, p. 053105, 2010.
22. M. C. Newton, S. J. Leake, R. Harder, and I. K. Robinson, *Nature Materials,* **9**, pp. 120–124, 2010.
23. E. Schlenker, A. Bakin, H. Schmid, W. Mader, S. Sievers, M. Albrecht, C. Ronning, S. Muller, M. Al-Suleiman, B. Postels, H. H. Wehmann, U. Siegner, and A. Waag, *Nanotechnology,* **18**, p. 125609, 2007.
24. Z. R. R. Tian, J. A. Voigt, J. Liu, B. McKenzie, M. J. McDermott, M. A. Rodriguez, H. Konishi, and H. F. Xu, *Nature Materials,* **2**, pp. 821–826, 2003.
25. A. K. Srivastava, M. Deepa, K. N. Sood, E. Erdem, and R.-A. Eichel, *Advanced Materials Letters,* **2**, pp. 142–147, 2011.
26. S. K. Panda, N. Singh, J. Hooda, and C. Jacob, *Crystal Research and Technology,* **43**, pp. 751–755, 2008.
27. C. Zollfrank, C. R. Rambo, M. Batentschuk, and P. Greil, *Journal of Materials Science,* **42**, pp. 6325–6330, 2007.
28. D. Y. Lee, J. E. Cho, N. H. Cho, M. H. Lee, S. J. Lee, and B. Y. Kim, *Thin Solid Films,* **517**, pp. 1262–1267, 2008.
29. M. Mader, J. W. Gerlach, T. Hoche, C. Czekalla, M. Lorenz, M. Grundmann, and B. Rauschenbach, *Physica Status Solidi-Rapid Research Letters,* **2**, pp. 200–202, 2008.
30. S. A. Morin, M. J. Bierman, J. Tong, and S. Jin, *Science,* **328**, pp. 476–480, 2010.
31. Z. W. Pan, Z. R. Dai, and Z. L. Wang, *Science,* **291**, pp. 1947–1949, 2001.
32. H. J. Fan, A. S. Barnard, and M. Zacharias, *Applied Physics Letters,* **90**, p. 143116, 2007.
33. C. Ronning, P. X. Gao, Y. Ding, Z. L. Wang, and D. Schwen, *Applied Physics Letters,* **84**, pp. 783–785, 2004.

34. Y. J. Kim, J. Yoo, B. H. Kwon, Y. J. Hong, C. H. Lee, and G. C. Yi, *Nanotechnology,* **19**, p. 315202, 2008.

35. W. L. Hughes and Z. L. Wang, *Applied Physics Letters,* **86**, p. 043106, 2005.

36. H. E. Unalan, Y. Zhang, P. Hiralal, S. Dalal, D. P. Chu, G. Eda, K. B. K. Teo, M. Chhowalla, W. I. Milne, and G. A. J. Amaratunga, *Applied Physics Letters,* **94**, p. 163501, 2009.

37. M. Willander, O. Nur, Q. X. Zhao, L. L. Yang, M. Lorenz, B. Q. Cao, J. Zuniga Perez, C. Czekalla, G. Zimmermann, M. Grundmann, A. Bakin, A. Behrends, M. Al-Suleiman, A. El-Shaer, A. Che Mofor, B. Postels, A. Waag, N. Boukos, A. Travlos, H. S. Kwack, J. Guinard, and D. Le Si Dang, *Nanotechnology,* **20**, p. 332001, 2009.

38. J. Alaria, P. Turek, M. Bernard, M. Bouloudenine, A. Berbadj, N. Brihi, G. Schmerber, S. Colis, and A. Dinia, *Chemical Physics Letters,* **415**, pp. 337–341, 2005.

39. I. Djerdj, G. Garnweitner, D. Arcon, M. Pregelj, Z. Jaglicic, and M. Niederberger, *Journal of Materials Chemistry,* **18**, pp. 5208–5217, 2008.

40. S. B. Zhang, S. H. Wei, and A. Zunger, *Physical Review B,* **63**, p. 075205, 2001.

41. F. Boccuzzi, G. Ghiotti, and A. Chiorino, *Journal of the Chemical Society-Faraday Transactions Ii,* **79**, pp. 1779–1789, 1983.

42. A. Janotti and C. G. V. de Walle, *Physical Review B,* **76**, p. 165202, 2007.

43. C. Gonzalez, D. Block, R. T. Cox, and A. Herve, *Journal of Crystal Growth,* **59**, pp. 357–362, 1982.

44. D. M. Hofmann, A. Hofstaetter, F. Leiter, H. J. Zhou, F. Henecker, B. K. Meyer, S. B. Orlinskii, J. Schmidt, and P. G. Baranov, *Physical Review Letters,* **88**, 2002.

45. S. B. Orlinskii, J. Schmidt, P. G. Baranov, C. D. Donega, and A. Meijerink, *Physical Review B,* **79**, 2009.

46. S. M. Evans, N. C. Giles, L. E. Halliburton, and L. A. Kappers, *Journal of Applied Physics,* **103**, p. 043710, 2008.

47. N. Y. Garces, L. Wang, L. Bai, N. C. Giles, L. E. Halliburton, and G. Cantwell, *Applied Physics Letters,* **81**, pp. 622–624, 2002.

48. Y. Hu, *Journal of Nanoparticle Research,* **10**, pp. 1371–1371, 2008.

49. Y. Hu and H. J. Chen, *Journal of Nanoparticle Research,* **10**, pp. 401–407, 2008.

50. K. Vanheusden, W. L. Warren, C. H. Seager, D. R. Tallant, J. A. Voigt, and B. E. Gnade, *Journal of Applied Physics,* **79**, pp. 7983–7990, 1996.
51. D. C. Look, R. L. Jones, J. R. Sizelove, N. Y. Garces, N. C. Giles, and L. E. Halliburton, *Physica Status Solidi a-Applied Research,* **195**, pp. 171–177, 2003.
52. S. Moribe, T. Ikoma, K. Akiyama, Q. W. Zhang, F. Saito, and S. Tero-Kubota, *Chemical Physics Letters,* **436**, pp. 373–377, 2007.
53. V. A. Nikitenko, *Journal of Applied Spectroscopy,* **57**, pp. 367–385, 1992.
54. P. V. B. Lakshmi, B. Ranganathan, N. Rajkumar, and K. Ramachandran, *International Journal of Modern Physics B,* **23**, pp. 3221–3229, 2009.
55. R. Laiho, D. S. Poloskin, Y. P. Stepanov, M. P. Vlasenko, L. S. Vlasenko, and V. S. Zakhvalinskii, *Journal of Applied Physics,* **106**, p. 7, Jul 2009.
56. R. Laiho, L. S. Vlasenko, and M. P. Vlasenko, *Journal of Applied Physics,* **103**, p. 123709, 2008.
57. M. Kakazey, M. Vlasova, M. Dominguez-Patino, G. Dominguez-Patino, T. Sreckovic, and N. Nikolic, *Science of Sintering,* **36**, pp. 65–72, 2004.
58. L. S. Vlasenko, *Physica B-Condensed Matter,* **404**, pp. 4774–4778, 2009.
59. V. Avrutin, N. Izyumskaya, U. Ozgur, D. J. Silversmith, and H. Morkoc, *Proceedings of the Ieee,* **98**, pp. 1288–1301, 2010.
60. T. Dietl, *Nature Materials,* **2**, pp. 646–648, 2003.
61. S. Q. Zhou, K. Potzger, Q. Y. Xu, G. Talut, M. Lorenz, W. Skorupa, M. Helm, J. Fassbender, M. Grundmann, and H. Schmidt, *Vacuum,* **83**, pp. S13–S19, 2009.
62. S. Q. Zhou, K. Potzger, G. Talut, H. Reuther, K. Kuepper, J. Grenzer, Q. Y. Xu, A. Mucklich, M. Helm, J. Fassbender, and E. Arenholz, *Journal of Physics D-Applied Physics,* **41**, p. 105011, 2008.
63. L. B. Duan, X. R. Zhao, J. M. Liu, T. Wang, and G. H. Rao, *Applied Physics a-Materials Science & Processing,* **99**, pp. 679–683, 2010.
64. N. Akdogan, A. Nefedov, K. Westerholt, H. Zabel, H. W. Becker, C. Somsen, R. Khaibullin, and L. Tagirov, *Journal of Physics D-Applied Physics,* **41**, p. 165001, 2008.
65. Q. Y. Xu, H. Schmidt, L. Hartmann, H. Hochmuth, M. Lorenz, A. Setzer, P. Esquinazi, C. Meinecke, and M. Grundmann, *Applied Physics Letters,* **91**, p. 092503, 2007.
66. P. Sharma, A. Gupta, K. V. Rao, F. J. Owens, R. Sharma, R. Ahuja, J. M. O. Guillen, B. Johansson, and G. A. Gehring, *Nature Materials,* **2**, pp. 673–677, 2003.

67. K. I. Hagemark, *Journal of Solid State Chemistry,* **16**, pp. 293–299, 1976 1976.

68. F. A. Kröger and H. J. Vink, *Solid State Physics,* **3**, 1956.

69. N. G. Kakazey, T. V. Sreckovic, and M. M. Ristic, *Journal of Materials Science,* **32**, pp. 4619–4622, 1997.

70. P. Erhart, K. Albe, and A. Klein, *Physical Review B,* **73**, p. 205203, 2006.

71. R.-A. Eichel, *Journal of Electroceramics,* **19**, pp. 11–23, 2007.

72. S. B. Orlinskii, J. Schmidt, E. J. J. Groenen, P. G. Baranov, C. de Mello Donega, and A. Meijerink, *Physical Review Letters,* **94**, p. 097602, 2005.

73. B. L. Yu, C. S. Zhu, F. X. Gan, and Y. B. Huang, *Materials Letters,* **33**, pp. 247–250, 1998.

74. A. Poppl and G. Volkel, *Physica Status Solidi a-Applied Research,* **125**, pp. 571–581, 1991.

75. A. Pöppl and G. Volkel, *Physica Status Solidi a-Applied Research,* **115**, pp. 247–255, 1989.

76. F. C. Rong, W. A. Barry, J. F. Donegan, and G. D. Watkins, *Physical Review B,* **54**, pp. 7779–7788, 1996.

77. S. B. Orlinskii, H. Blok, J. Schmidt, P. G. Baranov, C. D. M. Donega, and A. Meijerink, *Physical Review B,* **74**, p. 045204, 2006.

78. E. Erdem, H. C. Semmelhack, R. Bottcher, H. Rumpf, J. Banys, A. Matthes, H. J. Glasel, D. Hirsch, and E. Hartmann, *Journal of Physics-Condensed Matter,* **18**, pp. 3861–3874, 2006.

79. E. Erdem, K. Kiraz, M. Somer, and R.-A. Eichel, *Journal of the European Ceramic Society,* **30**, pp. 289–293, 2010.

80. E. Erdem, R. Bottcher, H. C. Semmelhack, H. J. Glasel, E. Hartmann, and D. Hirsch, *Journal of Materials Science,* **38**, pp. 3211–3217, 2003.

81. J. J. Schneider, R. C. Hoffmann, J. Engstler, A. Klyszcz, E. Erdem, P. Jakes, R. A. Eichel, L. Pitta-Bauermann, and J. Bill, *Chemistry of Materials,* **22**, pp. 2203–2212, 2010.

82. T. K. Gupta and W. G. Carlson, *Journal of Materials Science,* **20**, pp. 3487–3500, 1985.

83. R.-A. Eichel, E. Erdem, P. Jakes, A. Ozarowski, H. van Tol, R. C. Hoffmann and J. J. Schneider, *Functional Materials Letters,* **6**, p. 1330004, 2013.

Index